Lecture Notes in Mechanical Engineering

Lecture Notes in Mechanical Engineering (LNME) publishes the latest developments in Mechanical Engineering—quickly, informally and with high quality. Original research or contributions reported in proceedings and post-proceedings represents the core of LNME. Volumes published in LNME embrace all aspects, subfields and new challenges of mechanical engineering.

To submit a proposal or request further information, please contact the Springer Editor of your location:

Europe, USA, Africa: Leontina Di Cecco at Leontina.dicecco@springer.com
China: Ella Zhang at ella.zhang@cn.springernature.com
India, Rest of Asia, Australia, New Zealand: Swati Meherishi at swati.meherishi@springer.com

Topics in the series include:

- Engineering Design
- Machinery and Machine Elements
- Mechanical Structures and Stress Analysis
- Automotive Engineering
- Engine Technology
- Aerospace Technology and Astronautics
- Nanotechnology and Microengineering
- Control, Robotics, Mechatronics
- MEMS
- Theoretical and Applied Mechanics
- Dynamical Systems, Control
- Fluid Mechanics
- Engineering Thermodynamics, Heat and Mass Transfer
- Manufacturing Engineering and Smart Manufacturing
- Precision Engineering, Instrumentation, Measurement
- Materials Engineering
- Tribology and Surface Technology

Indexed by SCOPUS, EI Compendex, and INSPEC.

All books published in the series are evaluated by Web of Science for the Conference Proceedings Citation Index (CPCI).

To submit a proposal for a monograph, please check our Springer Tracts in Mechanical Engineering at https://link.springer.com/bookseries/11693.

Livan Fratini · Luigi Maria Galantucci ·
Luca Settineri
Editors

Selected Topics in Manufacturing

Emerging Trends from the Perspective of AITeM's Young Researchers

Springer

Editors
Livan Fratini
Department of Engineering
Università di Palermo
Palermo, Italy

Luigi Maria Galantucci
DMMM
Politecnico di Bari
Bari, Italy

Luca Settineri
DIGEP
Politecnico di Torino
Turin, Italy

ISSN 2195-4356 ISSN 2195-4364 (electronic)
Lecture Notes in Mechanical Engineering
ISBN 978-3-031-99500-2 ISBN 978-3-031-99501-9 (eBook)
https://doi.org/10.1007/978-3-031-99501-9

This Springer imprint is published by the registered company Springer Nature Switzerland AG
The registered company address is: Gewerbestrasse 11, 6330 Cham, Switzerland

About This Book

This book presents contributions drawn from a broader set of papers concerning a wide range of scientific and technological areas covered by AITeM (the Italian Manufacturing Association).

In particular, the first part, *AITeM Young Researcher Award 2025*, is written by young AITeM associates: their contributions overall reflect the wide nature of the research in manufacturing and, furthermore, highlights the most recent innovations and technologies proposed and discussed within the international scientific community. In this way, emerging technologies and established interdisciplinary connections with different technological areas are explored to advance the state of the art.

Additive manufacturing technologies, in particular, are shown to significantly impact not only simple product fabrication but also the development of a complex and intensely interconnected value creation processes ecosystem pursuing high-value-added products to compete globally.

In the book, the following topics are addressed: additive manufacturing, materials processing technology, assembly, disassembly and circular economy, manufacturing systems design and management, quality engineering and production metrology, process and system simulation, optimization and digital manufacturing.

An Editorial Committee composed by Prof. **Livan Fratini**, Università di Palermo (President), Prof. **Wilma Polini**, Università di Cassino, Prof. **Pierpaolo Carlone**, Università di Salerno, Prof. **Aldo Attanasio**, Università di Brescia, Prof. **Marco Sorgato**, Università di Padova, Prof. **Paolo Parenti**, Politecnico di Milano, Prof. **Flaviana Calignano**, Politecnico di Torino, Eng. **Olivier Terzo**, Fondazione Links, Eng. **Ciro Sinagra**, Laminazione Sottile S.p.A. peer reviewed and selected ten contributions among 22 papers proposed for the Award.

The second part, *White Papers*, includes three contributions on some emerging trends in manufacturing research. The contributions have been prepared by working groups that have formed around strategic research topics in the manufacturing sector, often related to emerging applications: adaptive processes for the manufacturing of welded profiles and tubes, sustainable and innovative production processes for composite materials, machining for aerospace. The papers here published aim to provide an overview of the new challenges posed by these frontier areas, demonstrating how only through the multidisciplinary and highly innovative approach that our community offers can these challenges be successfully addressed.

The *White Papers* underwent a reviewing process led by Prof. **Luigi Maria Galantucci** from the Politecnico di Bari to ensure their compliance with AITeM standards and to make their structure consistent.

Livan Fratini
Luigi Maria Galantucci
Luca Settineri

Contents

AITeM Young Researcher Award 2025

A Comparative Analysis of AA6061 and AlSi10Mg Processed via Friction Stir Extrusion

Sara Bocchi, Gianluca D'Urso, and Claudio Giardini

Abstract The aim of this study is to investigate the different behaviours of two aluminium alloys—AA6061, an age-hardening alloy, and AlSi10Mg, a casting alloy—under Friction Stir Extrusion processing. Friction Stir Extrusion (FSE), a sustainable solid-state recycling method, eliminates the need for melting and minimizes energy consumption, promoting efficient material reuse. The sustainability of this process is reinforced by the use of AlSi10Mg derived from recycled powder no longer suitable for Laser Powder Bed Fusion (L-PBF) due to granulometric changes post-printing. The experimental campaign analysed the influence of FSE parameters on microstructure and mechanical properties, revealing alloy-specific responses. These findings highlight the potential of FSE for manufacturing high-performance materials with reduced environmental impact, advancing the circular economy in metal recycling and additive manufacturing.

Keywords Sustainability · Aluminium alloys · Friction stir extrusion

1 Introduction

In the modern era, achieving sustainable development has become a key priority for industries and governments worldwide. The increasing demand for materials, powered by population growth and technological advancements, has led to a substantial rise in industrial production. However, this expansion comes with significant environmental costs, including high energy consumption, waste generation, and greenhouse gas emissions. The manufacturing industry contributes 36% of the global greenhouse gas emissions and due to the production of materials such as aluminium, steel, cement, and polymers, and this percentage keeps rising [1].

S. Bocchi (✉) · G. D'Urso · C. Giardini
Department of Management, Information and Production Engineering, University of Bergamo, Dalmine, BG, Italy
e-mail: sara.bocchi@unibg.it

L. Fratini et al. (eds.), *Selected Topics in Manufacturing*, Lecture Notes in Mechanical Engineering, https://doi.org/10.1007/978-3-031-99501-9_1

In the field of manufacturing, aluminium and its alloys definitely take the lead due to how well they combine strength, low weight, and corrosion resistance. Moreover, projections show that looking forward to 2050, the demand for aluminium will increase more than other essential materials at a rate of 2.6–3.5 times from 2005 [2]. Nonetheless, remelting–the primary method for recycling aluminium–has well-known drawbacks, including high energy consumption, emissions of harmful gases and contaminants, as well as smoke emission from remelting reactors, which contribute to environmental pollution [3]. In addition, several remelting cycles result in higher levels of impurities, which in turn lowers the mechanical properties of the recycled aluminium [4].

To overcome the challenges related to the remelting, the Welding Institute presented Friction Stir Extrusion (FSE) in 1993 as a new way to recycle metal waste, especially low-melting alloys of aluminium and magnesium [5–7]. This advancement allows for the primary recycling through the direct extrusion of new parts from scraps, metal chips or powder that stem from the machining process, thus lowering the required number of recycling processes these materials have to undergo. FSE operates as a low-energy, solid state recycling technique [8], designed for small-volume and low-mass components. What makes this process unique is the heat generated by friction between the tool and the workpiece, which produces enough heat for the process without requiring external heaters. It significantly reduces energy consumption; in fact, studies have shown that the process consumes less than 15% of energy required for conventional aluminium melting [9]. The material is softened because of the heat generated, which makes intense plastic deformation easier. The scrap is consolidated and bonded, resulting in extruded products with good mechanical properties and less probable additional working. In contrast to conventional fusion-based recycling, processing temperatures for FSE are about 400 °C, which makes direct recycling possible without having to melt the material. This prevents defects like dendritic structures, hot cracking, oxidation, and shrinkage that are related to casting [9]. The process has a rotating tool that goes into a hollow die, where the metal chips are compacted, stirred, and extruded into dense cylindrical rods. The final dimensions of these extrusions depend on the die size and the input material quantity, while the mechanical properties are affected by some set parameters of the process like rotational speed of the tool, tool descent feed and size of the chips.

The applicability of FSE extends across various industries, since a key advantages of FSE is its ability to process both metal powders and scrap into extruded rods, tubes, and wires [10–12]. The automotive world benefits from FSE because of its manufacturing of aluminium parts that reduce vehicle weight and fuel economy. The excellent thermal and electrical conductivity of aluminium allows FSEed components to be used in several applications in the field of electronics. The construction industry, on the other hand, uses FSE to develop aluminium profiles of high strength and toughness, finding their usage in building designing [13–15]. In spite of these advantages, a number of things to do with FSE need still further investigation, especially those focused on the influence of processing parameters on the properties of extruded materials. Indeed, as already mentioned, key process variables such as tool rotational speed, extrusion force, and feed rate directly impact the microstructure, mechanical properties, and energy efficiency of the final product [16–18].

These parameters govern material flow within the extruded section, which forms a complex three-dimensional helical structure [19] featuring stretched grains at the centre and recrystallized grains at the periphery [20, 21]. Tool speed determines the extent of heat generation and the pressure directs the amount of compaction. Tahmasbi and Mahmoodi showed that if the force applied was increased then there would be a more uniform and harder extrusion [18]. Similarly, studies by Hosseini et al. established that improper tool rotation speeds can lead to surface defects, such as cold cracks from insufficient heating at low speeds and porosity from excessive speeds [16].

Also material flow and strain evolution during FSE have been further explored through experimental work. Li et al. conducted strain analysis on FSE-produced aluminium wires, demonstrating that strain evolution plays a vital role in bonding between unbounded scrap material [22]. Similarly, Behnagh et al. investigated the recrystallization behaviour of magnesium wire using FEM simulations to predict microstructural changes during processing [17]. Flow tracing methods were also employed in the study of extruded AA6061 and AA2195 by Li et al. [23]. Baffari et al. investigated the influence of process parameters and initial material temperature on the mechanical performance of products extruded using FSE [15].

In his work, Sharifzadeh assessed how the manufacturing conditions influenced the microstructure and grain size of magnesium wires [24]. The FSE process with aluminium alloy AA6063-T52 has been investigated by Abu-Farha to recognize the changes in microstructure due to the different regions in which they are extruded [25].

FSE has been successfully applied to a wide range of aluminium alloys, such as AA1050, AA2024, and AA6061, as well as magnesium and copper-based alloys [26–28]. Research indicates that this technique produces extrusions with fine, homogeneous grain structures and a well-distributed second-phase precipitate network, which enhances mechanical performance and response to heat treatment.

Among aluminium alloys, the 6xxx series, particularly AA6061, is widely used in FSE due to its excellent formability, lower cost, and capability of being hardened through artificial aging after solution heat treatments [29–31]. The precipitation sequence of Mg and Si in AA6061 alloys contributes to its superior mechanical properties, making it a preferred choice for extrusion applications [32–34]. Additionally, dynamic precipitation phenomena activated during FSE can further enhance the alloy performance [35].

While the study involves the processing of two distinct aluminium alloys—AlSi10Mg (a cast alloy) and AA6061 (a wrought alloy)—using FSE, the core objective is not merely to explore the application of this technology to these materials. Rather, the primary aim is to demonstrate the feasibility of employing FSE for the recycling of waste powder derived from Laser Powder Bed Fusion (L-PBF) additive manufacturing processes. This focus addresses the need for sustainable pathways in metal recycling, particularly by reintroducing secondary raw materials into the production cycle.

A central innovation of this work lies in the investigation of AlSi10Mg, an alloy commonly used in foundry applications and additive manufacturing but not conventionally adopted in FSE. Successfully extruding this material via FSE—especially when using degraded or out-of-spec powder fractions (e.g., those affected by partial oxidation or failing to meet granulometric standards)—could offer a novel recycling route. This research seeks to validate such an approach, thereby broadening the scope of FSE beyond its traditional applications and contributing to a more sustainable and circular manufacturing ecosystem.

In addition to the overarching goal of feasibility demonstration, this study includes a comparative assessment of mechanical and microstructural properties of the extruded materials, process variability and its effect on quality, and optimization of processing parameters. It also explores differences between cast and wrought alloys in terms of recyclability, strength, and impact resistance.

From a practical standpoint, this method presents a promising pathway for scaling aluminium powder recycling within industrial environments. Given the increasing adoption of additive manufacturing (AM) technologies, the accumulation of unusable powder residues has become a growing concern for both economic and environmental reasons. Currently, such waste is often down cycled or discarded. The implementation of FSE as a post-AM recycling method offers a direct, energy-efficient solution by converting these powders into high-performance extruded profiles or components.

2 Materials and Methods

This study utilized gas-atomized AA6061 powder and AlSi10Mg powder as feedstock. AA6061 consists of spherical particles smaller than 106 μm, while AlSi10Mg consists of spherical particles greater than 30 μm, both stored in a dry air environment. The chemical compositions of both powders are reported in Tables 1 and 2.

The experimental work was carried out using the FE100 friction extrusion machine, developed by Bond Technologies, which is capable of applying forces up to 1000 kN, torques of 3500 Nm, and rotational speeds reaching 1000 rpm. Two main components were used in the experimental setup. The first component is a rotating powder container made of AISI 4140 steel featuring an internal diameter

Table 1 AA6061 powder chemical composition

Element	%	Element	%
Al	Bulk	Ti	0.01
Mg	0.83	Zn	0.01
Si	0.60	Ni	< 0.01
Cu	0.28	Sn	< 0.01
Fe	0.11	Zr	< 0.01
Mn	0.01	Cr	< 0.01

Table 2 AlSi10Mg powder chemical composition

Element	%	Element	%
Al	Bulk	Cr	0.01
Si	7.6	Ni	< 0.01
Fe	0.21	Zn	0.01
Cu	0.01	Ti	0.01
Mn	0.01	Sn	0.01
Mg	0.30	Zr	< 0.01

of 22 mm. The second component is a radial extrusion die made of AISI H13 steel, with a flat face and a die hole of 7 mm, expanding to 9 mm after a 5 mm bearing length. This configuration results in an extrusion ratio of 10.

Regardless of the aluminium alloy and the process parameters considered, the initial phase involved setting a rotational speed equal to the rotational speed of the process, combined with an applied force of 50 kN, to ensure full contact between the extrusion die and the material. Once the die advanced by 1 mm, further compacting and pre-heating the feedstock, the system transitioned to displacement-controlled friction extrusion through a gradual 5-s ramp.

Approximately 40 g of AA6061 and 35 g of AlSi10Mg powder were loaded into the container. Different powder quantities were used to ensure a similar occupied volume, considering the variation in powder grain size. A 1 mm-thick stopper was positioned over the die hole during the pre-compaction phase to prevent powder loss before extrusion, during the initial movement of the powder container. The material underwent a sequence of filling, tapping, and compression inside the container, with an initial hydraulic press load of 20, 50, 75 and 100 kN, reaching initial powder density within a range of 70–85% of the aluminium density.

For the execution of the experimental tests, two parameters were varied: the rotational speed of the container (S), ranging from a minimum of 200 rpm to a maximum of 1000 rpm in increments of 200 rpm, and the feed rate (F), ranging from a minimum of 1 mm/s to a maximum of 4 mm/s in increments of 1 mm/s. These configurations were chosen to represent the upper and lower boundaries of the processing range explored in previous research [19, 36]. Tables 3 and 4 present all the parameter combinations considered, along with the acronyms of the extruded sample. In each acronym, the first part (AA6 or AlSi) indicates the aluminium alloy type, the second number represents the rotational speed, and the last one corresponds to the feed rate.

The extruded samples were cut to obtain smaller specimens. Specifically, for all samples, two pieces with a length of 25 mm, designated as A1 and B1, and two pieces with a length of 10 mm, designated as A2 and B2, were extracted (Fig. 1). The A1 and B1 specimens were intended for microscopic analysis in the longitudinal direction, followed by hardness measurements. Conversely, the A2 and B2 specimens were used for microscopic analysis in the transverse direction and subsequent Vickers micro-hardness testing.

Table 3 AA6061 process parameters combinations and extruded acronyms

		F [mm/s]			
		1	2	3	4
S [rpm]	200	AA6-200-1	AA6-200-2	AA6-200-3	AA6-200-4
	400	AA6-400-1	AA6-400-2	AA6-400-3	AA6-400-4
	600	AA6-600-1	AA6-600-2	AA6-600-3	AA6-600-4
	800	AA6-800-1	AA6-800-2	AA6-800-3	AA6-800-4
	1000	AA6-1000-1	AA6-1000-2	AA6-1000-3	AA6-1000-4

Table 4 AlSi10Mg process parameters combinations and extruded acronyms

		F [mm/s]			
		1	2	3	4
S [rpm]	200	AlSi-200-1	AlSi-200-2	AlSi-200-3	AlSi-200-4
	400	AlSi-400-1	AlSi-400-2	AlSi-400-3	AlSi-400-4
	600	AlSi-600-1	AlSi -600/-2	AlSi-600-3	AlSi-600-4
	800	AlSi-800-1	AlSi-800-2	AlSi-800-3	AlSi-800-4
	1000	AlSi-1000-1	AlSi-1000-2	AlSi-1000-3	AlSi-1000-4

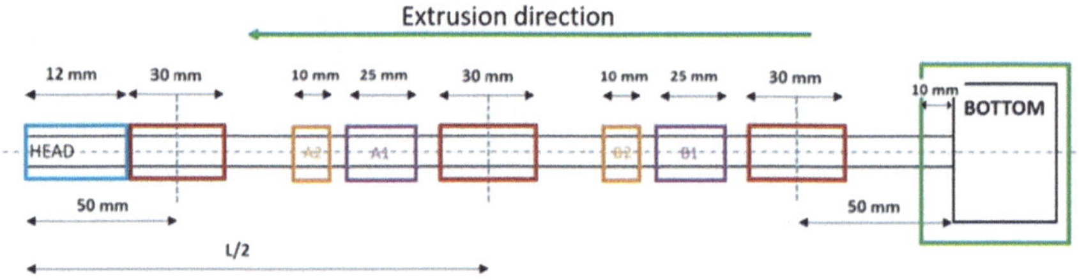

Fig. 1 Specimens cutting plan

Subsequently, the samples were embedded and polished using abrasive papers with progressively increasing grit numbers, concluding the polishing process with 1 μm alumina. The prepared samples were then used for hardness testing, metallographic etching, and observation under a digital optical microscope.

Vickers micro-hardness testing was conducted on each A2 and B2 specimen (Fig. 1) using an applied load of 100 gf. Each cross section of the samples was indented following a grid pattern measuring 17 × 17, giving a total of 289 contact points. Each grid was equally spaced apart at a distance of 0.5 mm. Consequently, the points satisfied the minimum distance requirements from the edges as specified by the relevant standard.

The final analysis was performed using electrochemical etching to reveal the grain structure. A masking template was placed over the polishing cell, which contains the cathode, to define the treated area. The sample, acting as the anode, was secured in place. Both the sample and the cathode were immersed in a Barker solution, which serves as the electrolyte to complete the circuit. A voltage of 25 V was applied for 60 s. After etching, the samples were analysed under a digital optical microscope to observe the samples microstructure.

3 Results and Discussion

For both alloys, for all the process parameters combinations, completely dense extruded wires were successfully obtained. At the end of the FSE process, the final length of the extruded parts was measured. The results are reported in Table 5.

When comparing the response of the two different alloys to the FSE process, both common behaviours and a fundamental difference were observed. Regarding similarities, both AA6061 and AlSi10Mg alloys exhibit three distinct zones in the bottom regions of the samples, characterized by different material densities and porosity levels (Fig. 2).

Table 5 Final extruded length for both the considered alloys

S [rpm]	F [mm/s]	Final extruded length [mm]	
		AA6061	AlSi10Mg
200	1	246	220
200	2	248	225
200	3	245	215
200	4	235	192
400	1	225	205
400	2	232	208
400	3	255	220
400	4	215	207
600	1	253	210
600	2	245	205
600	3	238	198
600	4	235	204
800	1	268	209
800	2	255	204
800	3	230	206
800	4	262	204
1000	1	254	205
1000	2	248	205
1000	3	255	215
1000	4	246	225

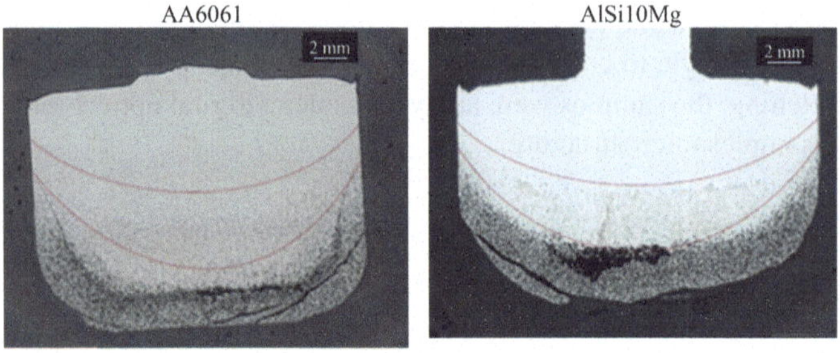

Fig. 2 Microscope images of the bottom part of sample AA6-600-1 on the left and AlSi-600-1 on the right. The red lines identify the three characteristic zones of each bottom

In the closest zone to the extruded wire, extending to a depth of approximately 5–7 mm at the centre of the bottom, the billet is nearly free of void. In the second zone, with a thickness ranging from 3 to 4 mm at the centre of the bottom, the number of voids increases, culminating in the third zone, which still contains residual starting powder.

In the FSE process, variations in temperature and pressure likely result from heat generation, friction due to contact with tooling, and mechanical compression. The process leads to considerable densification of the material, effectively eliminating initial porosity and compacting the structure. In the second zone, heat and pressure remain high but are insufficient to completely eliminate porosity, leaving some voids. In the third and outermost zone, being the farthest from the high-temperature and high-pressure zone created by the process, the effect is that very little compression and heating are experienced by the material. Therefore, this prevents densification, and the material retains its original powder state.

Another common characteristic is that, in B1 samples, following the cutting plan reported in Fig. 1, both alloys exhibit elongated grains oriented along the material flow direction. This grain elongation is likely influenced by the feed rate (Figs. 3 and 4).

The images indicate that, for a given rotational speed, an increase in the feed rate results in a larger region of elongated grains and, more importantly, reduces their

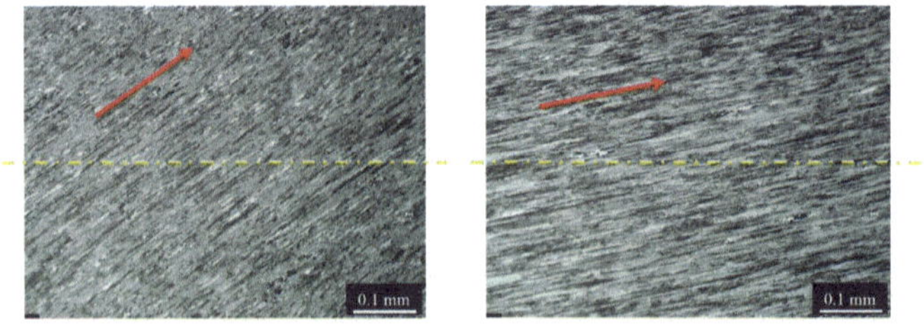

Fig. 3 Microscope images of the B1 part of sample AA6-200-1 on the left and AA6-200-4 on the right. The red arrows highlight the inclination angle of the grain with respect to the extrusion axis

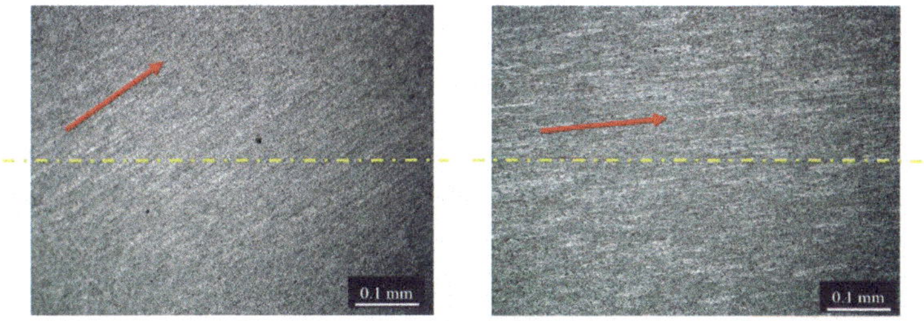

Fig. 4 Microscope images of the B1 part of sample AlSi-200-1 on the left and AlSi-200-4 on the right. The red arrows highlight the inclination angle of the grain with respect to the extrusion axis

inclination angle relative to the extrusion axis. Therefore, low feed rates seem to be preferable, as they allow for a nearly homogeneous extrusion. Specifically, low feed rates promote the formation of a larger region with equiaxed grains. Additionally, elongated grains exhibit a greater inclination relative to the extrusion axis, indicating that they were close to homogenization and were rotated by the force exerted by the rotational speed. This inclination of elongated grains indicates a state of near homogenization, where mechanical forces generated by the rotational movement of the die contribute significantly to grain rotation and deformation.

Moreover, although the orientation patterns of the microstructure appear comparable regardless of the alloy type, Figs. 3 and 4 highlight a notable difference in average grain size. Specifically, under identical processing parameters, the plastically deformed alloy exhibits significantly larger grain dimensions compared to the cast alloy, in both the longitudinal and transverse directions. This difference may be attributed to the higher intrinsic ductility and deformation response of the plastically deformed material, which facilitates grain growth and elongation during processing.

A third similarity between the AA6061 and AlSi10Mg is that, within each alloy, samples A1 and B1 exhibit different amounts of defects, especially cavities, visible as dark regions, which are presumed to be influenced by the rotational speed (Fig. 5).

Specifically, at low rotational speeds (Fig. 5a, b), A1 sample (Fig. 5a)—located closer to the head—exhibits defects, whereas B1 sample (Fig. 5b)—positioned near the bottom—is free of defects. Conversely, at high rotational speeds (Fig. 5c, d), A1 sample (Fig. 5c) is defect-free, while B1 sample (Fig. 5d) exhibits defects.

This observation is further confirmed by examining the head sections. The initial portion of the extruded wires is deliberately excluded, as it is affected by incomplete compaction, and only the correctly extruded material is considered. The head of sample AA6-200-4 (Fig. 6), processed at a rotational speed of 200 rpm and a feed rate of 4 mm/s, exhibits a significant number of defects. The head of sample AA6-600-4 (Fig. 7), processed at 600 rpm and a feed rate of 4 mm/s, exhibits very few defects.

The same behaviour is also observed for the AlSi10Mg alloy. In Fig. 8a, b, depicting sample processed at 200 rpm and at a feed rate of 1 mm/s, a defect reduction can be observed when transitioning from A1 (Fig. 8b) to B1 (Fig. 8a). Similarly, in Fig. 8c and Fig. 8d, corresponding to sample processed at 600 rpm and at a feed

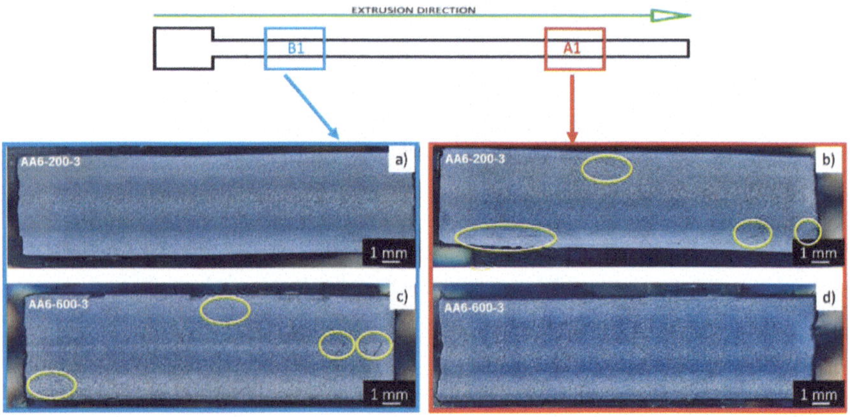

Fig. 5 Microscope images of the A1 and B1 samples of AA6061 extruded samples with the same feed rate (3 mm/s) but different rotational speed. **a** and **b** are referred to S = 200 rpm, whilst **c** and **d** to S = 600 rpm. The yellow profiles indicate the defects within the extruded product

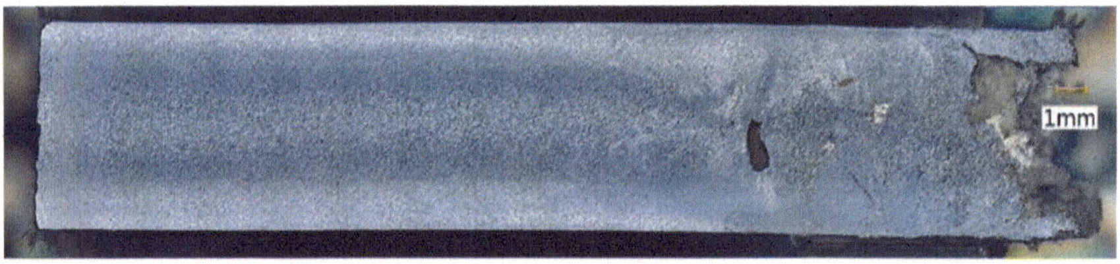

Fig. 6 Head of AA6-200-4

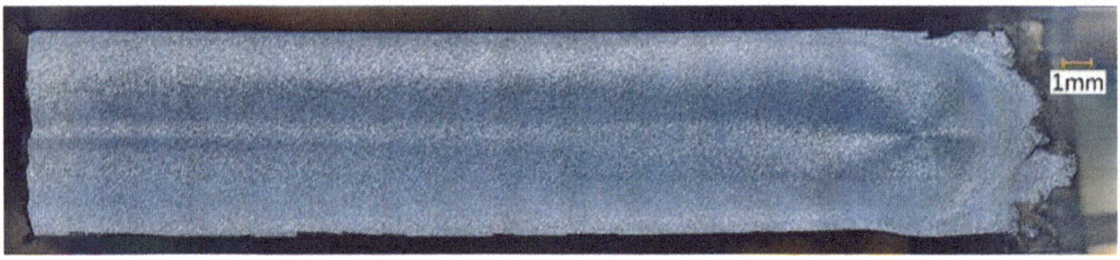

Fig. 7 Head of AA6-600-4

rate of 1 mm/s, fewer defects are present in sample A1 (Fig. 8d) compared to sample B1 (Fig. 8c).

To optimize the reduction of defects, a good strategy seems to be starting with higher rotational speeds to prevent defects in the initial part of the extrusion process and then gradually decrease the speed to avoid defect formation near the bottom.

The fundamental difference between the two alloys lies in the presence of circumferentially elongated grains in the AA6061 cross sections, a feature not observed in the AlSi10Mg samples. When examining samples A2 and B2 from all the specimens,

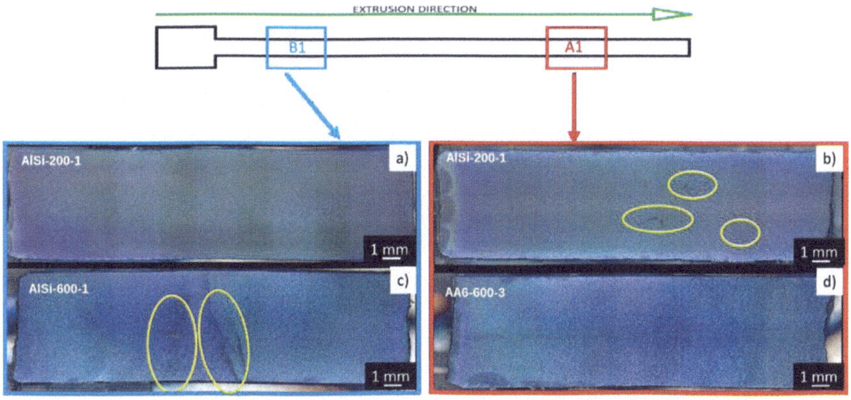

Fig. 8 Microscope images of the A1 and B1 samples of AlSi10Mg extruded samples with the same feed rate (1 mm/s) but different rotational speed. **a** and **b** are referred to S = 200 rpm, whilst **c** and **d** to S = 600 rpm. The yellow profiles indicate the defects within the extruded product

it is evident that in AA6061, the grains exhibit deformation due to the rotation of the container, regardless of the rotational speed considered (Figs. 9a and 10a).

In contrast, in the AlSi10Mg alloy, all grains appear equiaxed, indicating that the material has mixed more uniformly (Fig. 10a, b). This dissimilarity could be related to the different response of the considered alloys to the thermal input generated by the friction between the powder and the instrumentation. Indeed, AlSi10Mg is a cast alloy, which is not typically used for the machining processes.

Observing samples A2 and B2 of AA6061 and following the grains circumferentially elongated, it can be hypothesized that during extrusion, the material follows a spiral flow pattern (Fig. 11), as also reported in a previous simulative work [19].

The hypothesis regarding the spiral flow pattern of the material in AA6061 appears to be supported also by the of Vickers micro-hardness results (Fig. 12), where all

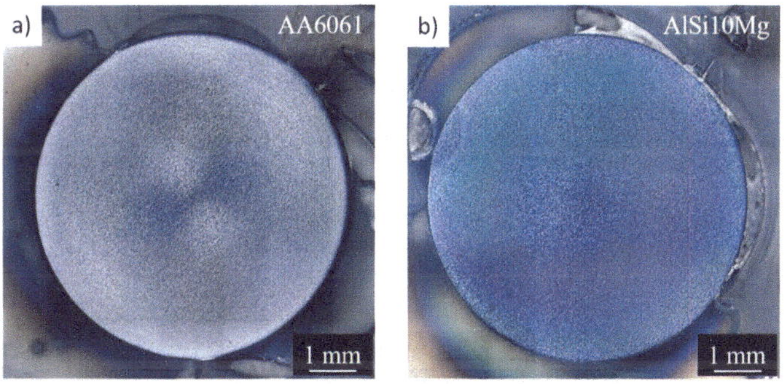

Fig. 9 Cross section images of AA6061 and AlSi10Mg samples extruded with S = 200 rpm and F = 1 mm/s

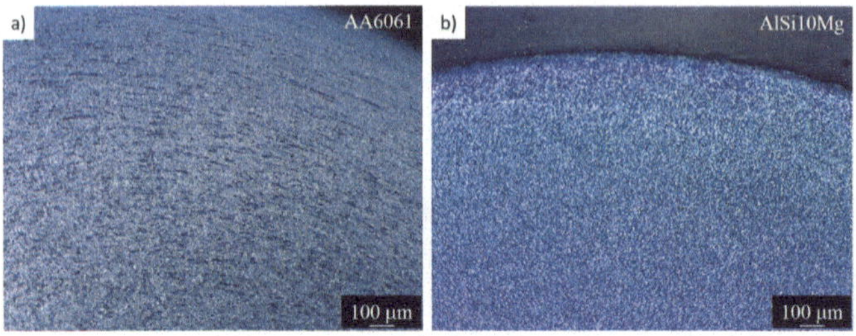

Fig. 10 A2 details of AA6061 and AlSi10Mg samples extruded with S = 600 rpm and F = 1 mm/s

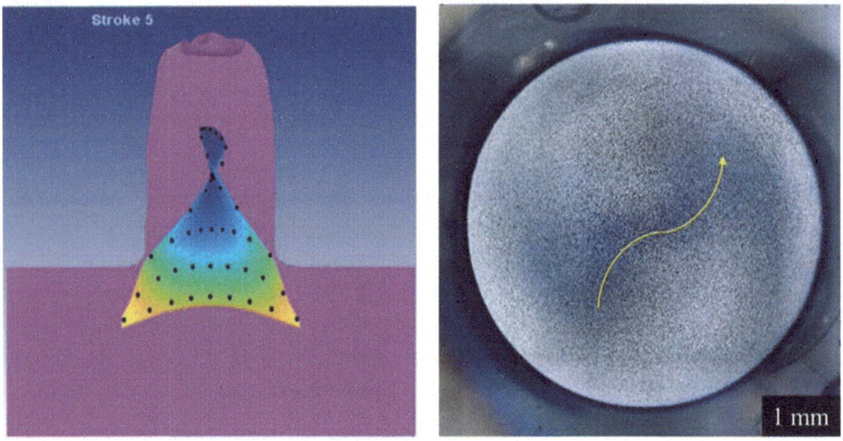

Fig. 11 On the left, the final position of the 45 considered nodes in the simulative model reported in Bocchi et al. [19]. On the right, the cross section of AA6-200-3

samples clearly exhibit two distinct areas: one with higher hardness and one with lower hardness.

It can therefore be concluded that the spiral flow of the material results in the formation of a higher-hardness zone that follows this helical movement.

4 Conclusions

This study demonstrates the technological feasibility of applying the Friction Stir Extrusion process not only to AA6061 but also to AlSi10Mg cast alloy derived from waste powder of the Laser Powder Bed Fusion process. The experimental results show that AA6061 and AlSi10Mg exhibit some similarities and a difference in their microstructural and mechanical responses to FSE.

In particular:

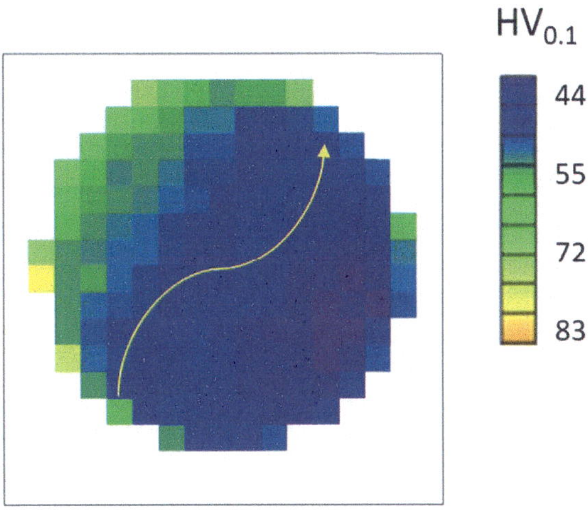

Fig. 12 Vickers micro-hardness map of AA6-600-1 cross section

- The bottom regions of both alloys exhibit three distinct zones, resulting from temperature and pressure variations induced by intense friction with the tool and mechanical compression.
- Metallographic studies show that in both alloys, elongated grains are aligned with the material flow direction along the extrusion axis. Feed rates have a significant influence on this grain elongation; lower feed rates result in a smaller diameter of the elongated grain region, a greater inclination of the grains relative to the extrusion axis, and, ultimately, a more uniform extrusion with a larger area occupied by equiaxed grains.
- The defect distribution varies along the length of the extruded material, with rotational speed being the influential factor. Lowering rotational speed during extrusion could reduce defect formation.
- The material flow during the extrusion process is helically oriented in a manner as revealed through metallographic tests and Vickers micro-hardness tests.
- The two materials were distinguished by the equiaxed grains in AlSi10Mg and the circumferentially elongated ones seen in AA6061 cross sections. This suggests that AlSi10Mg would be more uniform in material mixing during process time.

Acknowledgements The authors would like to thank Prof. Benjamin Klusemann, Dr.·Uceu Suhuddin and Dr. Lars Rath, from Solid State Materials Processing, Institute of Materials Mechanics, Helmholtz-Zentrum Hereon (Geesthacht, Germany) for the support in the production of the experimental samples used in this study.

References

1. Worrell E, Allwood J, Gutowski T (2016) The role of material efficiency in environmental stewardship. Annu Rev Environ Resour 41(41):575–598
2. Gutowski TG, Sahni S, Allwood JM, Ashby MF, Worrell E (2013) The energy required to produce materials: constraints on energy-intensity improvements, parameters of demand. Philos Trans R Soc A Math Phys Eng Sci 371(1986)
3. Baffari D, Buffa G, Campanella D, Fratini L (2017) Al-SiC metal matrix composite production through friction stir extrusion of aluminum chips. Procedia Eng 419–424
4. Bell S, Davis B, Javaid A, Essadiqi E (2006) Final report on effect of impurities in aluminum. p 18
5. Li J, Meng X, Li Y, Wan L, Huang Y (2021) Friction stir extrusion for fabricating Mg-RE alloys with high strength and ductility. Mater Lett 289:129414
6. Shunmugasamy VC, Khalid E, Mansoor B (2021) Friction stir extrusion of ultra-thin wall biodegradable magnesium alloy tubes—microstructure and corrosion response. Mater Today Commun 26:102129
7. Evans WT, Gibson BT, Reynolds JT, Strauss AM, Cook GE (2015) Friction stir extrusion: a new process for joining dissimilar materials. Manuf Lett 5:25–28
8. Baffari D, Reynolds AP, Masnata A, Fratini L, Ingarao G (2019) Friction stir extrusion to recycle aluminum alloys scraps: energy efficiency characterization. J Manuf Process 43:63–69
9. Manchiraju K (2012) Direct solid-state conversion of recyclable metals and alloys. Southwire Company, Golden, CO (United States)
10. Rath L, Suhuddin UFH, Klusemann B (2022) Comparison of friction extrusion processing from bulk and chips of aluminum-copper alloys. Key Eng Mater 926:471–480
11. Whalen S, Olszta M, Reza-E-Rabby M, Roosendaal T, Wang T, Herling D, Taysom BS, Suffield S, Overman N (2021) High speed manufacturing of aluminum alloy 7075 tubing by Shear assisted processing and extrusion (ShAPE). J Manuf Process 71:699–710
12. Li X, Zhou C, Overman N, Ma X, Canfield N, Kappagantula K, Schroth J, Grant G (2021) Copper carbon composite wire with a uniform carbon dispersion made by friction extrusion. J Manuf Process 65:397–406
13. Chan CYC, Suhuddin UFHR, Rath L, Bachiega FL, Klusemann B (2024) Grain structure evolution ahead of the die during friction extrusion of AA2024. In: Lecture notes in mechanical engineering, Springer Science and Business Media Deutschland GmbH, pp 422–428
14. Abdollahi H, Shafaei SZ, Noaparast M, Manafi Z, Niemelä SI, Tuovinen OH (2014) Mesophilic and thermophilic bioleaching of copper from a chalcopyrite-containing molybdenite concentrate. Int J Miner Process 128:25–32
15. Baffari D, Reynolds AP, Li X, Fratini L (2017) Influence of processing parameters and initial temper on friction stir extrusion of 2050 aluminum alloy. J Manuf Process 28:319–325
16. Hosseini A, Azarsa E, Davoodi B, Ardahani Y (2012) Effect of process parameters on the physical properties of wires produced by friction extrusion method. Int J Adv Eng Technol 3(1):2231–2963
17. Behnagh RA, Shen N, Ansari MA, Narvan M, Kazem M, Givi B, Ding H (2016) Experimental analysis and microstructure modeling of friction stir extrusion of magnesium chips. J Manuf Sci Eng Trans ASME 138(4)
18. Tahmasbi K, Mahmoodi M (2018) Evaluation of microstructure and mechanical properties of aluminum AA7022 produced by friction stir extrusion. J Manuf Process 32:151–159
19. Bocchi S, D'Urso G, Giardini C, Maccarini G (2022) A simulative method for studying the bonding condition of friction stir extrusion. Key Eng Mater 926:2333–2341
20. Baffari D, Buffa G, Fratini L (2017) A numerical model for wire integrity prediction in friction stir extrusion of magnesium alloys. J Mater Process Technol 247:1–10
21. Bocchi S, Negozio M, Giardini C, Donati L (2024) Prediction of the microstructure evolution during the friction stir extrusion of a AA6061 aluminum alloy. Mater Res Proc 678–687
22. Li X, Tang W, Reynolds AP, Tayon WA, Brice CA (2016) Strain and texture in friction extrusion of aluminum wire. J Mater Process Technol 229:191–198

23. Li X, Tang W, Reynolds AP (2016) Material flow and texture in friction extruded wire. In: Friction stir welding and processing VII. Springer, Cham, pp 339–347
24. Sharifzadeh M, Ansari MA, Narvan M, Behnagh RA, Araee A, Besharati Givi MK (2015) Evaluation of wear and corrosion resistance of pure Mg wire produced by friction stir extrusion. Trans Nonferrous Met Soc China 25(6):1847–1855
25. Abu-Farha F (2012) A preliminary study on the feasibility of friction stir back extrusion. Scr Mater 66(9):615–618
26. Rath L, Chan C, Suhuddin U, Buresch H, Ebel T, Klusemann B (2024) Processability of Mg-Gd powder via friction extrusion. In: Lecture notes in mechanical engineering, Springer Science and Business Media Deutschland GmbH, pp 431–441
27. Zhang H, Zhao X, Deng X, Sutton MA, Reynolds AP, McNeill SR, Ke X (2014) Investigation of material flow during friction extrusion process. Int J Mech Sci 85:130–141
28. Joshi VV, Jana S, Li D, Garmestan H, Nyberg E, Lavender C (2014) High shear deformation to produce high strength and energy absorption in Mg alloys. Magnes Technol Miner Metals Mater Soc 83–88
29. Cai M, Cheng GJ (2007) Microstructure-properties relationship in two Al–Mg–Si alloys through a combination of extrusion and aging. JOM 59(8):58–61
30. Murtha SJ (1995) New 6XXX aluminum alloy for automotive body sheet applications. SAE technical papers
31. Buha J, Lumley RN, Crosky AG (2006) Microstructural development and mechanical properties of interrupted aged Al–Mg–Si–Cu alloy. Metall Mater Trans A Phys Metall Mater Sci 37(10):3119–3130
32. Edwards GA, Stiller K, Dunlop GL, Couper MJ (1998) The precipitation sequence in Al–Mg–Si alloys. Acta Mater 46(11):3893–3904
33. Chrominski W, Kulczyk M, Lewandowska M, Kurzydlowski KJ (2014) Precipitation strengthening of ultrafine-grained Al–Mg–Si alloy processed by hydrostatic extrusion. Mater Sci Eng A 609:80–87
34. Wu Y, Liao H, Liu Y, Zhou K (2014) Dynamic precipitation of Mg_2Si induced by temperature and strain during hot extrusion and its impact on microstructure and mechanical properties of near eutectic Al–Si–Mg–V alloy. Mater Sci Eng A 614:162–170
35. Negozio M, Bocchi S, Rath L, Ghio E (2025) Finite element modeling of microstructure evolution and bonding during friction stir extrusion of AA6061 powder at different tool feed rates and rotational speeds. Mater Charact 219:114639
36. Bocchi S, D'Urso G, Giardini C (2023) Numerical modeling of a sustainable solid-state recycling of aluminum scraps by means of friction stir extrusion process. Materials (Basel) 16(12):4375

Multi-modal and Sensorized 3D-Printed Omnidirectional Snake Tongue Bioinspired Soft Actuator

Antonio Pavone and Gianluca Percoco

Abstract Traditional bioinspired soft actuators typically rely on a single actuation mechanism, limiting their movements to bending, twisting, or stretching, and often lack built-in sensing capabilities, reducing adaptability and interaction with external environment. The integration of multiple actuation mechanisms within the same actuator has recently gained popularity in the soft robotics 3D printing field. In the present paper, a novel 3D printed omnidirectional bioinspired soft actuator that mimics the snake tongue is presented. The multi-modal actuator monolithically fabricated using Multi-material Extrusion, seamlessly combining pneumatical actuation and embedded shape memory alloy spring actuators, with integrated 3D printed sensors. The integration of multiple actuation mechanisms and embedded sensing, enabled by cost-effective multi-material 3D printing, enhances soft robotics for precise, adaptive movements in biomedical, industrial, and surveillance applications.

Keywords Shape memory alloys · Additive manufacturing · Material extrusion · Soft robotics · Grippers · Embedded sensors · 3D printing

1 Introduction

Soft robotics has gained significant attention in recent years for its ability to replicate biomimetic movements [1–3]. Inexpensive Desktop Multi-material Extrusion (M-MEX), a class of 3D printing technology, has the capability to fabricate complex soft bioinspired structures offers notable advantages, including reduced costs, shorter production times, and reduction of manual assembly tasks [4–6].

A. Pavone (✉) · G. Percoco
Department of Mechanics, Mathematics and Management, Polytechnic of Bari, Bari, Italy
e-mail: antonio.pavone@poliba.it

Interdisciplinary Additive Manufacturing (IAM) Lab, TTEC Taranto Polytechnic, Polytechnic University of Bari, Taranto, Italy

Sustainable Materials for Additive Manufacturing for Soft Robotics (SMAM 4 SoRo), Polytechnic University of Bari, Bari, Italy

L. Fratini et al. (eds.), *Selected Topics in Manufacturing*, Lecture Notes in Mechanical Engineering, https://doi.org/10.1007/978-3-031-99501-9_2

From an actuation point of view, soft robotis fabricated through M-MEX are based on several actuation systems such as tendon driven [7], electromagnetic [8, 9], light-emission [10], pneumatic networks (PneuNets) [2, 11], shape memory alloys (SMA) [12, 13], shape memory polymer (SMP) [14], and multi-modal (hybrid) systems [15, 16]. Pneumatic actuators have been widely used in soft robotics, offering compliance, lightweight structures, and safe interactions with the environment [2, 11]. These 3D printed actuator systems are based on easy working principles: they are made of soft material, typically thermoplastic polyurethane (TPU), and consist of several inflatable and interconnected chambers, that can be expanded through compressed air. Indeed, unlike traditional actuators, PneuNets are capable of performing complex motion such as bending, twisting, and elongation. Herianto et al. [17] investigated the use of different commercial soft materials to obtain a repeatable and high-quality 3D printed soft PneuNet.

Despite their advantages, PneuNets have limitations, including the need for a precise pneumatic control system and the reliance on passive (rigid) layers to achieve a specific bending motion [18, 19]. On the other hand, Shape Memory Alloys (SMA) are smart materials, that undergo reversible phase under thermal activation, generating force and recover deformation autonomously. SMAs are widely used for soft fingers motion [13], and biomedical applications [20]. One limit of the SMAs is related to the need of one or more antagonist elements to move the SMA into the desirable elongation (before the activation), such as loads, opposite-side SMA, or tendons [21].

Both PneuNets and SMAs can move in a one-way direction: combining them, omnidirectional movements can be achieved [22–24]. Hybrid systems are studied to combine multiple actuation methods, resulting in multi-modal actuators that enhance motion complexity, improve response times, and increase adaptability in soft robotic applications [25, 26].

The proposed work aims to push the role of Desktop M-MEX for the fabrication of an omnidirectional multi-modal soft bioinspired actuator able to mimic complex movements of snake tongue (Fig. 1a), combining SMA springs and pneumatic actuation, the design enables complex movements while integrating a monolithically embedded piezoresistive sensor for real-time feedback. Despite the studied omnidirectional soft robots in the literature, which are often costly [27, 28], the authors propose a cost-effective approach using Desktop multi-material 3D printing to fabricate a soft robot without assembly tasks, and enhancing soft robotics applications in biomedical, gripper, and surveillance fields.

2 Aim of the Research

Omnidirectional soft robots capable of performing extremely complex tasks in terms of motions have always been researched, as it would enhance the role of soft robotics for practical applications, such as grippers, biomedical fields, and surveillance. At the state of the art, the most advanced omnidirectional soft robots require multiple

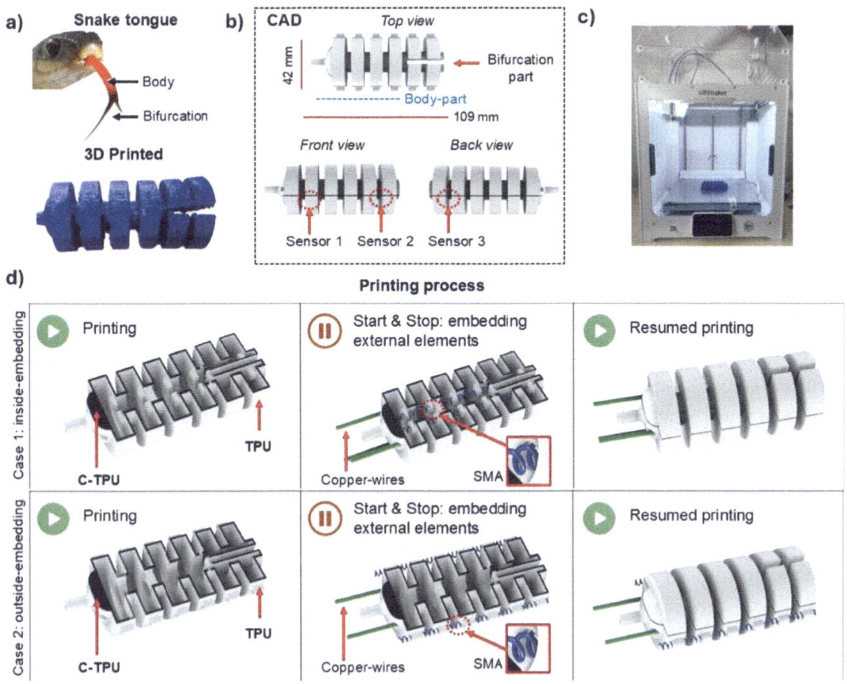

Fig. 1 Aim of the paper: **a** from snake tongue to 3D printed bioinspired soft actuator, **b** computer aided design (CAD) of the proposed actuator, **c** 3D printed machine used, and **d** start-and-stop printing process of the proposed case 1 and case 2 actuator

assembly steps and high-fabrication costs. The present research presents a novel approach to rapidly fabricate omnidirectional soft robots by leveraging a Desktop multi-material extrusion 3D printer. In particular, the latter is characterized by low cost (around 5000 euro) compared to other cost-effective Additive Manufacturing technologies such as Material Jetting (> 100,000 euro), capable of processing more than two polymers at the same time.

A framework based on the start and stop method is illustrated to create a snake tongue-inspired soft robot capable of moving across 3-Dimensions. A Design for Additive Manufacturing approach (DfAM) was used to create a soft robot based on two distinct actuation systems: pneumatic, and shape memory alloys. Additionally, the proposed soft robot fully exploited the capabilities of AM, as an embedded strain sensor was monolithically fabricated (by processing all electrically conductive polymer), with no assembly tasks (Fig. 1b). This research presents a novel approach to fabricate an inexpensive, assembly-free soft bioinspired robot with embedded actuation and sensing, capable of operating and providing real-time feedback immediately after its removal from the build-plate.

Table 1 Printing parameters

Material	Extrusion temperature (°C)	Nozzle size (mm)	Printing speed (mm/s)	Flow (%)
TPU	235	0.8	25	105
Conductive	240	0.8	20	135

3 Printing Process

A M-MEX machine Ultimaker S5 (Ultimaker, The Netherlands) was used to extrude at the same time both soft and conductive materials (see Fig. 1c). Furthermore, to integrate two shape memory alloys (Dynalloy, United States) and copper-wires for the 3D-printed sensors, the start-and-stop (S&S) [29] method was employed (such sown in Fig. 1d), optimizing the process within a one-shot manufacturing framework.

Soft pneumatical bioinspired actuator parts were fabricated with a shore A hardness of 80 Thermoplastic Polyurethane (TPU) (FiloAlfa, Turin, Italy) using printing parameters shown in Table 1, particularly a0.8mm nozzle was used in order to optimize the number of beads for improved airtightness of the chamber. Indeed, to prevent air leakage, three different parameters were studies: (i) number of continuous 3D printed beads, (ii) size of single bead, and (iii) printing orientation.

A significant thermal incompatibility issue arises when SMAs are embedded into 3D printed parts during the fabrication process. Flexible materials, such as used TPU, typically have a printing temperature (Tp) of around 230°C or higher, which is much higher than the activation temperature of Nickel-Titanium SMAs, which ranges from 70 to 90 °C. To achieve this, the SMP springs were integrated using *hole-structures* designed in the bioinspired actuator, with a height lower than that of the embedded sensors, such shown in Fig. 1d.

In order to sensorized the bioinspired soft actuator a strain gauge sensor was 3D printed using soft conductive Ninjatek Eel filament (NinjaTek, United States) according to the printer parameters shown in Table 1. The sensor was wired with embedded copper-wires during the S&S process.

4 Multi-modal Characterization

The complex movements that mimic the snake tongue behaviours were achieved combining both the SMA spring actuators and compressed air benefits. SMAs are able to generate significant force despite their small size without the use of bulky mechanical components (such as external motors) [30]. Additionally, their shape recovery property enables them to return to original shape when heated, shown high reliability and durability in order to increase the lifespan of the actuator structures wherein is employed. On the other hand, 3D printed pneumatic actuators (PneuNets) have been extensively studied in scientific literature demonstrating adaptivity and flexibility in soft robotics. Their easy and low-cost manufacturing process, combined with the possibility of integration with other actuation systems, offers innovative solutions for various engineering challenges [18, 26].

4.1 Pneumatical Actuation

The pneumatical actuator that embed the SMA springs and the 3D printed sensors, consists of three different pneumatical section in order to mimic the shape of snake tongue. The first part (body), in Fig. 1b, consists of 4 separate chambers that allows the elongation of the central body-part under compressed air inflation, and 2 cambers for the bifurcation-part. No rigid parts (passive layers) were used to achieve bending movements. Instead, the SMA springs replace the passive layer, generating an active and independent bending motion, characteristic of multi-modal actuation.

On the other hand, the final part of the body presents two small PneuNets that replay the bifurcation of snake tongue. Both body-part and bifurcation-parts are supplied by the same air channel, operating dependently. As a result, they undergo deformation simultaneously.

The dimensions of all 3D printed structure were 109 mm in length, and 42 mm in height. Each body-chamber can be elongate of 4.3 mm, and each bifurcation-chamber can be elongate of 2.6 mm.

As stated in Sect. 3, in order to characterize the design of pneumatical actuator parts, where studied: (i) number of continuous 3D printed beads, (ii) size of single bead, and (iii) printing orientation. After this, the displacement was characterized to define the motion of the pneumatical actuation system. First, according to Stano et al. [11], the horizontal printing direction was chosen to minimize leakage between the 3D-printed walls of the actuators, as showed in Fig. 2a. Additionally, a smooth internal chamber (Fig. 2b) design was implemented to allow for printing without the use of internal supports. To prevent air leakage during pneumatic actuation, reducing at the same times manufacturing time, a 0.8mm nozzle was used to print one continuous 0.8 mm bead for each chamber, following the process parameters outlined in Sect. 3. In contrast, using a 0.4 mm nozzle required two perimeter lines and resulted in increased infill density, which, in some cases, compromised air tightness due to excessive wall thickening and potential defects in interlayer adhesion. The 0.8mm nozzle, on the other hand, enabled a perimeter-only structure with no infill, which not only improved flexibility but also enhanced elongation performance during pneumatic activation and SMA bending. Furthermore, the horizontal printing orientation, ensured optima adhesion between different layers of the 3D-printed soft TPU, maintaining effective air tightness even when using a larger nozzle diameter (0.8mm).

Finally, to characterize the dynamic movements of the bioinspired actuator, the displacement was defined as: (a) elongation (displacement) for the body-part, and (b) elongation (displacement) for the bifurcation-part, when several pressure values were used. The results are shown in Fig. 2d, e. With input pressures of 1bar and

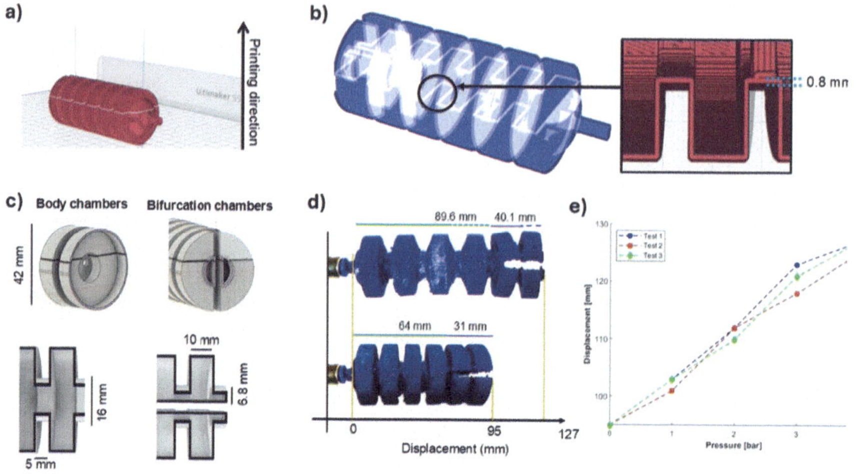

Fig. 2 Pneumatic actuation-mode characterization: **a** printing direction, **b** internal view of chamber walls, **c** body and bifurcation chambers dimension, **d** displacement characterizations, and **e** displacement tests at different air pressures

4bar, corresponding to the minimum and maximum values, the bioinspired actuator exhibits displacements of 8 and 32mm, respectively, with standard deviations of 0.94 and 0.89mm. The total elongation from the initial rest position corresponds to 33.68%. Additionally, the elongation of the body-part reaches 89.6mm, which is 10.65% bigger than the displacement of the bifurcation-parts (40.1mm), attributed to the increased number of actuator chambers.

4.2 *Shape Memory Alloys*

SMA actuator springs were purchased form Dynalloy Inc (Ca, United States) and shown the following properties: (i) austenitic starting temperature (TAs) equal to 90°C, (ii) external diameter of 3.45mm, and (iii) rest length equal to 45mm. Both used SMA springs were elongated to a total length of 84mm and crimped with a copper-wire using metal crimps. Two different SMA springs placements were tested: when embedded inside the pneumatic chambers (Fig. 3a), the bioinspired actuator exhibited lower bending performance compared to the configuration with externally embedded SMA springs. Therefore, the SMAs were embedded outside the chambers (Fig. 3b), resulting in enhanced bending performance.

After the PneuNet characterization, the SMAs were analysed studying: (i) input current, (ii) recovery time, and (iii) bending force (inside the actuator).

Before the embedding the SMA springs inside the actuator, the SMAs were characterized. Three different activation currents were tested: 3.2A according to Stano et al. [13], 3.4, and 3.6A, measuring for each input current the temperature of single the

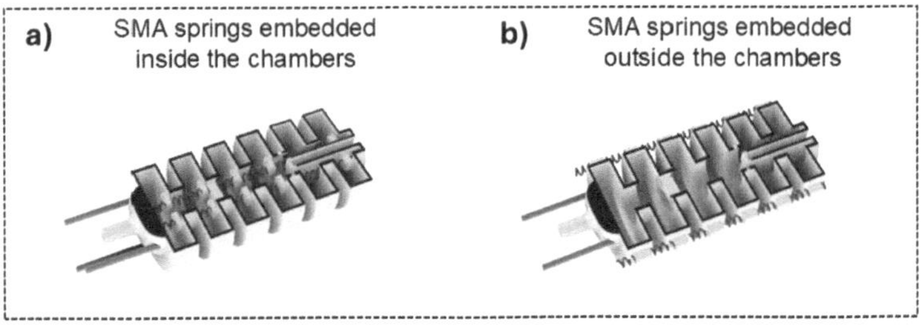

Fig. 3 Two different SMA springs placements: **a** inside-embedding placement, and **b** outside-embedding placement

SMA spring, the elongation, and the recovery time. For each current, the SMA started from a fixed rest position (elongation L of 45mm), under three different weights of 30, 40, and 50 g, respectively 0.3, 0.4, and 0.5N forces (see setup in Fig. 4a). A total of nine repetitions where performed. The results are shown in Table 3.

By applying current inputs, the temperature of the SMA spring ranges from 72°C to 96 °C, resulting in shape recovery times of 14.62±0.33, 16.03±1.20, and 23.02± 0.24 seconds, when 30, 40, and 50g loads were applied, respectively. The average recovery times at 3.4 and 3.6A decreased by 23.87% and 30.30%, respectively, when a 30g load was used. The 3.6A input current enables complete recovery of the SMA actuator spring, making it the optimal choice for maximizing recovery performance: under the maximum load (50g), the recovery time was the shortest compared to all tested input currents (32.97% faster than 3.2A). However, a current of 3.4A (with a recovery time 32.79% shorter than 3.2 A under 50g load conditions) achieves effective compression while reducing power consumption, offering a more energy-efficient alternative.

To characterize the bending performance of the SMA actuator springs, an ad hoc setup was used to evaluate the compliance factor of the embedded SMA springs: (a) an external two channels power supplied was used to increase the temperature of the SMAs (the current was fixed to 3.4A), (b) pneumatic air, and (c) digital camera to take pictures to calculate the displacement. Each SMA was tested three times alone and three times combined, for a nine total tests. The following test steps were performed: (i) activation of the SMA, (ii) vertical displacement evaluation, and (iii) SMA restoring using air pressure.

The bending performance were characterized for both inside- and outside-embedded (case 1 and case 2) SMA springs into the bioinspired actuator. As shown in Fig. 4c, for Case 1, the bending angles of SMA 1 and SMA 2 are 13.9°, 11.2°, and 9.8° (SMA 1) with 7.4° (SMA 2) when both springs are actuated. The maximum vertical displacement in Case 1 is 1.38mm when the SMA temperature reaches 110°C. Increasing the SMA temperature (with the same input current of 3.4 A) leads to an increasing of vertical displacement and bending angle of the actuator. For Case 2 (with SMA springs outside), the bending angles of SMA 1 and SMA 2 increase by 2.16% and 12.5%, respectively, due to the increase contraction of the SMA as it returns rest position (45mm). In Case 2, when both SMAs are actuated, the bending

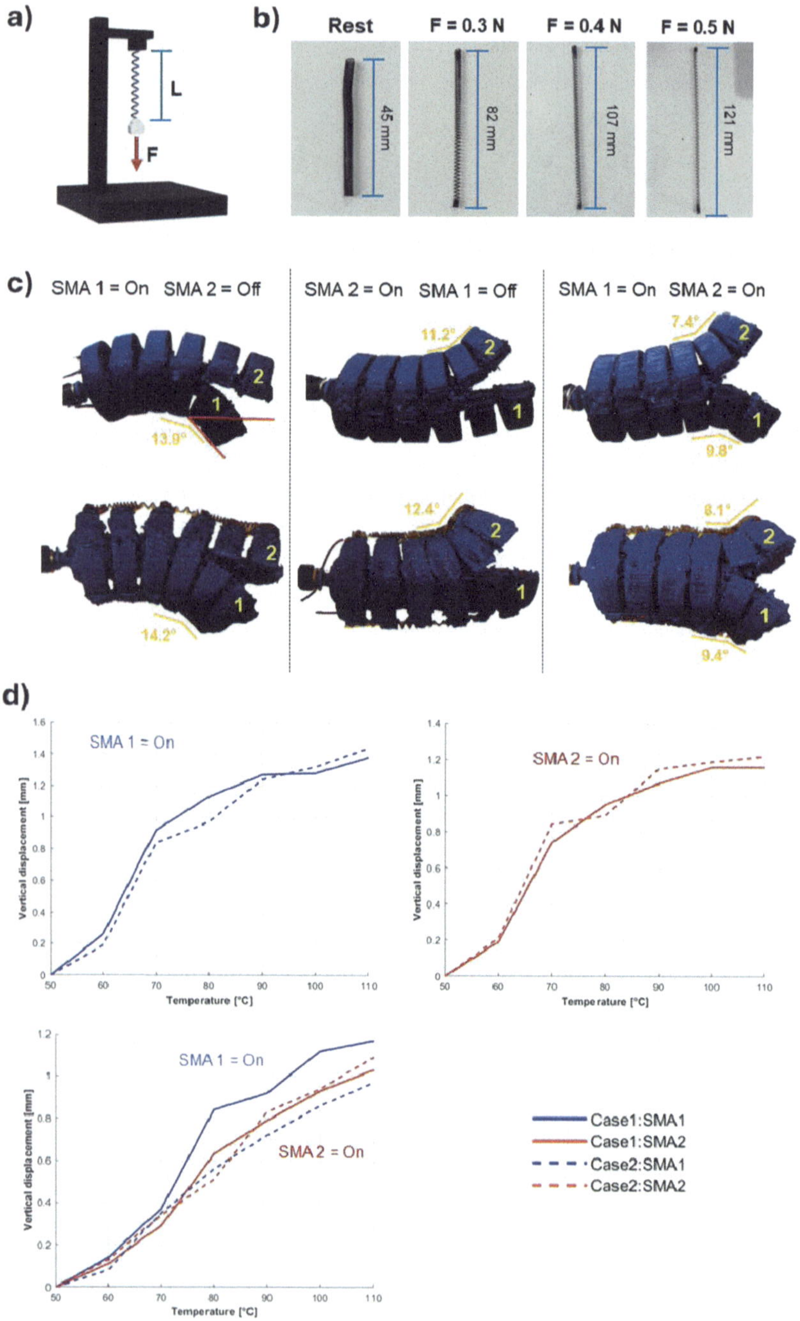

Fig. 4 SMAs characterization: **a** elongation test setup, **b** elongation under 30, 40, and 50 g load, **c** bending angle characterization of the actuator using SMA springs actuation (case 1 and case 2), and **d** vertical displacement of both cases 1 and 2

Table 3 SMAs characterization

Input current (A)	3.2			3.4			3.6		
Weight (g)	30	40	50	30	40	50	30	40	50
Recovery time (s)	14.62	16.03	23.02	11.13	13.27	15.47	10.19	12.73	15.43
Standard deviation	0.33	1.20	0.24	0.31	0.25	0.02	0.22	0.37	0.07
Temperature (°C)	83			97			117		
Elongation (mm)	82			107			121		

angle of SMA 2 increases by 9.46%, while that of SMA 1 decreases by 4.08%. The vertical displacement in Case 2 follows the same trend as the bending angle: when only SMA 1 is actuated, the displacement increases by 4.35%; when SMA 2 is actuated, the increase is 5.17%; and when both SMAs are actuated, the displacement increases by 5.83% for SMA 2, while SMA 1 shows a decrease of 17.09%. The results for vertical displacement are shown in Fig. 4d.

4.3 Multi-modal Actuation Systems

The multi-modal hybrid proposed approach offers several advantages consisting of a combination of pneumatic and SMAs actuation systems:

- Reducing recovery time of the SMAs without the need for antagonist elements, as the air itself acts as antagonist element.
- Increasing the level of complex motion, achieving movements that would not be possible with the use of a single actuation system.
- Omnidirectional complex movements.

To validate this, mobility (X and Y displacements) was performed testing several configurations of actuation (see Fig. 5a). Setting maximum tested pressure (4bar) while the springs are actuated, movements similar to the snake tongue is observed. The bifurcations gradually separate, while the body elongates toward its maximum expansion. Upon reducing the pressure, depending on the activated SMA springs, a reverse configuration is achieved. In this case, the actuator induces bending in the direction of the activated SMA. The SMA 2 exhibits a major banding than SMA 1, resulting in a displacement of 107 mm along the X-axis and 49 mm along the Y-axis (3 bar pressure, SMA 1 = Off, SMA 2 = On), as shown in Fig. 5b.

4.4 Embedded Sensing Systems

In order to obtain feedback during the motion of the bioinspired soft actuator, three independent 3D printed embedded strain gauges were monolithically fabricated.

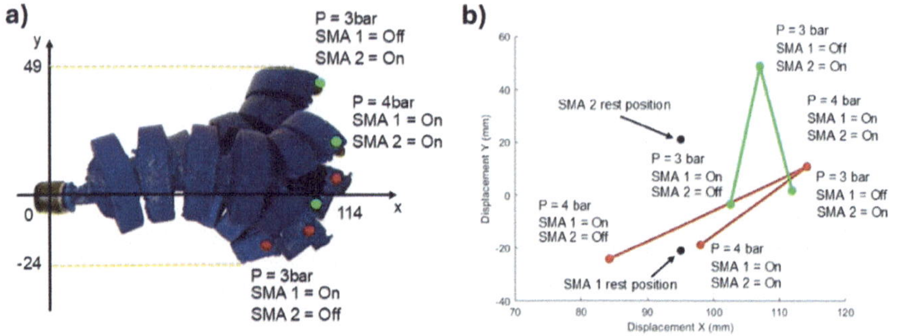

Fig. 5 Multi-modal approach characterization: **a** real complex movements, and **b** X and Y displacement using several configurations of pressure and SMA actuation

The sensors shape follows the perimeter of the TPU body-, and bifurcation-parts, with height of 0.6 mm (see Fig. 6a, b). In order to reduce the resistivity of the 3D printed sensors, two printing parameter were set: (i) single wall bead was 3D printed optimizing the nozzle path [13, 31], and (ii) the ironing printing parameter was employed [32, 33].

By utilizing the piezoresistive effect, the strain gauges exhibit a change in resistance in response to the mechanical strain (elongation and bending) applied during the actuation of the soft bioinspired actuator (see Eq. 1).

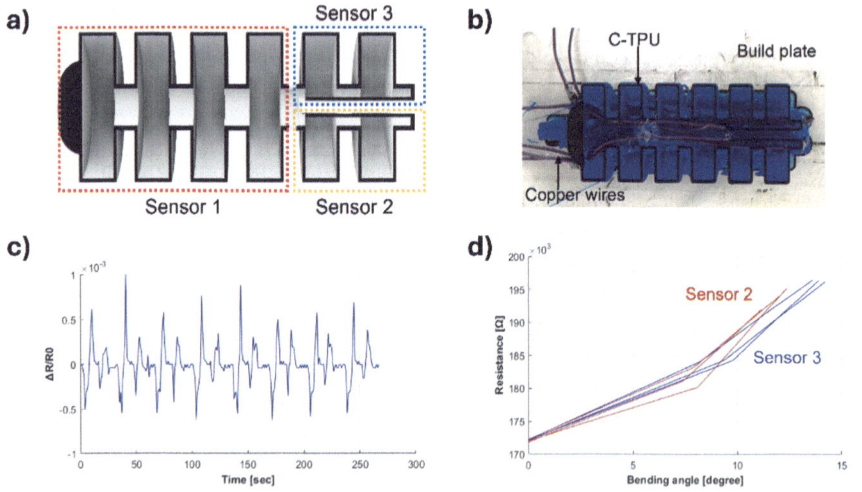

Fig. 6 Embedded 3D printed sensors: **a** CAD of three embedded sensors, **b** 3D printed actuator with sensors and copper wires (during the S&S), **c** sensor 1 characterization (during pneumatical actuation), and **d** sensor 2 and 3 characterization (bending angle vs. resistance)

$$\frac{\Delta R}{R_0} \tag{1}$$

The sensors feedback was characterized during both elongation and bending: the resistance variation, $\Delta R/R_0$, during elongation was calculated as the change in resistance when air pressure was insufflated into the chambers. In contrast, the resistance during bending was evaluated as the change in resistance occurring during the activation of the SMA springs.

As showed in Fig. 6c, the sensor 1 inside the body-part (pneumatic actuator), was tested suppling pressure air at 4 bar, then returning to 0 bar, for 230 s, completing a total of 8 activation cycles. The sensor's response remained almost constant, with a resistance R_0 of 159.48 kΩ and a normalized peak-to-peak amplitude of 4.6×10^{-4}.

It should be noted that increasing the pressure inside the chambers, the elongation of the 3D printed sensor 1 increases, and the resistance increased proportionally. An overview of the correlation between resistance and bending angle of the sensor 2 and 3 when the SMA springs were activated, is shown in Fig. 6d.

5 Conclusions

The present work proposes a cost-effective 3D printing approach for the fabrication of a novel omnidirectional soft robot with embedded sensors, based on two different actuation methods. An inexpensive Desktop Multi-Material extrusion 3D printer was used, zeroing post-processing and assembly tasks by recurring to the start-and-stop method through a custom.gcode script. The proposed omnidirectional soft robot is inspired by the snake tongue motion: pneumatic actuator and SMA springs actuator, allow complex motions of the robot, and a network of embedded 3D printed strain sensors provides real-time feedback.

The proposed framework paves the way for the fabrication of inexpensive, assembly-free, multi-actuation, and sensorized omnidirectional soft robots through common Desktop 3D printers. The proposed approach could be potentially used to fabricate soft robots for several fields, spanning from biomedical, to explorations allowing real-time feedback from the robots.

Funding The following research work was supported by the following funding sources: BRIEF—Biorobotics Research and Innovation Engineering Facilities—Missione 4,'Istruzione e Ricerca'—Componente 2,'Dalla ricerca all'impresa'—Linea di investimento 3.1,'Fondo per la realizzazione di un sistema integrato di infrastrutture di ricerca e innovazione', funded by European Union—NextGenerationEU, CUP: J13C22000400007.

National Recovery and Resilience Plan (NRRP), Mission 4 Component 2 Investment 1.3—Call for tender No. 341 of 15/03/2022 of Italian Ministry of University and Research funded by the European Union—NextGenerationEU. Award Number: PE00000004, Concession Decree No. 1551 of 11/10/2022 adopted by the Italian Ministry of University and Research, CUP D93C22000920001, MICS (Made in Italy—Circular and Sustainable).

Italian Ministry of University and Research under the Programme "Department of Excellence" Legge 232/2016 (Grant No. CUP—D93C23000100001).

References

1. Shintake J, Cacucciolo V, Floreano D, Shea H (2018) Soft robotic grippers. Adv Mater 30(29). https://doi.org/10.1002/adma.201707035
2. Al-Rubaiai M, Pinto T, Qian C, Tan X (2019) Soft actuators with stiffness and shape modulation using 3D-printed conductive polylactic acid material. Soft Robot 6(3):318–332. https://doi.org/10.1089/soro.2018.0056
3. Pavone A, Stano G, Percoco G (2023) One-shot 3D printed soft device actuated using metal-filled channels and sensed with embedded strain gauge. 3D Print Addit Manuf. https://doi.org/10.1089/3dp.2022.0263
4. Stano G, Percoco G (2021) Additive manufacturing aimed to soft robots fabrication: a review. Extrem Mech Lett 42:101079. https://doi.org/10.1016/j.eml.2020.101079
5. Yap YL, Sing SL, Yeong WY (2020) A review of 3D printing processes and materials for soft robotics. Rapid Prototyp J 26(8):1345–1361. https://doi.org/10.1108/RPJ-11-2019-0302
6. Walker S, Yirmibeşoğlu OD, Daalkhaijav U, Mengüç Y (2019) Additive manufacturing of soft robots. In: Walsh SM, Strano AP (eds) Materials. Woodhead Publishing, pp 335–359. https://doi.org/10.1016/B978-0-08-102260-3.00014-7
7. Khondoker MAH, Baheri N, Sameoto D (2019) Tendon-driven functionally gradient soft robotic gripper 3D printed with intermixed extrudate of hard and soft thermoplastics. 3D Print Addit Manuf 6(4):191–203. https://doi.org/10.1089/3dp.2018.0102
8. Stano G, Pricci A, Pavone A, Percoco G (2024) Electromagnetic assistance enables 3D printing of silicone-based thin-walled bioinspired soft robots. Addit Manuf 81:104028. https://doi.org/10.1016/j.addma.2024.104028
9. Peng H, Guimbretière F, McCann J, Hudson SE (2016) A 3D printer for interactive electromagnetic devices. In: UIST 2016—Proceedings of 29th annual symposium user interface software technology, no. October, pp 553–562. https://doi.org/10.1145/2984511.2984523
10. Zolfagharian A et al (2022) Silicon-based soft parallel robots 4D printing and multiphysics analysis. Smart Mater Struct 31(11):115030. https://doi.org/10.1088/1361-665X/ac976c
11. Stano G, Arleo L, Percoco G (2020) Additive manufacturing for soft robotics: design and fabrication of airtight, monolithic bending PneuNets with embedded air connectors. Micromachines 11(5). https://doi.org/10.3390/MI11050485
12. Bodkhe S, Vigo L, Zhu S, Testoni O, Aegerter N, Ermanni P (2020) 3D printing to integrate actuators into composites. Addit Manuf 35(January):101290. https://doi.org/10.1016/j.addma.2020.101290
13. Stano G, Ovy SMAI, Edwards JR, Cianchetti M, Percoco G, Tadesse Y (2022) One-shot additive manufacturing of robotic finger with embedded sensing and actuation. Int J Adv Manuf Technol 467–485. https://doi.org/10.1007/s00170-022-10556-x
14. Stano G, Pavone A, Percoco G (2024) 3D printing of shape memory polymers: embedding Nichrome-wires to enhance their performance BT. In: Carrino L, Galantucci LM, Settineri L (eds) Selected topics in manufacturing: emerging trends from the perspective of AITeM's young researchers. Springer Nature Switzerland, Cham, pp 3–18. https://doi.org/10.1007/978-3-031-41163-2_1
15. Yamamura S, Iwase E (2021) Hybrid hinge structure with elastic hinge on self-folding of 4D printing using a fused deposition modeling 3D printer. Mater Des 203:109605. https://doi.org/10.1016/j.matdes.2021.109605
16. Wang Z et al (2023) A magnetic soft robot with multimodal sensing capability by multimaterial direct ink writing. Addit Manuf 61(June 2022):103320. https://doi.org/10.1016/j.addma.2022.103320
17. Herianto, Mastrisiswadi H, Atsani SI, Sari WP, Tontowi AE (2024) Characterization of 3D printed multi-material soft pneumatic actuator. Int J Intell Robot Appl 0123456789. https://doi.org/10.1007/s41315-024-00395-2
18. Rossing L, Scharff RBN, Chömpff B, Wang CCL, Doubrovski EL (2020) Bonding between silicones and thermoplastics using 3D printed mechanical interlocking. Mater Des 186:108254. https://doi.org/10.1016/j.matdes.2019.108254

19. Zolfagharian A, Mahmud MAP, Gharaie S, Bodaghi M, Kouzani AZ, Kaynak A (2020) 3D/4D-printed bending-type soft pneumatic actuators: fabrication, modelling, and control. Virtual Phys Prototyp 15(4):373–402. https://doi.org/10.1080/17452759.2020.1795209
20. Sima Y et al (2024) An overview of 3D-printed shape memory alloys and applications in biomedical engineering. Adv Compos Hybrid Mater 7(5). https://doi.org/10.1007/s42114-024-00953-z
21. Van Humbeeck J (2018) Additive manufacturing of shape memory alloys. Shape Mem Superelasticity 4(2):309–312. https://doi.org/10.1007/s40830-018-0174-z
22. Yang X, Tan R, Lu H, Shen Y (2021) Starfish inspired milli soft robot with omnidirectional adaptive locomotion ability. IEEE Robot Autom Lett 6(2):3325–3332. https://doi.org/10.1109/LRA.2021.3062823
23. Hamidi A, Tadesse Y (2020) 3D printing of very soft elastomer and sacrificial carbohydrate glass/elastomer structures for robotic applications. Mater Des 187:108324. https://doi.org/10.1016/j.matdes.2019.108324
24. Alessi C, Bianchi D, Stano G, Cianchetti M, Falotico E (2024) Pushing with soft robotic arms via deep reinforcement learning. Adv Intell Syst 6(8). https://doi.org/10.1002/aisy.202300899
25. Rubies E, Palacín J (2020) Design and FDM/FFF implementation of a compact omnidirectional wheel for a mobile robot and assessment of ABS and PLA printing materials. Robotics 9(2). https://doi.org/10.3390/ROBOTICS9020043
26. Xavier MS, Tawk CD, Yong YK, Fleming AJ (2021) 3D-printed omnidirectional soft pneumatic actuators: design, modeling and characterization. Sens Actuators A Phys 332:113199. https://doi.org/10.1016/j.sna.2021.113199
27. Huang W, Xu Z, Xiao J, Hu W, Huang H, Zhou F (2020) Multimodal soft robot for complex environments using bionic omnidirectional bending actuator. IEEE Access 8:193827–193844. https://doi.org/10.1109/ACCESS.2020.3032983
28. Zou J, Lin Y, Ji C, Yang H (2018) A reconfigurable omnidirectional soft robot based on caterpillar locomotion. Soft Robot 5(2):164–174. https://doi.org/10.1089/soro.2017.0008
29. MacDonald E, Wicker R (2016) Multiprocess 3D printing for increasing component functionality. Science 353(6307). https://doi.org/10.1126/science.aaf2093
30. Song G, Kelly B, Agrawa BN, Lam PC, Srivatsan TS (2000) Application of shape memory alloy wire actuator for precision position control of a composite beam. J Mater Eng Perform 9(3):330–333. https://doi.org/10.1361/105994900770346006
31. Xiang D et al (2019) Enhanced performance of 3D printed highly elastic strain sensors of carbon nanotube/thermoplastic polyurethane nanocomposites via non-covalent interactions. Compos Part B Eng 176:107250. https://doi.org/10.1016/j.compositesb.2019.107250
32. Košir T, Slavič J (2023) Manufacturing of single-process 3D-printed piezoelectric sensors with electromagnetic protection using thermoplastic material extrusion. Addit Manuf 73(April):103699. https://doi.org/10.1016/j.addma.2023.103699
33. Stano G, Pavone A, Jafor MA, Matalgah K, Percoco G, Fleck TJ (2024) Enhancing the sensitivity of 3D printed sensors via ironing and void reduction. Virtual Phys Prototyp 19(1). https://doi.org/10.1080/17452759.2024.2331153

Dot-By-Dot Wire Arc Additive Manufacturing Finite Element Simulation for Thermal Cycle Prediction

Arturo Cioni, Gustavo H. S. F. L. Carvalho, and Niccolò Grossi

Abstract Wire arc additive manufacturing (WAAM) permits the production of high-performance, dense, metal structures of large size. Dot-by-dot WAAM is an innovative deposition strategy where short, interrupted weld beads (i.e., dots) are printed instead of the classic continuous bead to manufacture bars and lattice structures. This paper proposes a Finite Element simulation of the dot-by-dot WAAM process to accurately predict the heat-affected zone and interpass temperature, responsible for the geometry and quality of the product. The proposed approach, developed within the Simufact Welding environment, shows the influence of heat input, heat exchange conditions, material properties and idle time between dots. Multiple bars cases are investigated with variation on process parameters, dot size and idle time. Experimental validation on such cases proved the accuracy of the proposed approach in predicting the thermal cycle of the process and showed its impact on product quality.

Keyword Wire arc additive manufacturing (WAAM) · Thermography · Numerical modelling

1 Introduction

The current need to enhance engineering components, involving the refinement of material properties and the optimization of geometric configurations to achieve greater end-use performance, has necessitated the development and implementation of novel manufacturing processes and ad-hoc engineered fabrication methodologies. Wire Arc Additive Manufacturing (WAAM), a direct energy deposition (DED) technique, uses an electric arc to melt wire feedstock and build parts layer by layer [1]. WAAM offers key advantages over other metal AM methods, including higher deposition rates [2], lower part costs [3], and the ability to produce large-scale components [4]. It is widely applied in construction [5], shipbuilding [6], for build-up manufacturing [7] and repairs and custom alloy fabrication [8].

A. Cioni (✉) · G. H. S. F. L. Carvalho · N. Grossi
Department of Industrial Engineering, University of Florence, Firenze, Italy
e-mail: arturo.cioni@unifi.it

L. Fratini et al. (eds.), *Selected Topics in Manufacturing*, Lecture Notes in Mechanical Engineering, https://doi.org/10.1007/978-3-031-99501-9_3

A novel deposition strategy within WAAM technology is the so-called dot-by-dot (DbD) deposition. DbD is a WAAM technique that employs interrupted weld beads to achieve a vertical build. While productivity is generally lower than conventional WAAM, DbD WAAM allows for greater geometrical complexity such as lattice structures, which are unattainable with traditional WAAM [9]. Moreover, this deposition strategy allows greater control over the material's thermal cycles and its consequent properties. DbD retains the otherwise positive aspects of WAAM since it uses standard welding equipment and 100% of the stock material, reducing waste and lowering cost while being able to build very large structures.

However, DbD presents challenges in geometry control due to its sensitivity to thermal cycles, which are difficult to model because of changing geometries and heat transfer conditions. Furthermore, the rapid nature of the process limits thermal stabilisation, distinguishing it from conventional WAAM and highlighting the need for further research.

One of the main leading parameters of the entire process is the interpass temperature. It is defined as the layer temperature between depositions, i.e. the temperature of the preceding layer before the deposition of the subsequent layer. Unless the process is carefully modelled or monitored, heat accumulation leads to rapid increases in the interpass temperature, which leads to melting and build failure [10]. It is also fundamental in determining the thermal cycle and the resulting microstructure. If not considered by carefully selecting dot idle time (the time between each dot deposition) [11], it might lead to undesirable mechanical properties or even internal stresses [12]. Moreover, the interpass temperature plays a major role in the final geometry of the components due to the effect of temperature on the dimensions of the dot [13].

Industry standards for geometry prediction consist mostly of trial and error. Interpass temperature measurement can prove challenging as thermocouples are limited to substrate measurements, and are effective only for a limited number of layers and large idle times, such that the deposited material and the substrate can reach thermal equilibrium [14, 15]. Only non-contact methods can be employed to measure the just deposited material without interrupting the process, but there is little data on the temperature dependence of material emissivity, which determines how effective the material is at emitting thermal radiation involved in non-contact heat transfer, along with additional effects coming from the arc and from the melted weld pool [16]. Radel [17] developed a closed-loop system using a CCD camera and slicer to guide a robotic torch based on CAD comparison. Kozamernik [18] used a torch-mounted pyrometer to control interpass temperature in air-cooled layers, demonstrating its impact on part quality. Vazquez [19] applied thermography to dynamically adjust idle time, reducing build duration and defects.

An interesting solution for the study of this complex and often challenging-to-acquire information, and its correlation with the idle time, lies in its prediction and analysis via finite element method (FEM) simulation. This is generally done at the macro-scale level with thermal models being used to select optimal parameters by observing the time–temperature behaviours of the model, the biggest approximation is on the heat transfer between the electrical arc and material, which is modelled with an appropriate heat source [20]. Farias [21] conducted a sensitivity analysis on idle time in classical WAAM, using Artificial Neural Networks to generate a process map. Montevecchi [11] proposed a FEM-based idle time scheduler to maintain constant interpass temperature and prevent collapse. Current research efforts on FEM simulation of WAAM processes focus on the traditional continuous deposition approach. At the same time, there are very few works on DbD, and most aim to characterise the final products of the process instead of modelling the process itself, as in the work of Silvestru [22]. Regarding FEM model for DbD deposition, Wu [23] simulated thermal behaviour in aluminium DbD to estimate idle time, including a case study on a "Y" intersection. Mao [24] investigated how varying arc waveforms and inclinations affect thermal cycles and microstructure. However, these models lacked validation and calibration, limiting their ability to reduce experimental trials.

A significant gap is therefore found in studies regarding DbD WAAM deposition, in particular concerning calibrated and validated FEM modelling. Temperature data is essential for the accurate and precise parametrisation and prediction of the process and is crucial for improving and spreading the DbD technique, enabling the fabrication of increasingly complex components. The use of FEM obviates a series of experimental trials, but most existing models are either unvalidated or do not take into consideration the unique challenges of the DbD deposition, such as the comparatively extensive non-stable arc phase, the rapid thermal cycles and the difficulties in predicting part geometry. This paper proposes a model calibration process, which can be applied to geometries obtained through a DbD process, and an experimentally validated FE simulation of the dot-by-dot WAAM manufacturing of straight bars and an X shaped joint, with the aim of predicting thermal history and interpass temperature, crucial parameters for deposed material geometry, toolpath generation and quality of the part.

2 Proposed Approach

This work presents a simulation of the dot-by-dot WAAM process, including the procedure to calibrate the required parameters. Indeed, several input parameters are needed for implementing such model in the specific case: geometrical parameters, material thermal properties, heat source and heat transfer parameters. The latter are generally unknown, and a dedicated procedure is required for their identification. This work proposes a two-step calibration procedure to retrieve simulation parameters for a specific case to be applied to several scenarios. Using the calibrated parameters, the proposed approach is validated on nine different case studies of straight bars varying the dot diameter and idle time. Although such validation is essential to show the model's accuracy, the application of the proposed simulation model becomes even more crucial when a complex structure that includes inclined bars and intersections needs to be studied. To show the model's potential in such a scenario, its application to a structure composed of four inclined bars intersecting at one point is investigated. Figure 1 provides the general overview of the proposed approach, which will be detailed in the next section.

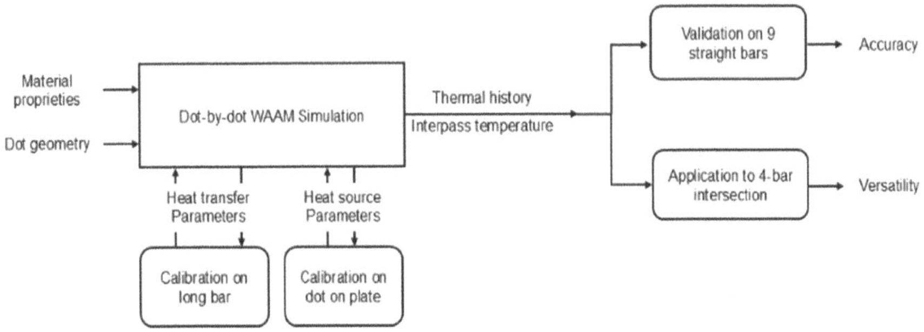

Fig. 1 Scheme of the proposed approach

The most adopted solution for thermal simulation of the WAAM process is based on the finite element method. This work will use the same methodology with some modifications to adapt the workflow to the DbD process. The software used for the simulation is Hexagon's Simufact Welding. The direct energy deposition suite was used, implementing a non-linear macro-scale thermal analysis. As in most other simulation software, the user must define the work geometry and meshes, the materials which can be imported through a library, the trajectories of the welding torch, the heat transfer coefficients, boundary conditions and the welding parameters. In this work, the deposition of dots was performed keeping the torch axis along the bar. This strategy leads to inclined layer lines on the geometry and as such, to guarantee a mesh that leads to reliable element activation, the geometry should be prepared beforehand by cutting along the plane defined by the dots' surfaces.

2.1 *Heat Source Model and Welding Parameters*

In order to model the effect of the electrical arc in a WAAM process, a Goldak double ellipsoid heat source is used as it is commonly employed in similar works [20]. The DbD process, which resembles a point welding process in the sense that there is no torch travel speed; therefore, the double ellipsoid shape of the Goldak model collapses in a hemisphere ($a = C_f = C_r$), as shown in Fig. 2.

All the geometrical characteristics must be defined, as well as the heat distribution function, which is assumed to be a Gaussian distribution. In Simufact, it is governed by the Gaussian parameter "M", which defines the steepness of the Gaussian bell curve and, therefore, the energy per area. Higher values will make the Gaussian bell curve sharper, while a value of 0 will give an all-out even energy distribution. The welding parameters consist of the power input and the velocity of the Goldak model. The power input is assumed to be a product of three factors: current, voltage and efficiency.

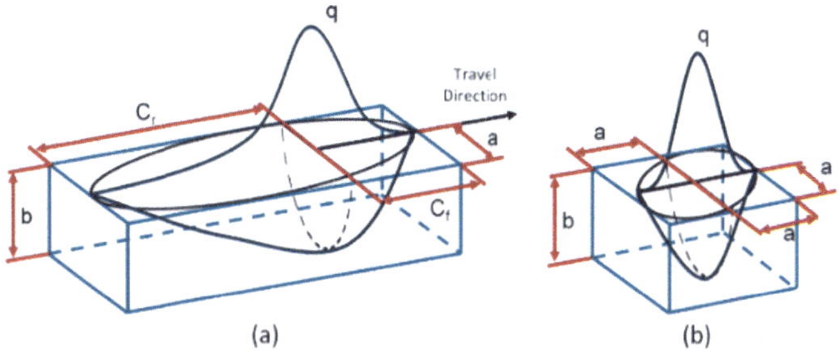

Fig. 2 Goldak heat source model **a** traditional WAAM, **b** DbD WAAM

Power input dictates how the process begins and evolves and must be input accurately. Unfortunately, this proves to be a challenge in the DbD process as the deposition time, where the arc is active, is short for each dot, and in turn, the time where the electrical arc is considered stable is even shorter. This means that for a considerable duration of the deposition time, a transient arc, with questionable power delivery capability, takes part in the material deposition. Therefore, typical efficiency values found in the literature (e.g. 0.8) do not represent the reality in DbD but are more suited to a continuous process, during which the transient part of the arc found at the beginning and at the end of the long deposition can be ignored. Hence, the user must calibrate the model to find the optimal efficiency value. Current and voltage, on the other hand, are characteristic of the process and must be obtained from experimental data.

The velocity is tied to the trajectories the heat source must follow during a Simufact simulation. Since the welding torch does not physically move during a DbD deposition, a fictitious trajectory at least two orders of magnitude smaller than the dot diameter can be implemented without a significant difference in results. The travel speed of the heat source must therefore be set in order to match the desired welding time. For example, if a welding time of 1 s is required and the fictitious trajectory is defined as 0.001 mm long, then the velocity must be set at 0.001 mm/s.

2.2 *Trajectories*

Heat source trajectories are defined through point coordinates and an associated unit vector for torch orientation. This information can be imported from a ".csv" file. To avoid undesired element activation, the trajectory should lie on the plane identified by the torch orientation vector and the corresponding point coordinate. Each trajectory has an associated welding time, defined by trajectory length and heat source velocity, and an associated idle time, which can be defined either manually or by editing the program's ".csv" files. During the welding time, the heat source is active, heat is input into the model and elements get activated. The idle time governs how long the system waits between welding times while the model cools down according to the thermal properties of the material, the heat transfer coefficients and the boundary conditions.

2.3 Heat Transfer Coefficients

The heat transfer coefficients assigned to the model are fundamental to the final results of the simulation. They consist of two parameters: the convective heat transfer coefficient (which should be modelled as temperature dependent), and the emission coefficient (which governs radiative heat transfer). The heat transfer coefficient's temperature dependency was chosen in accordance with Incropera [25] with Eq. (1) valid for natural laminar flow for vertical cylinders and Eq. (2) for rectangular plates:

$$h_{dots} = \alpha * (T - T_{amb})^{\frac{1}{4}} \tag{1}$$

$$h_{sub} = \beta * (T - T_{amb})^{\frac{1}{4}} \tag{2}$$

where "h_{dots}" and "h_{sub}" are the convective heat transfer coefficients, "T" and "T_{amb}" identify the nodal and ambient temperature, respectively, and both "α" and "β" are coefficients that can be modified to obtain the best-fit curves to experimental data to choose final parameters after calibration. The emission coefficient must be obtained in the same manner.

3 Materials and Methods

3.1 Experimental Setup

The calibration and validation experiments were conducted on a DMU 75 monoBLOCK 5-axis machine equipped with a custom-made holder for a Fronius 320i welding torch. Deposition was performed using the GMAW process with Cold Metal Transfer (CMT) technology in spot-welding mode, which was enhanced by a gap-bridging feature. Shielding gas was a mix of 82% Argon and 18% CO_2 at 15 L/min, with pre-gas and post-gas times set to 0.5 and 2 s, respectively.

Temperature data was collected using an Optris PI 640i infrared thermal camera and managed via Optris PIX Connect software. Arc activation signals from the welding equipment were also recorded to precisely determine deposition times. The camera was positioned horizontally for calibration and at a 45° angle for validation.

The deposition material was 1 mm diameter AWS FILCORD ER70S-6 low-carbon steel wire, while the substrate consisted of 5 mm thick EN10025:2 S235JR low-carbon steel plates, suspended to guarantee convective heat dissipation. Metallographic samples were sectioned to the dot centre and prepared according to ASTM E3 standards. Etching was performed with 2% Nital, and microstructural analysis was carried out using a Nikon Eclipse LV150 optical microscope.

3.2 Preliminary Analysis

A preliminary analysis was carried out in the simulation environment to understand the qualitative impact of the different parameters on the simulation results. Two datasets were analysed; the first is the maximum temperature, which is reasonable to assume is taken from elements close to the deposition surface for a bar-shaped deposition. The second dataset is the substrate temperature. In the case of a bar substrate, the observed node lies at a 9 mm distance from the top of the substrate, in the middle of the bar. Such a node was chosen because experimentally it was observed that a distance closer to the top of the substrate would be affected by slag in such a manner as to modify material emissivity. Figure 3 shows the effects of changing the simulation's parameters in terms of substrate temperature. It can be observed from Fig. 3 how α, which determines the "h_{dots}" values, has noticeable effects on the developed process for final cooling behaviour, while it does not impact the initial dots. The efficiency value, which governs effective heat input into the simulation elements, significantly affects temperatures for the entire process, as expected. The layer height, which determines the position of the heat source and the slicing of the geometry, noticeably affects the substrate temperature. Since the layer height determines the geometry input into the simulation, correctly predicting the layer height is important in obtaining an effective simulation process. The emissivity value "*irr*" does not have a significant impact. The preliminary analysis shows that, once the geometry is correctly included, the most critical parameters in the calibration study are the convective heat coefficient and the efficiency value. The identification of such parameters will be detailed in the next subsections.

3.3 Calibration

In order to set up the simulation model, several parameters must be identified. This procedure is not trivial since the parameters are often inter-dependant and most of them are temperature dependent. In this work, a calibration procedure is proposed, composed of two experiments: "Dots on Bar" test and "Dot on Plate" test.

Dots on Bar: The "Dots on Bar" process involves depositing new "hot" dots onto a pre-existing bar at room temperature, created using the same method. As shown in Fig. 4 he bars are sufficiently long to neglect substrate effects during new dot deposition; therefore, "h_{sub}" does not influence the temperatures, leaving the heat flow to be

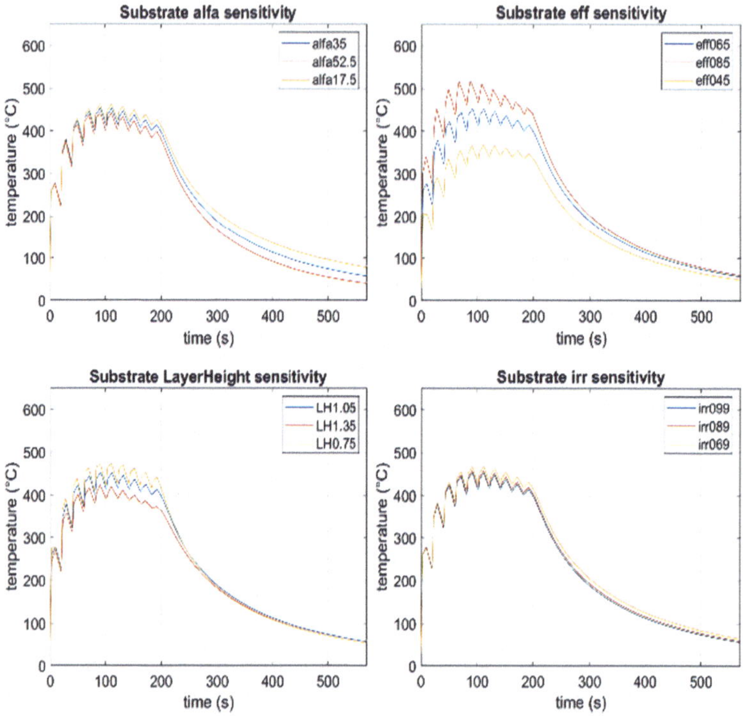

Fig. 3 Sensitivity analysis on Simufact input parameters. Layer height in mm

controlled only by "α", the efficiency and the emissivity, which is defined as the ratio of the thermal radiation emitted from the body against the thermal radiation emitted from an ideal black body. A simulated model emulates the process, and thermal parameters are obtained by fitting the simulated and experimental temperature of the substrate (the bar, in this case). A single geometry is chosen with a dot diameter of 6 mm, as detailed in Table 1, with the description of plate dimensions. The resulting parameters will be used in other configurations in the validation section.

The experiment was carried out on a single deposited geometry configuration with a 6 mm dot diameter. Temperatures are measured on the substrate (i.e., bar), opportunely matte-black painted, at a 9 mm distance from the substrate's top surface. The thermal camera is positioned in a horizontal position and perpendicular to the measurement point, as recommended by the manufacturer. This approach was selected to obtain a robust and accurate measurement of the temperature because the emissivity of the bars is not easy to set up since it is influenced by roughness and temperature, while matte-black paint allows for accurate measurements using a 0.99 emissivity value, which has been confirmed through preliminary tests. Table 2 presents the experiment parameters.

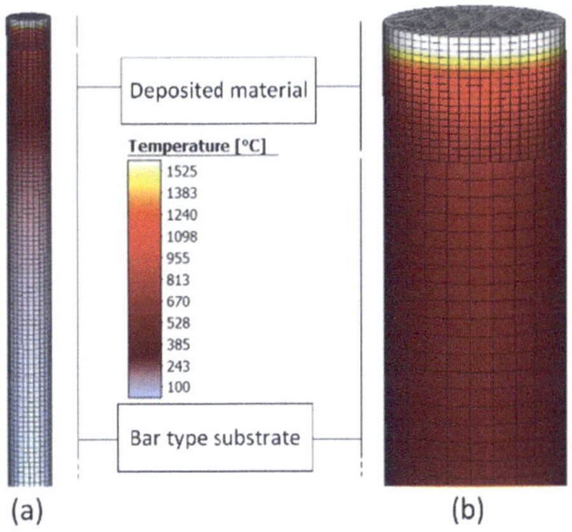

Fig. 4 Dots on bar calibration process representation. **a** 6 mm bar, **b** deposition zone close-up of 6 mm bar

Table 1 Calibration deposition geometry data

Calibration geometry	Geometry	Number of dots	Height (mm)	Diameter/width (mm)	Length (mm)
Dots on bar	Dot	10	10.5	6	–
Dots on bar	Bar	–	–	179.55	6
Dot on plate	Plate	–	5	72	100

Table 2 Calibration experimental test parameters

Geometry	Number of dots	Nominal welding time (s)	Effective welding time (s)	Nominal idle time (s)	Wire feed speed (m/min)
Dots on bar	10	0.6	0.74	40	6

The meshes used for the simulation used the parameters shown in Table 3. Note that all the meshes have been defined based on convergence analysis for the investigated results.

The materials used experimentally have an equivalent in Simufact's library of materials: SG1-JMP-MPM was chosen for all the deposited geometry and the long cold bar substrate, and S235-SPM was chosen for the rectangular square substrate.

Table 3 Meshing data for calibration tests

Geometry	Mesh type	Element order	Element size (mm)
Dots	Hexa	1	0.35
Bar	Hexa	1	0.8
Rectangular plate	Hexa	1	1

Table 4 Goldak calibration experimental tests parameters

Geometry	Number of dots	Welding time (s)	Idle time (s)	Wire feed speed (m/min)
Dot on plate D4	1	0.32	20	4
Dot on plate D6	1	0.74	20	6
Dot on plate D8	1	1.15	40	8

Dot on Plate: The geometrical parameters of the Goldak heat source model were calibrated using the "Dot on Plate" test. In this experiment, dots of three different diameters, D4, D6, and D8, were deposited onto a rectangular plate. Microstructural analysis was used to identify the heat-affected zones (HAZ), which indicate the maximum temperatures reached during deposition. The model then simulated the HAZ extent, and a fitting procedure was applied by adjusting the heat source's geometrical and Gaussian parameters to match the experimental HAZ profiles. The results of these tests are summarised in Table 4, and the final parameter set was selected based on the best fit across all dot sizes.

3.4 Validation

Once the thermal parameters and Goldak heat source model were fully calibrated, the simulation was validated across nine distinct test cases. These involved depositing new straight bars onto the same type of rectangular plate substrate used during calibration. Each bar varied in process parameters, dot diameter, and number of dots. For each configuration, three different idle times were tested, resulting in nine validation cases detailed in Table 5. The same experimental setup as in the calibration phase was used, except for the position of the thermal camera, which was set to be able to measure the interpass temperature accurately (Table 5). For each bar, the relative calibrated Goldak geometry and the calibrated thermal parameters were used. As a first approximation, the substrate geometry convective heat transfer coefficient "h_{sub}" was assumed equal to the convective heat transfer coefficient "h_{dots}". Table 5 summarises the geometrical parameters for the simulation.

Table 5 Validation tests geometry, simulation and process parameters

Validation geometry ID	Ø (mm)	Dots	Total height (mm)	Idle time (s)	Welding time (s)	WFS (m/min)	Mesh type	Element order	Element size (mm)
A1-A2-A3	4	100	51	10–15–20	0.32	4	Tetra	2	0.35
B1-B2-B3	6	60	63	20–30–40	0.74	6	Tetra	2	0.35
C1-C2-C3	8	60	87	40–30–50	1.15	8	Tetra	2	0.5
Substrate	–	–	5	–	–	–	Hexa	1	1

4 Results

4.1 Calibration

Dots on Bar: Calibration was essential to determine the optimal thermal parameters for process simulation. As shown in Fig. 5, the calibrated model closely replicates experimental temperature behaviour, particularly during the initial heating and early cooling phases. This indicates that the selected heat source efficiency accurately represents the arc's thermal input. Table 8 summarises the optimal parameter set, which will be used for all the following simulations**.** It can be argued that while the chosen "α" parameter defines a "h_{dots}" that adequately models the thermal behaviour, a difference between the experimental and simulated curve can be observed during the final cooling part. This leads to the conclusion that the temperature dependency of the heat exchange parameter is not perfectly accurate to reality, most likely due to the imperfect temperature dependence that was chosen for the "**α**" parameter, but for the proposed use case it is deemed acceptable. The more extreme spikes and valleys in the simulated model results can be explained by the fact that while the simulation uses nodal measurements, the experimental values are derived from a small average area around the maximum temperature point identified by the thermal camera.

Dot on Plate: The fitting process led to the choice of the geometrical and Gaussian parameters shown in Table 6.

Figure 5 presents a comparative analysis of the HAZ obtained through microstructural examination and the simulated temperature distribution isotherms, visually represented by colour areas. The delineation of HAZ boundaries in the microstructural analysis was estimated on a correlation between observed microstructural transformations and established approximate critical temperatures derived from the respective chemical composition in the Fe–C phase diagram. For instance, the lower critical temperature, i.e. the initiation of austenite formation upon heating at approximately 723 °C, the approximate temperature for complete austenitisation upon

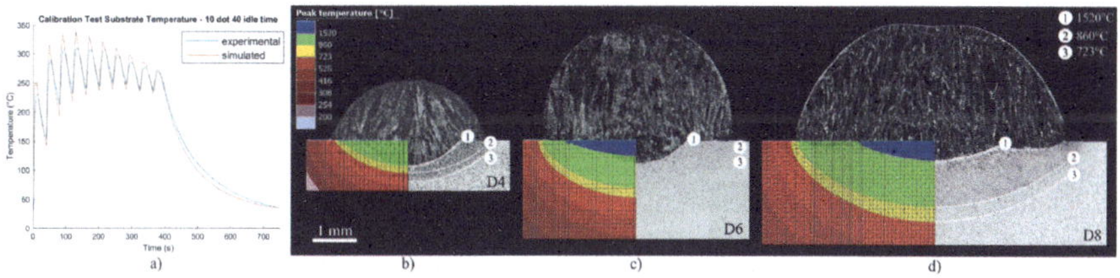

Fig. 5 **a** Calibration results in terms of interpass temperature and comparison between the peak temperatures obtained from the simulation (coloured regions) and the HAZ actual microstructure. The white lines delineate the different critical regions within the HAZ microstructure. **b** 4 mm dot, **c** 6 mm dot, **d** 8 mm dot

Table 6 Final Goldak parameters and associated HAZ results

Geometry	$a = C_f = C_r$ (mm)	b (mm)	Gaussian parameter (mm)	> 723 °C radius (mm)	> 723 °C depth (mm)	> 860 °C radius (mm)	> 860 °C depth (mm)	Average error (%)
Dot on plate D4	2.1	1.3	1	1.7	0.75	1.6	0.6	– 24.3
Dot on plate D6	6	1.2	1	2.3	1.3	2.2	1	+ 9.8
Dot on plate D8	5	2.5	1	3.54	1.90	3.32	1.50	+ 16.6

heating, and the melting point. These empirical observations were subsequently juxtaposed with the corresponding HAZ simulations.

The simulation provided a reasonable approximation of the HAZ extent for the smallest dot diameter, though with limited precision. Accuracy improved significantly with larger dots, where simulated HAZ boundaries—based on critical temperature isotherms—closely matched microstructural observations, confirming the model's predictive strength. This improvement is due to the greater thermal predictability of larger dots. In contrast, smaller dots are more complex to model accurately because of their lower mass relative to the substrate, which increases variability in cooling rates and thermal cycles. Despite these challenges, the model showed high fidelity in replicating experimental results for larger dots, demonstrating its effectiveness in predicting HAZ formation in dot-by-dot welding.

4.2 *Validation*

Experimental results of the temperature distribution acquired by the thermal camera were compared with simulation results. Since the goal of the model is to predict interpass temperature, results in terms of this variable are analysed. Figure 6 shows the validation results. Table 7 reports the root mean square percentage error (RMSPE) between simulated and experimental interpass temperature, acquired in the steady part at the end of the test, where the temperature reaches a stable value for all the tests.

Looking at the experimental results, all the tests exhibit the same trend, as expected. Three zones can be highlighted: (i) a slow increase of temperature, (ii) a constant temperature increase till approaching the maximum value, (iii) a steady zone where interpass temperature does not change significantly. The first zone is highly influenced by the substrate that drains the heat generated. In the second zone, the constant idle time is insufficient to level interpass temperature, while in the third zone, the balance between heat generated and dissipated is achieved, and a steady process begins.

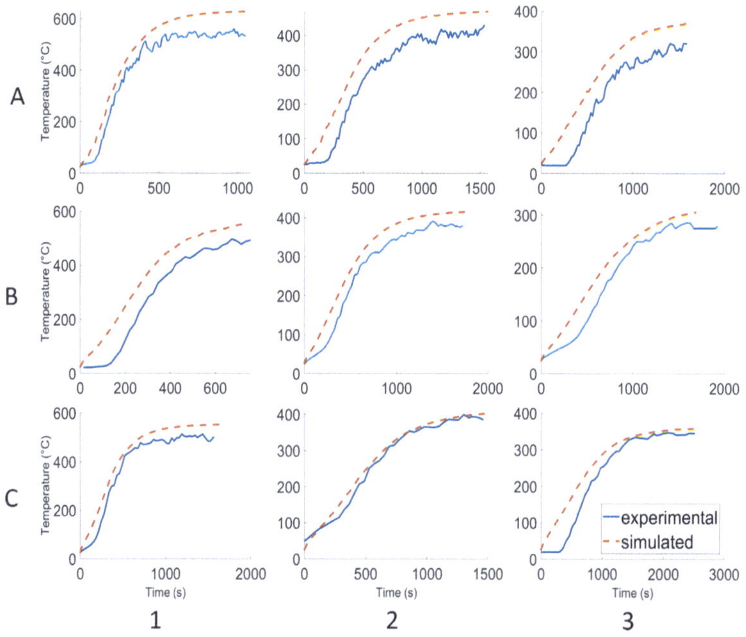

Fig. 6 Validation test of the interpass temperature results

Table 7 Validation tests steady interpass temperature and root mean square percentage error

Steady interpass temperature	A1	A2	A3	B1	B2	B3	C1	C2	C3
Experimental (°C)	534.0	431.4	303.9	492.1	380.7	277.9	502.8	385.7	346.2
Simulation (°C)	630.4	470.9	373.6	554.1	416.6	306.7	553.5	403.7	359.5
RMSPE (%)	12.6	8.06	11.23	8.91	6.69	7.97	7.04	1.25	2.65

The simulation effort successfully follows the experimental interpass temperature trend in all proposed test cases, independently of varying dot diameter or idle time between depositions. Good accuracy is achieved in the steady zone with an error below 10% for almost all the tests on steady Interpass Temperature prediction. The most significant difference between the simulated and experimental curves can be observed in the process's initial phase. Here, the substrate heat exchange modelling plays a crucial role. The assumption that the heat exchange coefficients are the same between the vertical dots and the horizontal plate substrate may not be accurate enough to represent the initial part of the process adequately. A more comprehensive model of the substrate could take into consideration this aspect, even if, in the first phase of the process, the substrate does not participate in the heat exchange process in its entirety: a considerable portion of the substrate's surfaces remain close to ambient temperature meaning that the surfaces interested in heat exchange do not resemble a rectangular plate. Some of the discrepancies could also be related to the thermal camera measurements since the emissivity of the hot deposited material is not easy to estimate.

In addition, the extraction of interpass temperature is not trivial since it relies on the zone analysed. While this is true of all the tested cases, the effect is more pronounced in the experiments with smaller diameters, as observed from the higher average RMSEP of the A1-A2-A3 cases. This could be explained by the smaller available surface area to extrapolate infrared data, making it more challenging. Furthermore, minor defects due to the physical process, such as a variation in effectively deposited dot diameter or the effective layer height, play a much more impactful role on smaller geometries, where the percentage error is higher than on larger geometries. This could explain why the best results of the validation process are obtained for the largest geometry investigated, specifically the tests C1, C2, and C3, each with 8 mm diameter dots. Another aspect that could be explored is the effect of modelling pre and post-gas in the simulation. The heat exchange models proposed assume natural laminar flow, but this is not entirely representative of the heat change effects caused by a gas impingement present before, during, and after the welding phase. An adequate modelling of such a phenomenon will likely lead to lower simulated interpass temperatures.

5 Application to Interpass Temperature

Having observed the simulation model's performance on simple straight bars, a more complex configuration that could be implemented as a practical application in the manufacturing of lattice structures derived from interconnected bars is proposed. The structure consists of the joint of two inclined bars that meet forming an "X" shape joint.

The bars were deposited and simulated using the parameters of the validation "B" bar, and the same rectangular substrate was used. Table 8 summarises the parameters, where "$\boldsymbol{\alpha}$" and "eff" have been Table 8obtained through calibration efforts, while "irr" has been taken from existing literature regarding emissivity of matte black painted surfaces.

The constant idle time does not allow the process to have a constant interpass temperature, leading to worse control over the resulting geometry. Therefore, three

Table 8 Simulations coefficients

Coefficient	Description	Source	Value
α	Parameter governing convective heat exchange	Calibration	8.5
eff	Heat input efficiency	Calibration	0.6
irr	Emissivity	Literature	0.99

idle time sets have been used for different sections of the process. Up until the joint, each bar is independent; as such, they are similar to the B1 case: idle time is set as 20 s per bar. When the bars start to join, they are treated like a single bar: dots are deposited along the vertical direction with a 20-s idle time between deposition. When the joint section is complete and the bars start to diverge, a more conservative approach to avoid melting is preferred: 20 s of idle time is imposed between each dot, which means 2 consecutive dots on the same bar are deposited 40 s apart from one another, more like B3. Table 9 shows the data used in the X joint case simulation.

The simulation will use the thermal and heat source parameters found in the calibration process. Figure 7 shows the results of the "X" joint case in terms of Interpass Temperature.

The interpass temperature behavior reveals four distinct phases. Initially, it mirrors the B1 case: two independent bars formed by 6 mm dots with 20-s idle intervals. The only difference is a warmer substrate due to dual-bar deposition. As the bars begin to join, thermal behavior shifts—heat from a single deposition now dissipates through both bars, causing a rapid temperature drop. Once the joint is complete and the bars diverge, the temperature declines more gradually, with the joint acting as a heat reservoir. Eventually, the bars become thermally independent again, and the interpass temperature stabilizes at a lower level due to effectively doubled idle time.

The simulation closely follows the interpass temperature trend during the initial and final stages. In the mid-section where the joint forms and the bars diverge, the behaviour becomes complex. Experimentally, a temperature drop is observed

Table 9 Geometry, meshing and simulation data for X joint application

	Diameter (mm)	Number of dots	Layer height (mm)	Mesh type	Element order	Element size (mm)	Idle time (s)	Welding time (s)
X joint	6	139	1.1	Tetra	2	0.5	20–40	0.74

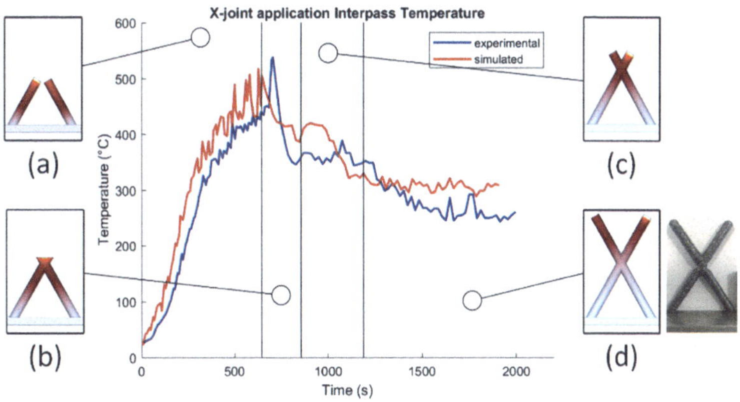

Fig. 7 "X-joint" interpass temperature comparison

immediately after the joint is completed, followed by a second dip once the bars are far enough apart to behave independently. This lower temperature is due to the increased idle time per bar.

The simulation captures this trend but predicts the second dip earlier than observed. This suggests the model underestimates the joint's thermal influence duration. A more detailed geometry and thermal formulation could improve accuracy in this region. Nonetheless, the model effectively reflects experimental trends and values, making it a reliable tool for idle time optimisation and toolpath planning. It can support the development of idle time control routines to enhance interpass temperature control, part geometry, and overall productivity.

This procedure can be extended to complex alloys where interpass temperature must be controlled to avoid property deterioration. This is critical for alloys like titanium (reactive with oxygen at high temperatures), duplex stainless steel (risk of sigma phase formation), or high-hardenability low-alloy steels (risk of cold cracking from martensite formation).

6 Conclusions

Dot by Dot (DbD) is a specialized form of Wire Arc Additive Manufacturing (WAAM) that uses spot welding to deposit discrete material drops rather than continuous beads. This study introduces an experimentally validated simulation framework to predict the thermal behaviour of the DbD process, with a focus on interpass temperature—the temperature just before each new dot is deposited—which significantly influences the final component geometry. A calibration method identifies key simulation parameters, which are then validated across various test cases with differing geometries, dot counts, and idle times. The model consistently predicts interpass temperatures with acceptable accuracy. A complex application involving an X-shaped joint formed by intersecting bars is shown as an application test case. While the intersection presents challenges due to intricate heat transfer dynamics, the simulation captures the transition between converging and diverging bars, demonstrating its potential for idle time optimisation and toolpath planning. The following conclusions can be drawn:

- Heat exchange parameters, particularly the convective heat exchange parameter, have a dominating role in determining the interpass temperature profile and must, therefore, be accurately modelled.
- The simulation environment is a useful tool to predict Interpass Temperatures of DbD WAAM depositions and can be applied even in complex cases.
- Future works regarding lattice structures obtained through joined bars can use the proposed simulation framework to anticipate joint thermal behaviour and select adequate idle times to improve productivity and part quality.

- The simulation of interpass temperature provides valuable insights regarding applications where interpass temperature control is imperative, such as for titanium alloys, duplex stainless steels, and nickel alloys.

Funding Part of the research activities were conducted within the CIRCULAR WELD PRIN 2022PNRR project funded by the European Union—Next Generation EU, Mission 4 Component 1 CUP B53D23027320001.

References

1. Singh SR (2021) Wire arc additive manufacturing (WAAM): a new process to shape engineering materials. Mater Today Proceed 44:118–128
2. Wang C (2021) The effect of wire size on high deposition rate wire and plasma arc additive manufacture of Ti-6Al-4V. J Mater Process Technol 288
3. Dias M (2022) Economic and environmental potential of wire-arc additive manufacturing. Sustainability 14(9)
4. Jafari D (2021) Wire and arc additive manufacturing: opportunities and challenges to control the quality and accuracy of manufactured parts. Mater Design 202
5. Sian IEJ, Wang (2023) A preliminary study of deposition rate, material property and stability of WAAM stainless steel plates. Adv Steel Const 19(1):23–30
6. Nota C (2023) A first feedback on manufacturing and in-service behaviour of a WAAM-made propeller for naval application. Welding in the World
7. Campatelli G (2021) Design and testing of a WAAM retrofit kit for repairing operations on a milling machine. Machines 9(12)
8. Rodrigues TA (2022) Steel-copper functionally graded material produced by twin-wire and arc additive manufacturing (T-WAAM). Mater Design 213
9. Arrè L (2023) Wire-and-Arc Additive Manufacturing for lattice steel structures: overview of the experimental characterization on dot-by-dot rods. Proceed Civil Eng 6(3–4):757–762
10. Montevecchi F (2018) Heat accumulation prevention in wire-arc-additive-manufacturing using air jet impingement. Manufact Lett 17:14–18
11. Montevecchi F (2018) Idle time selection for wire-arc additive manufacturing: a finite element-based technique. Addit Manuf 21:479–486
12. Penot C (2023) Interpass temperature effects on WAAM ER316L stainless steel corrosion using potentiostatic pulse tests. npj Mater Degrad 7(1)
13. Williams S (2024) Interpass temperature impact on bead geometry of mild steel in wire-arc additive manufacturing. Manuf Lett 41:992–997
14. Ding J (2011) Thermo-mechanical analysis of Wire and arc additive layer manufacturing process on large multi-layer parts. Comput Mater Sci 50(12):3315–3322
15. Baier D (2022) Thermal process monitoring and control for a near-net-shape wire and arc additive manufacturing. Prod Eng Res Devel 16(6):811–822
16. da Silva L, Joao, R, Ruham P, Scotti A (2019) The potential of IR pyrometry for monitoring interpass temperature in wire + arc additive manufacturing. Evol Mech Eng 3(1)
17. Radel S (2019) Skeleton arc additive manufacturing with closed loop control. Addit Manuf 26:106–116
18. Kozamernik N (2020) WAAM system with interpass temperature control and forced cooling for near-net-shape printing of small metal components. Int J Adv Manuf Technol 110(7–8):1955–1968
19. Vazquez L (2023) Maximising the deposition rate of 5356 aluminium alloy by CMT-twin-based WAAM while reducing segregation-related problems by local IR thermography. Metals 13(11)
20. Sampaio RFV (2023) Modelling of wire-arc additive manufacturing—a review. Adv Ind Manuf Eng 6

21. Farias FWC (2021) Prediction of the interpass temperature of a wire arc additive manufactured wall: FEM simulations and artificial neural network. Add Manuf 48
22. Silvestru V-A (2021) Performance under tensile loading of point-by-point wire and arc additively manufactured steel bars for structural components. Mater Design 205
23. Wu B (2021) Robotic skeleton arc additive manufacturing of aluminium alloy. Int J Adv Manuf Technol 114(9–10):2945–2959
24. Mao H (2025) Effects of arc waveforms on morphology, microstructure, and properties of unsupported bars in wire and arc additive manufacturing. J Mater Process Technol 337
25. Frank P Incropera DPD, Bergmann, TL, Lavine AS (1996) Fundamentals of heat and mass transfer. Wiley

Micro-Geometrical Evolution of Titanium Surfaces by Dry Electropolishing

N. Burattini, L. Bottini, and A. Boschetto

Abstract Dry electropolishing is a metal finishing process that employs an ion transport mechanism through co-polymers particles retaining a small amount of sulfonic or metanosulphonic acid. It offers advantages over traditional liquid electropolishing, such as reduced chemical waste, more uniform polishing, and no part shape variation. The tuning of the process parameters is still challenging, and an in-depth investigation is necessary to evaluate their effect on the surface morphology. The work analyzes the micro-geometrical evolution of the profiles through a particular alignment of periodical surfaces taking into account different starting morphologies and investigating the effect of the processing current times. Morphological analysis is addressed to the understanding of the crucial role played by the initial roughness on the electropolishing capability, especially in the case of Additively Manufactured surfaces characterized by strong anisotropy. A statistical modeling of the roughness percentage reduction is developed to determine a relationship with the investigated factors. Sets of favorable combinations of processing parameters are found on simple geometry and applied to a case study characterized by curved and intricated internal surfaces.

Keywords Electro-physical and chemical processes · Surface analysis · Dry electropolishing

1 Introduction

Electropolishing (EP) is a non-contact and damage free process widely employed to treat metal and alloy-based parts by an anodic dissolution process, in which the material is removed ion by ion from the surface asperities [1]. The part serves as the anode, and it is immersed into an electrolytic bath together with the cathode. Power activation produces an electrical current that moves from the anode to the cathode, resulting in the oxidation of the metal surface and in the removal of surface

N. Burattini (✉) · L. Bottini · A. Boschetto
Sapienza University of Rome, Via Eudossiana 18, ITA, Rome, Italy
e-mail: burattini.1962781@studenti.uniroma1.it

L. Fratini et al. (eds.), *Selected Topics in Manufacturing*, Lecture Notes in Mechanical Engineering, https://doi.org/10.1007/978-3-031-99501-9_4

irregularities, which are dissolved in the electrolyte and diffuse through the film to the cathode at a controlled rate [2]. This process is widely used for its flexibility, productivity [3] and applicability to very complex geometries such as parts fabricated by Additive Manufacturing (AM) [4]. Belonging to the same family of EP, a new process called Dry Electropolishing (DEP) was developed by GPAInnova®. The process is similar to the EP with the main difference in using solid electrolyte retaining acidic groups that creates an electro erosion in contact points between them and the metallic surfaces. No liquid agent is required for the entire process, only a small quantity of distilled water is used to humidify the electrolyte solid particles. Applying pulsed current, the process passivates and depassivates the surface, solubilizing the created oxides and catching the metal ions [5]. The solid electrolyte guarantees electrical contact between the cathode and only the peak asperities of the processed surfaces permitting the roughness reduction without part shape modifications [6]. Moreover, this dry process permits to overcome other negative aspects of the EP such as reduction of operator and environment safety problems eliminating the use of liquid acids and aggressive electrolytes; reduction of chemical deformation and corrosion of the treated part; less chemical waste generation [7]. The process is suitable for some materials such as stainless steel, titanium, copper, WC–Co and specific electrolytes should be used for each one [8].

Two mass transport mechanisms were reported in the literature to explain the electropolishing process for titanium i.e. "the product mass transport-limiting salt film mechanism" and "the acceptor mass transport-limiting mechanism". The former regards the titanium ions generation on the anode surface that cannot diffuse fast enough into the electrolyte generating the preferred dissolution only on the surface peaks where the current density distribution is higher. The latter regards an acceptor molecule required to chelate the titanium ions and remove them from the surface that affects the velocity of the process [9]. These mechanisms are influenced by the voltage and the time values in the following workflow. A positive voltage is applied on the processed part for a so-called anodic time t+: during this period ions diffusion occurs and the surfaces are polished. A process time break tp+ allows for cationic interchanges between particles: diffusion and homogenization is reached with the aid of mechanical movements and vibrations. A polarity inversion for a cathodic time t− is characterized by no material removal and provides part surfaces depassivation. Lastly, another pause interval tp− provides the same homogenization. The sum of these four intervals constitutes a process cycle.

Only some works in the literature explored the polishing operation by DEP and the effect of the process parameters on the obtainable surface quality. In [9] the effects of the voltage and the anodic time t+ of a DEP process applied on titanium rods characterized by different initial surface roughness obtained by rough machining, fine machining and hand polishing were investigated. The results showed that both parameters have influence on DEP process efficiency: in particular, more is the voltage and the anodic time more relevant is the roughness reduction. In [10] the authors used an electrical parameters (including voltage and anodic/cathodic times) matrix to study the effect on the surface quality improvement on WC–Co specimens. The results showed that the best electrical parameters for achieving an optimal surface finish in terms of elimination of defects, average roughness reduction and planarity between constitutive phases, are the pulse/pulse reverse waveform by the simultaneous application of voltage and negative pulse time higher than the positive one. Additionally, the authors assessed that the material removal mechanism may be described as a combination of electrochemical and mechanical action where the latter is produced by the contact of the polymeric particles with the oxides generated on the surface of the WC–Co, whereby the material is anchored to the surface of the solid active-polymeric media. In [11] the DEP process was applied on the cemented carbide end-mill to improve the geometry of tools cutting edges. The investigated parameter was only the DEP processing time, and the evaluated output were the evolution of cutting-edge parameters and the resulting surface finishing. Findings demonstrated enlargement of the cutting-edge radius, a decrease in surface roughness, and the mitigation of defects induced during previous manufacturing stages. Moreover, a direct correlation between DEP time and primary cutting edge micro-geometry parameters was found. In [12] the DEP of a dental bridge realized in Co-Cr by Laser Powder Bed Fusion (LPBF) was reported. The authors analyzed the efficiency of the process in terms of roughness reduction: a decrease of about 73% and 84% was observed for Rz and Ra respectively. No variation of process parameters was applied but three different parameters sets were used in sequence to reach these results. In [4] Ti6Al4V sheets were processed by DEP to investigate the changes of the surface morphology and the electrochemical properties. The authors assessed that the formation of the thickening passivation film on the titanium surface and the transfer rate of anions to the surface are in competition. The results showed that the roughness of the polished specimens decreased but some local corrosion pits appeared after 1 h of DEP process. The process parameters were fixed for the entire experimentation, and no variation was considered. In [13] the DEP process was applied to finish 316L specimens fabricated by LPBF. Surface morphology and roughness evolution under different DEP conditions were investigated. The results displayed that the top surface roughness was reduced by over 91% while the side surface by over 93%.

As evidenced by the literature, the effect of DEP process parameters and initial roughness morphology on the micro-geometrical evolution is not cleared and limited to average roughness. This work promotes a new method considering these aspects and extending the analysis to a variety of parameters.

2 Materials and Methods

2.1 Specimens Fabrication

The selected geometry for investigating the DEP was parallelepiped $20 \times 15 \times 6\ mm^3$. Specimens were fabricated by LPBF technology using Ti6Al4V material. The powder was supplied by EOS GmbH. The chemical composition is reported in Table 1.

Table 1 Chemical composition of the used powders

Al	V	O	N	C	H	Fe	Y	Ti	Other
5.50–6.75	3.50–4.50	0.20	0.05	0.08	0.015	0.30	0.005	Balance	0.1 max

The used machine is an EOSINT® M290 equipped with a 400 W Ytterbium fiber continuum laser, a beam diameter of 100 μm and wavelength 1075 nm, a building chamber of 250 × 250 × 325 mm^3. The processing parameters were set to the standard values provided by the equipment manufacturer and powders supplier: 280 W laser power, 1200 mm/s scan speed, 30 μm layer thickness and 140 μm hatch distance. The fabrication was executed in an argon inert atmosphere with less than 0.1% oxygen. The layers were manufactured by adopting the stripe scanning strategy with a maximum length of 5 mm and no overlap. A hatch rotation of 67° was applied between consecutive layers to minimize part distortions and delaminations. After the fabrication, no thermal treatment was performed, and the parts were detached from the building platform by an abrasive cutting performed on a Hitech Europe 400XL, with a Ø 400 mm carborundum wheel working at 19.6 m/s cutting speed and 18 mm/s feed speed. Since the surface morphology of a LPBF manufactured part is chaotic, one of the vertical surfaces of the specimen was subjected to a milling operation. This way, a regular and deterministic profile allows a basic understanding of the profile evolution under the DEP process. An Alcor universal milling machine and a Kennametal 8%Co flat end mill 8 mm in diameter were employed. The spindle was inclined by an angle α of 10° to avoid the multiple cycloid scratches and the cutting parameters were set as following: 200 rpm rotational speed; feed of 0.05, 0.1, 0.2 mm/min/tooth respectively for the tracks v040, v080 and v160 as shown in Fig. 1a. The cutting depth was set to 0.1 mm to preserve the LPBF contour microstructure.

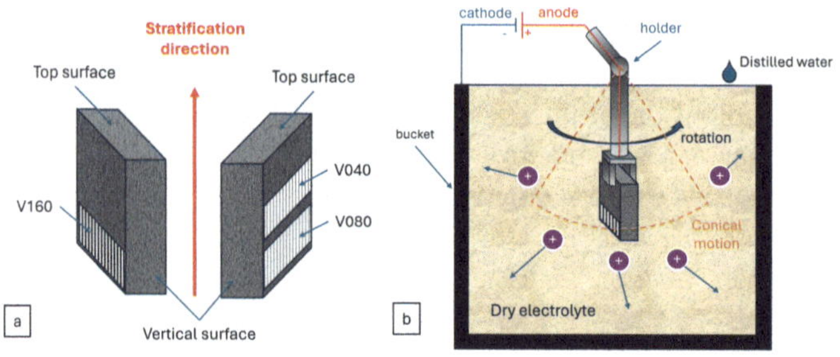

Fig. 1 Specimen surfaces specifications (**a**) and dry electropolishing schematic (**b**)

2.2 DEP Process

The DEP process was realized by a D-Lyte Desktop PRO machine. It is characterized by a bucket containing the electrolyte which is negatively charged and the workpiece fixed by a holder which acts as the anode (Fig. 1b). An electronically controlled motor moves the holder according to a conical motion to provide a homogeneous mixing of the granular material. Additionally, the bucket is subject to a vibration to help the homogenization of the charge surrounding the under conditioning surfaces.

The selected electrolyte is the Ti01 MIX MSA PLUS-S provided by GPA Innova ®. It is specifically designed for the processing of titanium alloys. It is composed of a mix of two different size/type spherical particles of copolymer material with a macroporous structure: the type 1 particles retain metanosulphonic acid (MSA) and account for 80% wt of the total charge; the remaining 20% is composed of the type 2 characterized by smaller diameter retaining sulfonic (S) acid. The granulometry of these two particle types was determined by using a Dino-Lite AM7515MZT digital microscope equipped by polarized lens. The sampling procedure was characterized as follows: 4 min mixing through the machine equipment to establish the process condition; syringe extraction directly within the bucket; high resolution photographs taken under the microscope; a total of 114 images were analyzed via a designed for the purpose digital image processing code developed in Wolfram Mathematica 14.

In Fig. 2a representative image of the particles is reported: the type 2 particles can be recognized because their natural color is dark orange, the bigger white particles are of the type 1. The elaboration through digital image processing allows for determining the area and the equivalent particles diameter. The distribution for each type is graphed in Fig. 2b. The equivalent disk diameters of type 2 assume a symmetric distribution with a mean value of 290 μm and a standard deviation of 18 μm. The type 2 is more scattered and the 95% of the equivalent diameters lie between 373 μm and 683 μm. The composition fraction calculated on the volumes resulted in 22.7% of type 2 which is in accordance with the value declared by the manufacturer datasheet.

The DEP parameters can be classified into two groups: mechanical and electrochemical. The former ones include the positioning of the sample in the holder, the speed and the motion strategy of the motor, and the vibration of the bucket.

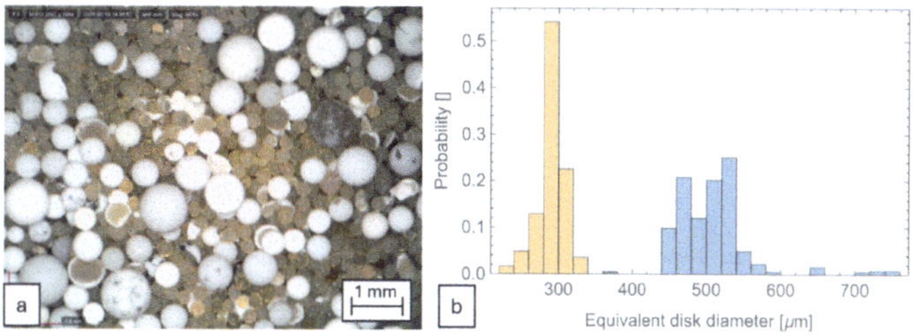

Fig. 2 Electrolyte used in the experimentation (**a**); granulometric distribution of the type 1 in violet and type 2 in orange (**b**)

To establish uniform conditioning of the specimen faces, the rotation direction was set alternate with an inversion step of 16 cycles. The rotational speed was set at 48 rpm. The electrochemical parameters are the voltage, the anodic and cathodic times and the pauses between them. Additionally, in the preliminary experimentation, the humidity of the charge was a critical factor to establish the electrochemical condition. This last parameter is not directly controlled by the machine, hence direct and indirect measurements were performed during the experimentation. A Testo 605i humidity probe was employed to detect the relative humidity (RH) inside the charge as shown in Fig. 3a. The data were acquired through a data logger and achieved as a function of the time. The equipment's electronics allows for an instant measurement of the anodic and cathodic currents. By using digital image acquisition the machine display was recorded and processed by using a Tesseract text recognition method for acquiring the current trends. In Fig. 3b the cathodic and anodic currents are reported for 30 min processing at a constant RH of 71.7% and 24 °C. For the first minutes a mirrored behavior is observed, and, after a settling period of 10 min, the currents reach a plateau. The processed current revealed a high sensitivity to the RH variation as reported in Fig. 3c. Notwithstanding an accurate relative humidity (RH) measurement obtained by sealing the bucket and waiting 60 min for the stabilization of the environment, the overall precision for these measurements was about ±0.5% RH. As a demonstration of this variability the residual analysis carried out on the linear regression of the data exhibited a standard deviation of 0.46% RH.

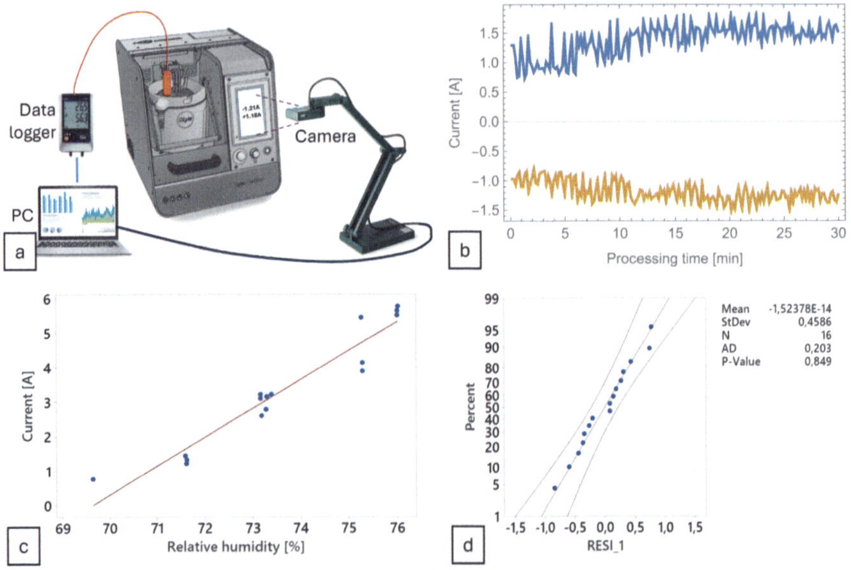

Fig. 3 Current and humidity measurements setup (**a**), current measurement trend for 30 min processing (**b**), scatter plot of measured current as a function of the measured humidity and linear regression (**c**), probability-probability plot of residuals (**d**)

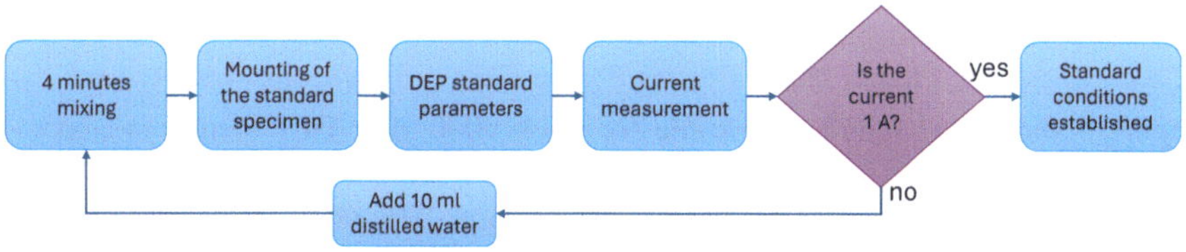

Fig. 4 Workflow of the standard working conditions monitoring

Hence, the relationship between these factors cannot ensure a good repeatability of the DEP process. For this purpose, a different approach was employed. The humidity was used only for monitoring whilst the standard working conditions were defined by the procedure reported in Fig. 4. The charge at a humidity reasonably near 71% RH is mixed for 4 min. A standard specimen (a LPBFed parallelepiped $20 \times 15 \times 6\ mm^3$) is mounted and DEP processed. The current measurement is checked to be 1 A. If true, the standard conditions are set and another specimen with different processing parameters are employed. Conversely, if the current is insufficient, 10 ml distilled water is added, and the procedure starts again with the mixing. This procedure allowed the best repeatability of the DEP process.

A full factorial plan was designed to study the effects of some electropolishing process parameters on the roughness reduction and the removal mechanism of this process. The chosen parameters were the anodic time t+ and the cathodic time t− varied on two levels: 10 and 20 μs for the t+ , 30 and 60 μs for the t−. The other parameters were maintained fixed: the motor speed was set on "medium" and the motor direction on "alternate"; the vibration was activated; the voltage was set on 25 V; the t_p+ and t_p− were both set on 10 μs.

2.3 Roughness Measurements

The specimens were measured before the polishing and after 30, 60 and 120 min processing time. The measurements were acquired on the 3 tracks v040, v080 and v160 obtained by milling, on a vertical LPBF surface and on the horizontal one for a total of five measurements for each specimen. They were performed by a Mitutoyo Surftest SJ-412 according to [14]: the sampling length and the evaluation length were set to 2.5 and 12.5 mm, respectively. The profiles were filtered by a spline profile filter with a cut-off λc and a short wavelength cut-off λs of 2.5 mm and 8 μm, respectively [15].

3 Results

Figure 5 shows the roughness profile for track v040 after DEP setting t+ and t− at 10 μm and 60 μm respectively. These values were chosen by giving a longer time for depassivation than the polishing operation. The v040 at the beginning is characterized by a periodic trend as expected from a milling operation (Fig. 5a). The mean spacing parameter RSm exhibits a value of 188 μm which is close to the set feed ratio value of 180 μm per tooth. The measured total and average roughness (Rt and Ra) comply with the predicted values of the sharpened edge formula reported in Eq. (1):

$$R_t = \frac{feed}{\frac{1}{\tan(\alpha)} + \frac{1}{\tan\left(\frac{\pi}{2} - \alpha\right)}} \tag{1}$$

As further confirmation of the profile quality, the shape parameters describe a symmetrical and platykurtic distribution, namely the skewness Rsk is close to 0 and the kurtosis Rku is less than 3. In particular, a triangular profile is characterized by a Rsk and Rku of 0 and 1.9 respectively.

The Amplitude Density Function (ADF) shown on the right of the profile is characterized by fluctuations due to a slight variability of the peaks and the pits. The root mean square of the local slope Δ_q which is related to the surface reflectivity is 0.36°. After 30 min DEP (Fig. 5b) the Ra and Rt are halved because of a marked rounding of the peaks. This is confirmed by the ADF and the shape parameters: the profile is now asymmetric with a negative Rsk which indicates surfaces with rounded peaks and sharp valleys. This confirms the assessment that the DEP does not alter the pits [6]. A further rounding of the peaks is observed at 60 min working time (Fig. 5c). The Ra decreases below 2 μm with the same values for the shape parameters. At 120 min the distribution tends to be symmetrical, (Rsk is 0.16). Notwithstanding the Ra is only 0.52 μm the periodicity is still evident and no modification of the RSm is provided even at this working time (Fig. 5d). This behavior is well evident by observing the power spectrum shown in Fig. 6a. Before the conditioning the principal frequency is detected at the value 5.51 mm^{-1} which corresponds to a spacing of 181.5 μm very close to the machining feed. Other secondary frequencies, multiple of the principal one, can be observed. As the DEP takes place the frequency peaks are progressively reduced to 2 h processing when only the principal frequency remains. This is another evidence that the pits are not affected by the DEP. Therefore, an interesting approach can be considered by aligning the Firestone Abbot (F-A) curves to the lowest cumulative point. Figure 6b shows the aligned curves for different working times: it is well evident how the kernel roughness depth Rk is totally changed by the processing since a very big content of the profile is affected. The areas between the curves represent the material removed by the processing. The area to working time ratios are reported within the graph: it is well evident that these values progressively reduce probably because of the different profile morphology.

t+	t-	Track	Ra [μm]	Rt [μm]	Rsk []	Rku []	Δq []	RSm [μm]
10 μs	60 μs	v040	6.08	26.6	0.04	1.89	0.36	188.558

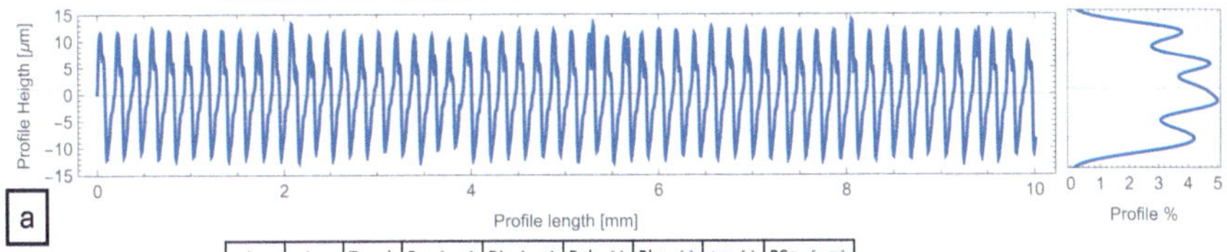

t+	t-	Track	Ra [μm]	Rt [μm]	Rsk []	Rku []	Δq []	RSm [μm]
10 μs	60 μs	v040	3.27	13.5	-0.57	2.04	0.19	186.442

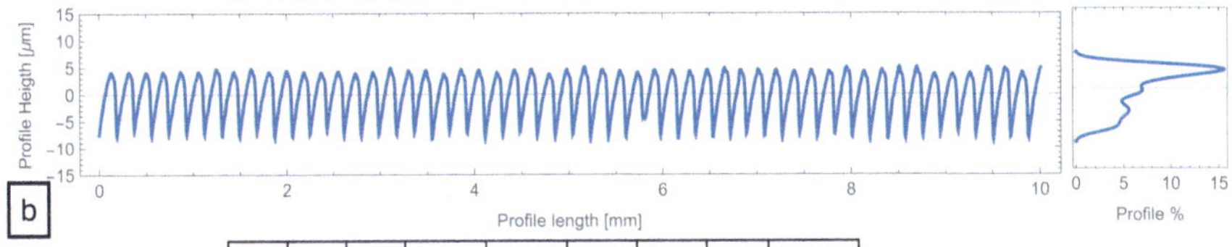

t+	t-	Track	Ra [μm]	Rt [μm]	Rsk []	Rku []	Δq []	RSm [μm]
10 μs	60 μs	v040	1.94	9.3	-0.6	2.17	0.11	188.173

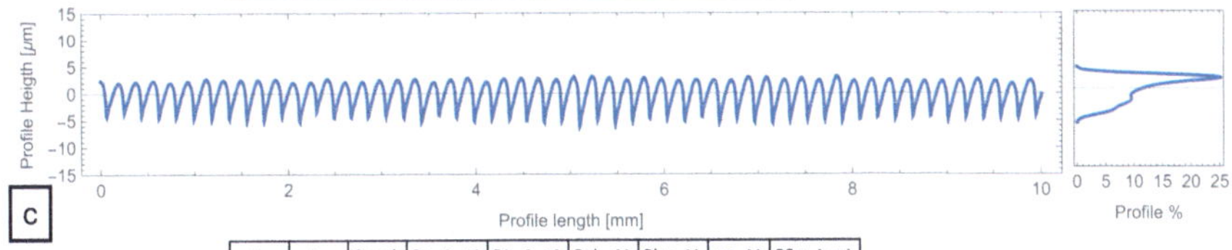

t+	t-	Track	Ra [μm]	Rt [μm]	Rsk []	Rku []	Δq []	RSm [μm]
10 μs	60 μs	v040	0.52	2.8	-0.16	1.96	0.02	186.731

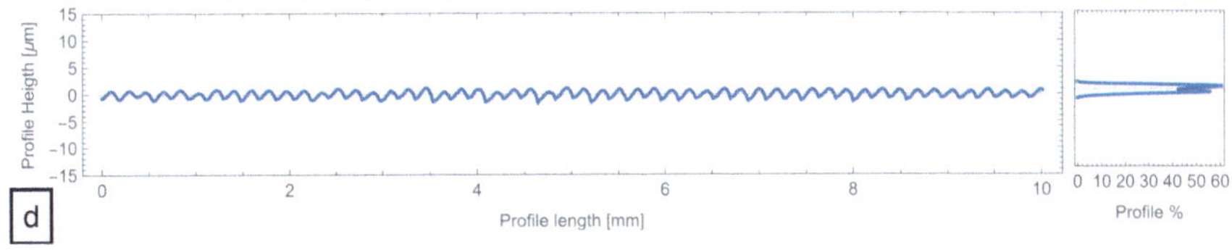

Fig. 5 Roughness profiles, roughness parameters and ADF for as is component (**a**), after 30 min (**b**), after 60 min (**c**) and after 120 min (**d**) DEP processing

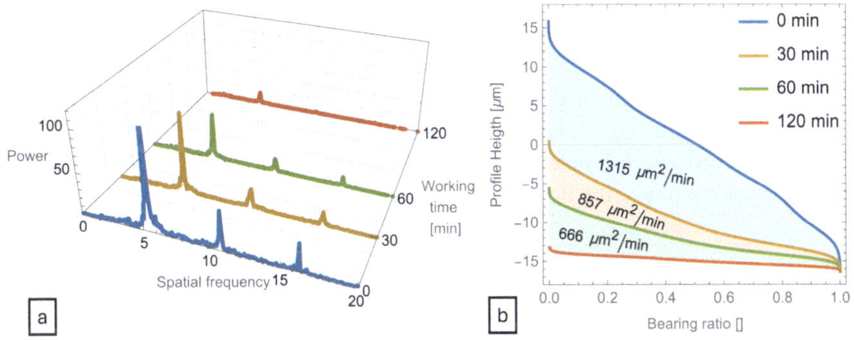

Fig. 6 Power spectrum (**a**) and aligned F-A curves of roughness profile for different working times (**b**)

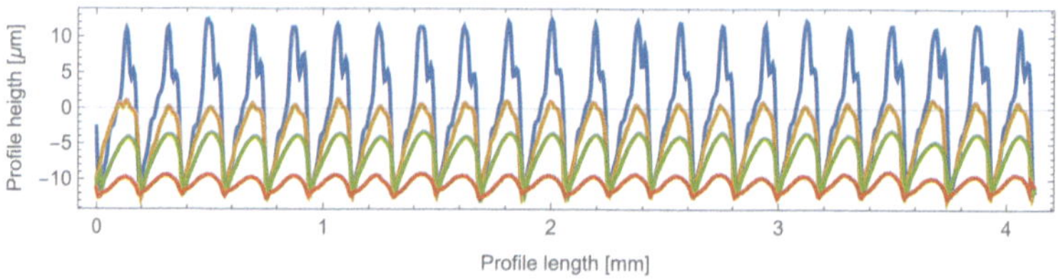

Fig. 7 Roughness profiles enlargement at different working times for v40 specimen aligned by using the F-A curves

By using the displacements found for the F-A curve alignment, it is possible to align the entire roughness profiles at different working times. Figure 7 shows these outcomes for the first 4 mm for better describe the peak and valley shape. At the beginning, the profile peaks are sharpened thus a pronounced current is expected. At higher processing times, a marked peaks rounding takes place.

To compare the effects of the investigated DEP processing parameters, the aligned profiles before and after 1 h are considered. In the previous case, characterized by t+ and t− set of 10 μs and 60 μs respectively, the Ra after 60 min is reduced by 69.4% (Fig. 8a). With the aim to improve the DEP effectiveness the anodizing time was maintained at 10 μs and the cathodic time was halved. As a result, the cycle time passed from 90 μs to 60 μs hence the t+ to cycle time ratio passed from 11.1% to 16.7% and the t− to cycle time ratio passed from 66.7% to 50.0%. Notwithstanding this depassivation time reduction, the profile was well rounded (Fig. 8b). Considering that the initial profile in these conditions was not very close to the previous ones, the Ra percentage reduction was only slightly improved.

This way, another combination of the current times was set by maintaining t− at 30 μs and by increasing the t+ at 20 μs, aiming an improvement of the dissolution process. The Fig. 8c confirms the goodness of this set: the profile periodicity almost disappeared, and Ra decreased down to 0.3 μm. As a completion of the factorial plan, the combination t+ and t− at 20 μs and 60 μs respectively was investigated. This set is expected to provide a less efficient cycle since the dissolution to total time ratio is reduced. The result shown in Fig. 8d goes against this supposition and exhibits an outcome comparable with the previous case.

The previously commented scenarios claim that the selection of the processing times is not easy since the outcome strongly depends upon the initial profile characteristics. In order to evaluate the efficiency of the DEP for the different tracks at different working times, the Ra percentage reduction was considered as following:

$$R_a^{\%R} = \left(1 - \frac{R_a^{(t_i)}}{R_a^{(t_0)}}\right)100 \tag{2}$$

where $R_a^{(t_i)}$ and $R_a^{(t_0)}$ are the Ra values at i-th working time t_i and t_0. The values of $R_a^{\%R}$ are plotted in Fig. 9. The abovementioned comments led at 60 min are now extended to the time evolution trends. The v040 tracks are scarcely affected by the

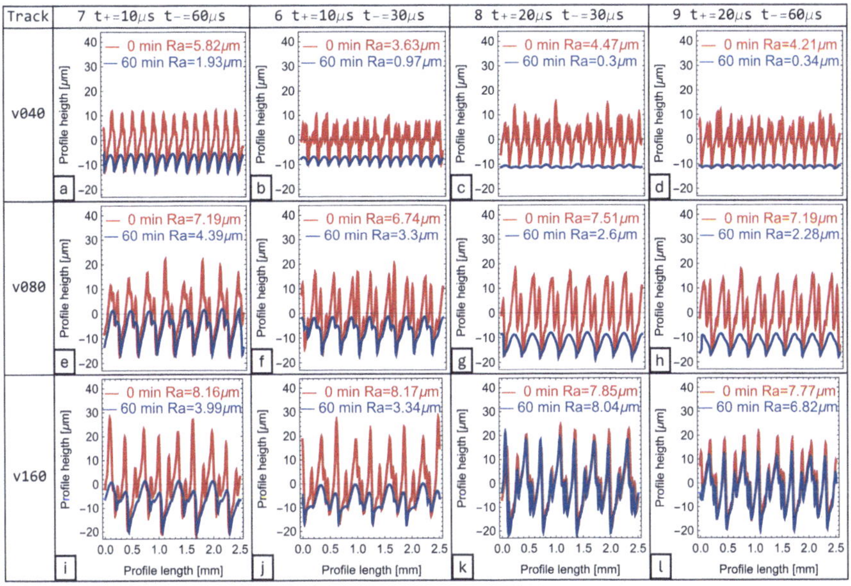

Fig. 8 Aligned roughness profiles at different working times for the investigated tracks before the DEP (line red) and after 60 min (line blue)

change in t− values (Fig. 9a) whilst the contribution of the t+ is crucial: at low t+ the trend is increasing and the plateau is not reached at 120 min, denoting a slower action with respect to the employment of high t+ . A different result is obtained for v080 track (Fig. 9b): the worst result is at high t+ with an ineffective t−, and at low t+ the t− plays a role; in fact lesser the t− better the improvement. A totally different outcome is measured for v160 track as shown in Fig. 9c. Now high t+ are partially or totally unaffecting the Ra and low t+ are exhibiting good parallel trends: at low t− and at high t− 75% and 68% Ra reduction are obtained respectively.

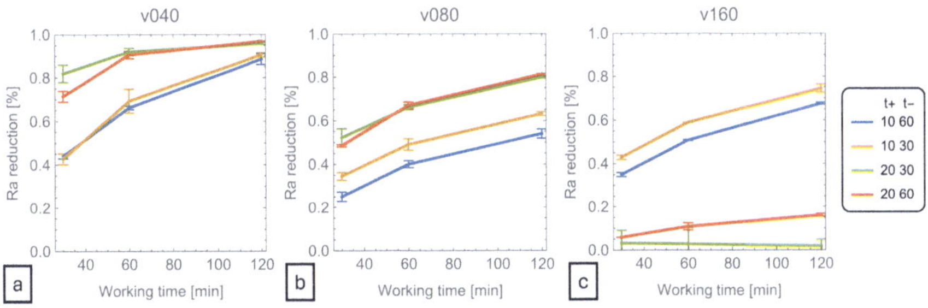

Fig. 9 Mean and standard deviation trends of $R_a^{\% R}$ as a function of the DEP processing time for track v040 (**a**), v080 (**b**) and v160 (**c**) at the investigated anodic and cathodic times

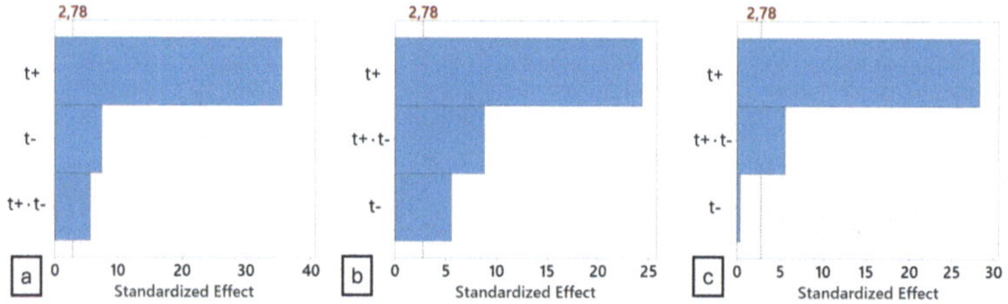

Fig.10 Pareto plot of standardized effect of factors t+ , t− and interaction t+ ·t− on $R_a^{\% R}$ for track v040 (**a**), v080 (**b**) and v160 (**c**)

In order to model the complex behavior of the factors and their interaction, a Design of Experiments analysis was carried out. The $R_a^{\%R}$ was modeled by considering the complete formulation of Eq. 3 for each track type:

$$R_a^{\%R}|_{track} = \mu + \alpha \cdot t_+ + \beta \cdot t_- + \gamma \cdot t_+ \cdot t_- \quad (3)$$

The regression analysis revealed that the interactions in all the models are significant with a confidence level of 99% as well as most of the single factors. As a report summary, the Pareto plots of the standardized effects adopting T-Student statistics is considered. Figure 10a reports the track v040: the largest effect is given by the t+ factor and the t− and interaction have a confidence level greater than 95% which corresponds to a T-Student value of 2.78. Similarly, the v080 is mainly affected by the t+ followed by the interaction and the t− which are significant (Fig. 10b). The t− term in v160 is not significant but it will not be excluded from the model since the interaction is significant (Fig. 10c).

The found models are reported in Eqs. 4 to 6.

$$R_a^{\%R}|_{v040} = 0.7039 + 0.01197 \cdot \mathrm{t}_+ - 0.005086 \cdot \mathrm{t}_- + 0.000235 \cdot \mathrm{t}_+ \cdot \mathrm{t}_- \quad (4)$$

$$R_a^{\%R}|_{v080} = 0.6540 - 0.00183 \cdot \mathrm{t}_+ - 0.009683 \cdot \mathrm{t}_- + 0.000533 \cdot \mathrm{t}_+ \cdot \mathrm{t}_- \quad (5)$$

$$R_a^{\%R}|_{v160} = 1.4667 - 0.07810 \cdot \mathrm{t}_+ - 0.00945 \cdot \mathrm{t}_- + 0.000644 \cdot \mathrm{t}_+ \cdot \mathrm{t}_- \quad (6)$$

The analysis carried out on the residuals assessed they follow a normal distribution since the Anderson–Darling test returned a p-value not less than 0.388 for all the models. The R^2 values of all models are higher than 99% which indicates the models fit very well the data. The adj-R^2 are very close to the R^2 indicating that no overfitting is present. Finally, the pred-R^2 denotes a high ability to predict the response for new observations since all these values are higher than 97%. A representation of the models is reported in Fig. 11. The contours reveal a differently pronounced effect of the interaction at different levels of t− which enables the following conclusions: v040 track can be processed by using t+ set at 20 μs and indifferently setting the t−

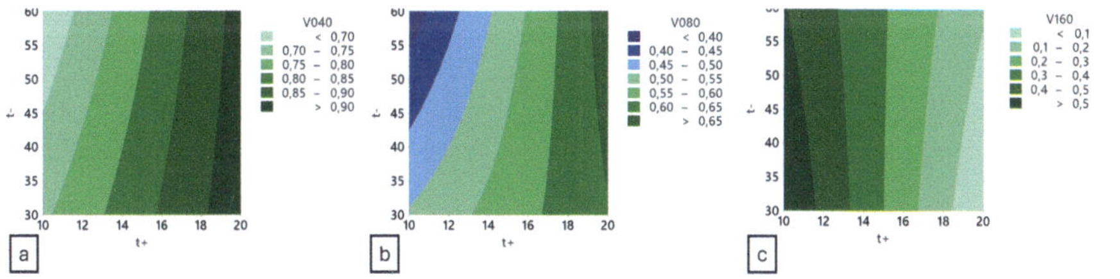

Fig. 11 Contour plots of v040 (**a**), v080 (**b**) and v160 (**c**) track models

(Fig. 11a); from Fig. 11b it is well evident that the best combination is on the top right, i.e. {20 μs; 60 μs}; finally the v160 is well processed at {10 μs; 30 μs}.

As a conclusive remark, no combination enables the possibility to optimally work surfaces characterized by different starting quality. Since the AM part quality spans over a comparable range, at least two combinations, namely the {20 μs; 60 μs} and the {10 μs; 30 μs} must be investigated. The analysis of the AMed top and vertical surfaces are reported in Figs. 12 and 13 before and after 120 min DEP. The top ones are typical of the LPBF process (Fig. 12a and b): a slight periodicity can be noticed and the RSm is in accordance with the 140 μm hatch spacing projection by 25°, namely the hatch rotation of the last layer. The profile amplitude distribution shows a slightly asymmetric and leptokurtic shape, as indicated by the Rsk and Rku. After the DEP, the combination {10 μs; 30 μs} flattens the profiles and clears most of the LPDF tracks (Fig. 12c). As the set {20 μs; 60 μs} is applied, a marginally better result is obtained (Fig. 12d) and the profile heights almost follow a normal distribution (Rku near 3).

Figure 13 shows the results gained on the vertical surfaces. The starting height distribution is close to a Gaussian confirming a chaotic behavior as expected for these surfaces (Fig. 13a and b). Also, the scattering is relatively pronounced. The

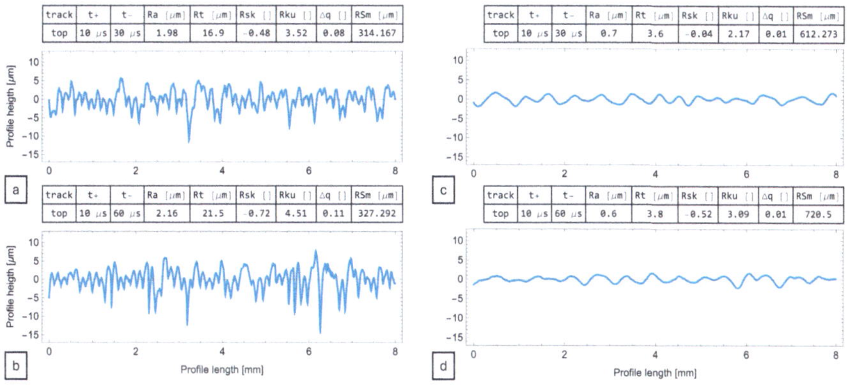

Fig. 12 Top surface before (**a**) and after (**c**) Dep processing at {10 μs; 30 μs} and before (**b**) and after (**d**) DEP processing at {20 μs; 60 μs}

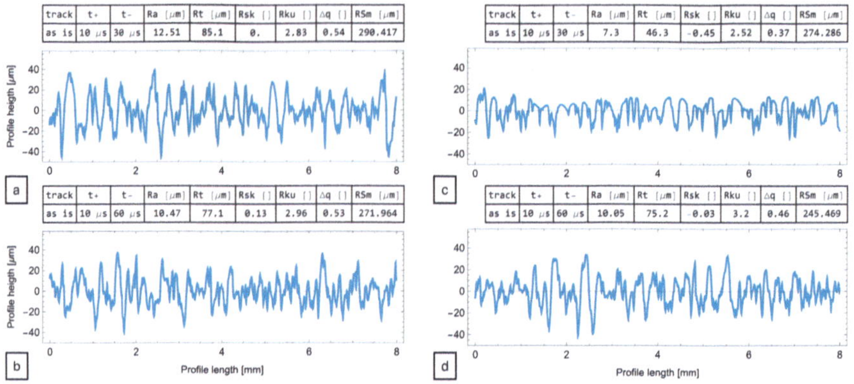

Fig. 13 Vertical surface before (**a**) and after (**c**) Dep processing at {10 μs; 30 μs} and before (**b**) and after (**d**) DEP processing at {20 μs; 60 μs}

combination {10 μs; 30 μs} allows reaching 7.3 μm Ra which corresponds to a reduction of 41.6% (Fig. 13c). The surface appears smoother as confirmed by the decreasing of the Rku. As expected, the conditioning by adopting the combination {20 μs; 60 μs} does not provide satisfactory results (Fig. 13d): the reduction is negligible, and the shape parameters are almost the same.

In order to test the DEP on an intricate geometry, a Triply Periodic Minimal Surface gyroid lattice was selected for the hard accessibility of the feature. It was fabricated and processed by using the set {10 μs; 30 μs}. Figure 14a shows the parts before and after the conditioning for 120 min. To determine how the DEP works the internal features, the part was halved by using a Struers Labotom-5 metallographic cutter, and roughness measurements were taken on internal and external curved surfaces. In Fig. 14b the two halved parts are reported together with the measurements. Since most of the surfaces are curved the slope is indicated by an interval; only the surface colored in green is exactly orthogonal to the stratification direction. This last feature is well conditioned since the initial Ra decreases from 3.2 to 1.2 μm accounting for a 62.5% reduction, in accordance with the previously results. The curved horizontal surface is a combination of many slopes and deeply differs from the top surface: in fact, the initial Ra is about 12 μm for the external and internal surfaces. As the DEP is applied, a different reduction is observed between the external and internal surfaces probably attributable to the hard accessibility of these features. The vertical zones show a similar trend: the difference in reduction is at the order of 10%.

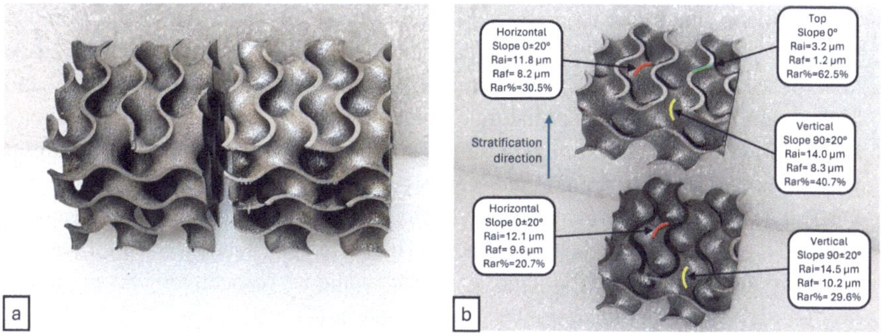

Fig. 14 Lattice structures before and after the DEP processing for 120 min (**a**) and roughness measurements on the halved part (**b**)

4 Conclusions

This paper investigates the improvement via DEP process of Ti6Al4V surface quality fabricated by LPBF. Through a profilometric analysis different periodical surfaces were treated by DEP with different combinations of the time parameters. The analysis strived to determine the evolution of the profiles along the working time by adopting a specific alignment method. The statistical modeling of the outcomes allowed to judge the effects of the time parameters on the average roughness reduction. Based on the achieved and discussed results, the conclusions can be highlighted as in the following:

- The study of the treated periodical profiles allowed to describe the morphological modification provided by the DEP. A marked peak rounding is observed without altering the deepest valleys.
- An alignment of the F-A curves allowed the evaluation of the material removed which gradually reduces along the working times.
- The change in t+ and t− affected the evolution of the periodical profiles in contrasting ways when different profiles are considered. The low roughness profiles were markedly improved at high t+ and, low depassivation time is preferred. At an intermediate level of the initial profiles, a less marked enhancement is observed due to the larger area below the peaks. Also in this case, the higher the t+ value, the more effective the rounding of the peaks. On the contrary, the high initial roughness is better improved at low t+ and the DEP improvement is limited to 60% in 60 min. Hence, the initial roughness plays a crucial role in establishing good processing conditions.
- The statistical analysis carried out on the roughness percentage reduction confirmed the morphological considerations and the regression models exhibited a strong dependance upon the single factors as well as their interaction. The models were characterized by high goodness of fit and predictability.

- The analysis undertaken on the AMed surfaces denotes a marked difference between the top and vertical surfaces, as expected for LPBF process. Hence two DEP processing conditions were set in accordance with the models for low and high roughness levels. The polishing action showed a very good agreement with the outcome observed and modeled on the periodical profiles notwithstanding it was applied to highly irregular and scarcely repetitive profiles associated with the vertical and top surfaces respectively. Both the selected combination of t+ and t− allowed a good result on the top surfaces. The vertical surfaces were improved by 30% only by employing 10 μs and 30 μs for t+ and t− respectively. As predicted no modification is obtained with the combination {20 μs; 60 μs}.
- The considered case study provided additional considerations. Since the gyroid lattice structure is composed of curved surfaces the roughness measurements were characterized by varying slopes. However, a comparable reduction is observed for external flat top, curved horizontal and curved vertical surfaces. The correspondent internal features were processed to a lesser extent because of the very difficult accessibility.

Further development of this research will be focused on the improvement of the DEP on high roughness profiles. A way can be the modification of the working voltage. Also, the employment of the mechanical parameters, namely the rotational speed and the inversion time, can lead to better homogeneity between internal and external surfaces.

Funding Financed by the European Union—NextGenerationEU (National Sustainable Mobility Center CN00000023, Italian Ministry of University and Research Decree n. 1033–17/06/2022, Spoke 11-Innovative Materials & Lightweighting). The opinions expressed are those of the authors only and should not be considered as representative of the European Union or the European Commission's official position. Neither the European Union nor the European Commission can be held responsible for them.

References

1. Han W, Fang F (2019) Fundamental aspects and recent developments in electropolishing. Int J Mach Tools Manuf 139:1–23
2. Yang G, Wang B, Tawfiq K, Wei H, Zhou S, Chen G (2016) Electropolishing of surfaces: theory and applications. Surf Eng 33(2):149–166
3. Piotrowski O, Madore C, Landolt D (1998) The mechanism of electropolishing of titanium in methanol-sulfuric acid electrolytes. J Electrochem Soc 145(7):2362–2369
4. Cheng Y, Wang L, Yu S, Min R (2021) Studies on the electro-mechanical polishing effects of titanium alloy with ion-exchange resin solid particles. J Alloy Compd 854:157269
5. Millet S (2020) Method for smoothing and polishing metals via ion transport via free solid bodies and solid bodies for performing the method. US 2020/0270761 A1 Patent
6. Min R, Wang L, Cheng Y, Hu X (2021) Dry electrochemical polishing of copper alloy in a medium containing ion-exchange resin. RSC Adv 11:35898
7. Woo HG, Kim JS, Ryu JY, Lee JS, Lee SJ (2020) Dry electropolishing of an additively manufactured spacer grid. In: Transactions of the Korean nuclear society virtual spring meeting, July 9–10–2020

8. Millet S (2021) Use of H_2SO_4 as an electrolyte in processes for smoothing and polishing metals by ion transport via free solids US 10,975,491 B2 Patent
9. Simeunovic S, Jung C, Mory D, Seiler D, Wild M (2021) Investigating dry electro-chemical polishing of titanium structures: novel polishing methods for the MedTech industry. Curr Directions Biomed Eng 7(2):77–80
10. Riu-Perdrix G, Llanes L, Roa JJ (2025) Optimizing and understanding dry-electropolishing process on WC-Co cemented carbide material. Int J Refract Metal Hard Mater 128:106997
11. Riu-Perdrix G, Valencia-Cadena A, Llanes L, Roa JJ (2024) Cemented carbide end-mill edge preparation using dry-electropolishing. J. Manuf. Mater. Process. 8(1):28
12. Rautio T, Jaskari M, Päkkilä J, Järvenpää A (2023) Dry electropolishing of laser powder bed fusion manufactured cobalt-chrome. in: 4th international conference on materials research and innovation. In: IOP conference series: materials science and engineering, p. 1280
13. Bai Y, Zhao C, Yang J, Fuh JYH, Lu WF, Weng C, Wang H (2020) Dry mechanical-electrochemical polishing of selective laser melted 316L stainless steel. Mater Des 193:108840
14. ISO 21920-3:2021 Geometrical product specifications (GPS) Surface texture: Profile
15. ISO 16610-22 (2015) Geometrical product specifications (GPS) filtration Part 22: linear profile filters: spline filters. International Organization for Standardization (ISO), Geneva

Femtosecond Laser Cutting of Current Collectors: A Study for High-Quality Li-Ion Battery Fabrication in Electric Vehicles

Caterina Angeloni, Dongkyoung Lee, and Alessandro Fortunato

Abstract The average power of ultrashort laser sources has steadily increased, alongside significant investments by battery companies in advanced technologies, which necessitate solutions to improve ablation efficiency and quality in laser cutting of Li-Ion electrodes. Although burst mode processing has been primarily investigated in micro-processing, its application to cutting remains rarely explored. This paper introduces a simplified efficiency model that provides, with minimal computational cost, rapid insights into process dynamics of femtosecond laser cutting of 8 μm copper and 12 μm aluminum current collectors both in the single pulse and burst mode. The model results are then compared with actual cutting quality under constant conditions of 40 W average laser power and 200 fs pulse length. Results demonstrate that burst mode triples scanning speed, albeit with heat accumulation in aluminum, while MHz processing enables scalpel-like cuts in copper with a kerf half the spot size.

Keywords Laser cutting · Zero defects · Burst mode

1 Introduction

As the dawn of third-generation electric vehicles, the competition for high-energy–density batteries is becoming increasingly intense. Next-generation vehicles must achieve a mileage of 500 km minimum on a single charge, necessitating the development of high-performance lithium-ion batteries (LIBs) with exceptional capacity, rate capability, stability and long lifespan [1]. Thus, relevant research is being conducted through material and technological innovation to improve performance and quality

C. Angeloni (✉) · A. Fortunato
Department of Industrial Engineering (DIN), University of Bologna, Bologna, Italy
e-mail: caterina.angeloni2@unibo.it

D. Lee
Department of Mechanical and Automotive Engineering, Kongju National University, Cheonan, South Korea

L. Fratini et al. (eds.), *Selected Topics in Manufacturing*, Lecture Notes in Mechanical Engineering, https://doi.org/10.1007/978-3-031-99501-9_5

along with the manufacturing costs. In this respect, current collectors (CCs) made of aluminum (Al) and copper (Cu), are critical in LIB systems as they constitute ~15–50% of the weight, hence impacting significantly the price, weight, and energy density. CCs in automotive applications are thin, 6–30 μm thick foils which undergo various manufacturing processes, starting with trimming, followed by coating, and then notching to prepare them for the assembly and welding of battery connections. Laser beam is the state-of-the-art technology for cutting metal foils, in both trimming and notching, thanks to its ability to concentrate the high energy of the beam within a small spot, its rapid processing time, its wide range window of process parameters and its contact-free process [2]. Trimming (uncoated metal foil cutting) is performed in a "roll-to-roll" configuration with web speeds of up to 3 m/s, while laser spot velocities can reach 20 m/s for complex geometries. As a thermal process, laser cutting introduces thermal defects like spatter, dross, and heat-affected zone, which must be minimized for optimal battery performance [3].

Most studies in the literature have optimized laser cutting using a Pulsed Wave (PW) nanosecond lasers in IR or green wavelength. For example, PW nanosecond laser with a 532 nm wavelength has achieved good results in terms of kerf width and speed for 10 μm copper cutting with a direct interference pattern [4].

However, high-quality continuous laser beams (CW) recently available in the market, with spot dimension of 60 μm and gaussian beam shapes are increasing their applicability in industrial environment. Lee et al. [5] developed a numerical model for CW laser cutting of CCs at speeds up to 5 m/s, while Angeloni et al. [6] drew a thorough comparison of CW and PW nanosecond lasers for high-speed cutting (≤28 m/s) of Al and Cu CCs demonstrating how these are affected by laser irradiance (the ratio of laser power to spot area) and fluence (the product of irradiance and contact time). The key finding was that, in all studied settings, CW single mode fiber lasers allowed for improved cut quality, except at ≤4 m/s for copper foils. Even with optimized process parameters, thermal defects occur because ablation is mainly due to melt ejection and evaporation rather than direct solid–vapor transition, as in ultra-short pulsed laser systems [7].

Ultrashort pulsed lasers (pulse length <10 ps) minimize thermal effects, achieving nearly dross-free cut edges by ending pulses before thermal diffusion occurs. The problem of ultrashort laser power limitation or efficient ablation evolved gradually over time: the average power doubles every three years, with pulse repetition rates (PRRs) reaching tens of megahertz [8]. Even at such high pulse repetition rates, the resulting fluence exceeds the optimal one. A practical solution to reduce pulse energy (and fluence) without sacrificing the average power is temporal pulse division, which creates *pulse bursts*. These bursts typically feature intra-burst repetition rates in the MHz-GHz range, significantly enhancing ablation efficiency and reducing surface roughness [9]. However, the use of such techniques for cutting Li-ion battery electrodes remains limited, being applied intensively in micromachining. The primary challenges for industrial implementation up to now include relatively low processing speeds, high investment costs and the delicate requirements of the equipment and environment (e.g., a controlled atmosphere). Pico-second lasers in burst mode have improved cutting speeds of Cu CCs up to 3.8 m/s and reduced heat-affected zones

HAZ [10]. Heidari Orujo et al. [11] optimized burst fluence in 10 ps laser cutting of CCs, achieving peak speeds of 4.7 m/s and 2.1 m/s for Al and Cu, respectively. It was not observed any improvement when varying the burst shape, namely equal pulse (EP) and increasing pulse (IP). Recent studies are exploring GHz-burst mode for electrode femtosecond cutting, with Audouard et al. [12] demonstrating increased speed and metal-dependent cutting times.

To the best of author knowledge, no studies have been published on femtosecond laser cutting of Li-ion current collectors only, likely due to the challenges previously mentioned. As industries are becoming more prone to invest in advanced technologies capable of achieving dross below 1 μm when cutting in a single pass at high cutting speeds, a critical question emerges: can quality be improved without compromising power and productivity?

This study investigates femtosecond laser cutting of CCs, analyzing speed and frequency effects on the final quality. Both MHz and GHz burst mode were tested with adjustable pulse numbers (N and P indexes) from 1 to 10. Results showed that higher pulse splitting increased maximum speed. Hence, with P and N fixed at 10, five frequency levels were studied (burst fluence 0.6~2.8 J/cm^2), identifying max speed for each. As a result, ablation efficiency and quality were assessed to strike a physical explanation of the results obtained in both the single mode and burst mode (Fig. 1).

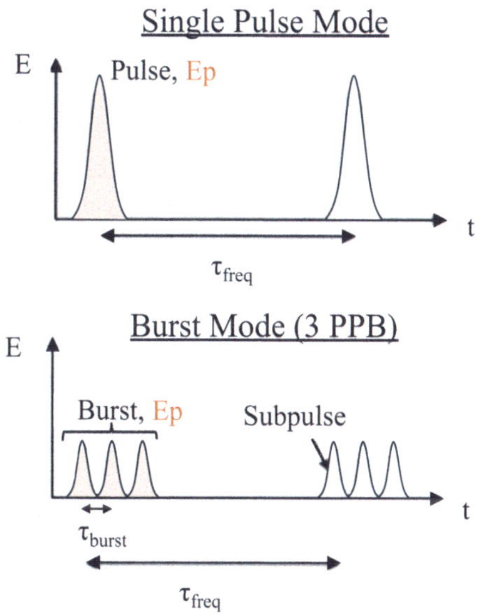

Fig. 1 Schematic representation of burst mode (PPB: pulses per burst)

2 Materials and Methods

2.1 Materials, Laser Source and Cutting Setup

A simple metal foil geometry and linear cutting path were considered to characterize and optimize the laser trimming process used for current collectors. Strips of copper Cu (CuAg ≥ 99.8%) and aluminum Al (1100 alloy, wt.% Al ≥ 99.0%, Fe ≤ 0.5%, Si ≤ 0.5%, Cu ≤ 0.2% other elements ≤ 0.05%) foil, 8 μm and 12 μm thick respectively, were cut from rolls and placed on a vacuum chuck shown in Fig. 2 to prevent bending and ensure the foil remained in the same focal plane during the single-pass cutting. Each strip was 20 mm × 60 mm in size, with parallel laser cuts performed over a length of 20 mm in each case. To avoid thermal interaction and local deviation of the material from the focal plane, the distance between subsequent cuts was 5 mm, while a maximum of 10 cuts were made on each strip. Since the Rayleigh distance is very small (0.69 mm), to ensure the consistency of our results, each process parameter combination was repeated three times.

Laser system employed for experiments was a solid-state laser (Carbride, Light Conversion) with a variable pulse duration in the range of 200 fs–10 ps and radiating at the light wavelength of $\lambda = 1030$ nm. It can produce a linearly polarized Gaussian beam with the diameter of 4.3 mm, pulse energies up to 400 μJ (40 W max) and repetition rates from a single pulse up to 1 MHz. During all experiments the pulse duration of 200 fs and an average power of 40 W were fixed. This laser source can work also in burst mode, as Fig. 1 explains. Two values of intra-burst repetition rate could be chosen: 66.7 MHz and 2.5 GHz for a respective τ_{burst} of 15 ns and 400 ps.

A galvanometer scanner (SG8220, Sino Galvo) and F-theta lens with a focal distance of 100 mm were used to scan and focus the laser beam into a spot of 30 μm. Figure 2 reports the specifications of the laser system. The experimental campaigns were conducted in a controlled environment: temperature was maintained between 15 °C and 30 °C, and relative humidity was kept below 80% (non-condensing).

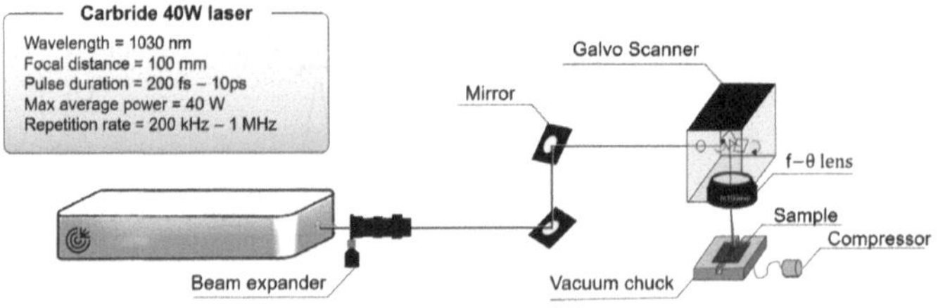

Fig. 2 Experimental set-up: ultrafast laser pulses guided via mirrors towards a galvanometer scanner and focus on the metal foil through the f-theta lens

Table 1 Process parameters for single pulse (SP) and burst mode (BM) campaigns

	Scanning speed (m/s)	Frequency (kHz)	P	N
SP	0.1–0.2–0.4–0.7–0.9	200–400–600–800–1000	1	1
BM	0.4–0.8–1.6	200–600–1000	1–10	1–10

2.2 Design of Experiments

For the Single Pulse *SP* experimental campaign, five levels of pulse fluences (5.7–28.3 J/cm^2) and five levels of speeds (0.1–0.9 m/s) were tested. The process parameters were chosen so that the no cut condition was found for both aluminum and copper. With the burst mode, thanks to the pulse splitting, the pulse fluences decreased up to 0.57 J/cm^2 ensuring the cutting condition. Maximum 40 W power and minimum 200 fs pulse length were kept constant to exploit at full potential the laser source. The burst mode *BM* campaign was conducted to evaluate the effect of intra-burst separation time, ranging from 400 ps (2.5 GHz) to 15 ns (66.7 MHz), on both quality and efficiency. Additionally, the number of pulses per burst was also varied from 1 to 10: P index for GHz mode and N for the MHz mode.

For each level of frequency, the maximum speed was found, both for BM and SP processing. Table 1 summarizes the combinations of process parameters investigated to assess fs laser cutting quality vs efficiency across SP and BM regimes. All tests were performed three times on both Aluminum and Copper.

2.3 Modeling of Ablation and Efficiency Analysis

Femtosecond laser ablation has been widely used in micromachining for its precision and minimal thermal defects by promoting direct sublimation. This results in cleaner cuts and reduced heat affected zones, making it ideal for defect-free processes such as the cutting of thin metallic foils. Hence, this study characterizes the relationship between laser pulse fluence, thermal defects, and kerf geometry to optimize ablation efficiency and resulted quality.

By adopting the approach in Ref [13] the laser beam at the entrance of the scanner is characterized by a two-dimensional Gaussian shape with a fluence distribution described by:

$$F(r) = F_{\max} e^{-\frac{2r^2}{w^2}} \left[\frac{J}{\mathrm{cm}^2}\right] \quad (1)$$

$$F_{\max} = \frac{2E_p}{\pi w^2} \left[\frac{J}{\mathrm{cm}^2}\right] \quad (2)$$

With $F_{\max}$ [J cm^2] being the laser peak fluence in the center, r is the radial distance from the beam center, w is the laser spot radius and $E_p[J]$ is the laser pulse energy, described as $E_p = P/f$.

In ultrashort pulse lasers not all the energy is efficiently used to remove the material, part remains in the latter as residual heat and part of the energy is emitted from the surface with the hot liquid/gaseous matter [14]. To ensure efficient material removal, laser has to work near its optimum fluence, which is a function of the material's fluence threshold. Threshold fluence F_{th} is the value of local fluence $F(r)$ after which material ablation occurs. For cutting applications, theoretical (TH) threshold fluence is determined experimentally by measuring the kerf width *D*, which corresponds to the spot size where the local fluence equals F_{th}. This is expressed as:

$$F_{th}^{TH} = F(D/2) = F_{\max} e^{-\frac{D^2}{2w^2}} \tag{3}$$

Energy losses after kerf formation and heat accumulation due to pulse overlapping has to be taken into account. This introduces two additional efficiency factors: thermal η_t and energy η_e efficiency. The real threshold fluence for cutting is given by:

$$F_{th}^{RE} = F_{th}^{TH} \eta_t \eta_e \tag{4}$$

The ratio between the real threshold value and its respective pulse fluence is called "cut efficiency" η_{cut} as reflects how effectively the laser energy is utilized to achieve material ablation under given process parameters.

Thermal efficiency η_t

High power and high repetition rates regimes in femtosecond laser processing lead to heat accumulation in the material, resulting in detrimental effects [15]. Thermal efficiency quantifies the extent of this thermal accumulation. The interaction between fs pulses and the metal foils is governed by two characteristic times: the pulse-to-pulse interval $\tau_{freq} = 1/f$ and the thermal relaxation time τ_{rel}. Hence, thermal efficiency is defined as:

$$\eta_t = 1 - e^{-\frac{\tau_{freq}}{\tau_{rel}}} \tag{5}$$

The thermal relaxation time quantifies the time required for heat to dissipate within the material over a specific distance, such as the laser spot radius w. Thermal relaxation time is given by:

$$\tau_{rel} = \frac{w^2}{4\alpha} \tag{6}$$

where α is the thermal diffusivity, defined as $\alpha = k/(\rho C_p)$, with *k* being the thermal conductivity, ρ the density and C_p the specific heat capacity. Material properties for both Aluminum and Copper were taken from standard sources [16] and the calculated τ_{rel} are 0.577 μs and 0.489 μs, respectively.

When the time between two subsequent pulses τ_{freq} is shorter than τ_{rel}, thermal accumulation occurs, leading to localized residual heat. Typically, the residual heat generated during the ablation process by a single ultra-short laser pulse is comparably small and avoids detrimental effects on the cut edge. However, as the repetition rate increases, residual heat may not dissipate fast enough through heat conduction into the workpiece, leading to material damage around the cut edge. For the SP processing frequencies tested in this study, this condition occurs only for aluminum at the maximum frequency of 1 MHz. This corresponds to a minimum τ_{freq} of 1 μs, which is less than, yet very close to τ_{rel} for both Cu and Al. As a result, heat spreads rapidly from the region heated by the laser pulse, preventing melting in copper but not in aluminum due to its lower thermal conductivity. When BM is activated, different mechanisms of laser ablation come into play, requiring a more attentive selection of optimal laser conditions. The influence of the number of sub pulses and their time interval is described in the *Burst Mode* section.

Energy Efficiency η_e

This factor describes the fraction of energy effectively used for ablation and it is defined as:

$$\eta_e = \frac{E_{v,\min}}{E_v} \tag{7}$$

where $E_{v,\min}$ is the minimal volumetric energy density required to open the kerf. It is calculated for each level of frequency tested at its maximum speed. For higher overlapping, it is supposed that part of the beam energy doesn't contribute fully to vaporize the material, and it is therefore considered as "lost". Hence, the volumetric energy density for each combination of speed and frequency is modelled as:

$$E_v = \frac{E_p}{V}\left[\frac{\mu J}{\mu\text{m}^3}\right] \tag{8}$$

where E_p is the energy of a single laser pulse $[\mu J]$ and V is the removed material between subsequent pulses defined as:

$$V = \frac{tDv}{fN_b}[\mu\text{m}^3] \tag{9}$$

where t is the foil thickness (8 μm for copper and 12 μm for aluminum), D is the kerf width and N_b is the number of pulses within a burst.

Burst Mode

When burst mode is activated, the calculation of the ablated volume per pulse V and the total pulse energy E_v must account for the division of total pulse energy among N_b

sub-pulses within a single burst. Additionally, thermal accumulation within the burst has to be modelled. Since the time interval between consecutive sub-pulses is shorter than the material's thermal relaxation time, the energy from successive sub-pulses accumulates because the heat does not fully dissipate between them. Consequently, this thermal accumulation necessitates a reformulation of the energy distribution within a single burst, as described by:

$$F_{\max,burst} = \frac{F_{\max}}{N_b} \sum_{i=1}^{N} e^{-\left(\frac{(i-1)\tau_{burst}}{\tau_{rel}}\right)} \tag{10}$$

The first sub impulse (i = 1) contributes with its full energy, while the following ones (i > 1) overlap thermally, each with a contribution reduced by an exponential factor. In other words, sub pulses within a burst do not act independently: their energy accumulates based on residual heat from previous pulses.

When the sub-pulse separation $\tau_{burst} \ll \tau_{rel}$, the exponential decay term in the fluence summation becomes negligible and heat does not dissipate significantly, leading to a near-complete energy accumulation. In this case, the peak burst fluence $F_{\max,burst}$ closely matches the single-pulse peak fluence and remains nearly unaffected by τ_{burst}. Conversely, when $\tau_{burst} \leq \tau_{rel}$, heat dissipates efficiently between sub-pulses, reducing thermal accumulation and lowering $F_{\max,burst}$. The latter is then utilized in Eq. (3) for the subsequent determination of F_{th}^{RE}.

2.4 Sample Characterization

Digital microscopy (DM, Leica DVM6) was used to provide preliminary assessment of cut quality and to identify defects for each combination of process parameters in terms of uncompleted cuts, dross, melted edges, spatter, and presence of a HAZ (see Fig. 3). Furthermore, the kerf width was determined for all tests with the mean value calculated based on the three data points collected along the entire cut.

SEM-FEG microscopy (Tescan Mira3 with a Schottky emitter) was then used to examine the representative cuts on both the front surface (the area directly subject to laser irradiation) and the underside of each foil. Each cut was observed via SEM at the start, middle, and end of the laser path.

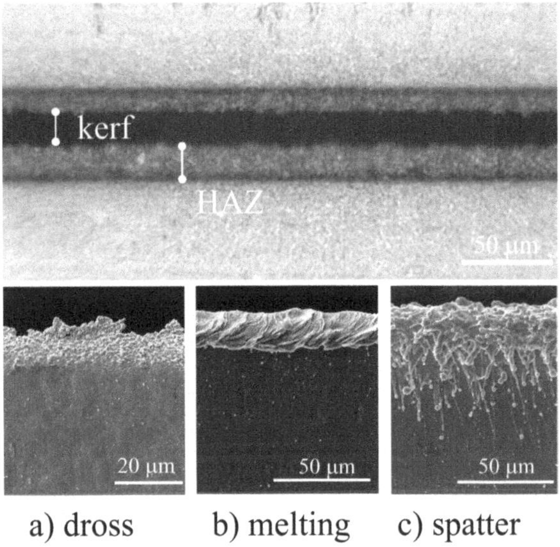

Fig. 3 Defects classification and definition of kerf and HAZ width

3 Results and Discussions

3.1 Ablation Efficiency

The influence of scanning speed v and laser pulse repetition rate *PRR* on cutting efficiency, defined as the maximum portion of energy utilized for ablation (as described in Sect. 2.3), was investigated.

Experimental results show a linear decrease in kerf width with increasing speed ($R^2 = 98\%$) and PRR ($R^2 = 80\%$). At higher speeds, maximum energy efficiency is achieved because a big portion of incident energy contributes to material removal, with minimal heat accumulation from pulse overlap. Moreover, the standard deviation of mean efficiency at each speed level decreases fivefold at lower speeds, indicating that the cutting process becomes less sensitive to frequency variations, as shown in Fig. 4.

Aluminum and copper registered a maximum efficiency of 0.79 and 0.86, respectively, at 400 kHz. As frequency increases, part of the incident energy in aluminum is retained within the material, reducing overall efficiency. Another effect of heat accumulation is the decrease in the SP threshold fluence F_{th}^{RE} at higher frequencies, leading to increased cutting speeds. For aluminum, the SP maximum cutting speed jumps from 0.6 m/s for 200 kHz $\left(F_{th}^{RE} = 14.9J/\text{cm}^2\right)$ up to 0.8 m/s for 1000 kHz $\left(F_{th}^{RE} = 3.5J/\text{cm}^2\right)$. This effect is less pronounced in copper, which, due to its higher thermal conductivity, is less affected by heat accumulation and the maximum cutting speed is set constant to 0.9 m/s. On average, efficiency values in copper are 24% higher than in aluminum.

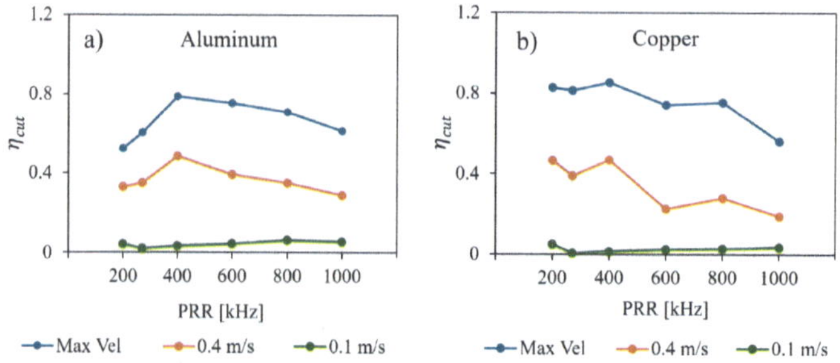

Fig. 4 Single pulse cut efficiencies for **a** aluminum and **b** copper

The concept of a progressive decrease in the ablation threshold with increasing frequency, leading to higher material absorptivity due to residual heat from previous pulses, becomes more prominent in the burst mode. Micromachining literature has long-established that dividing a pulse's energy into a series of sub-pulses (in MHz and GHz range) minimize thermal damage and mechanical stress, enabling higher processing speeds. Huang et al. [10] demonstrated that maximum speeds could be achieved with the highest degree of pulse splitting. Hence, the maximum speeds for each frequency level were identified for both materials in the MHz and GHz modes when N = 10 and P = 10. The ablation thresholds F_{th}^{RE} attained for both Al and Cu were lower for all burst regimes in comparison to the single-pulse processing.

In MHz mode, aluminum reaches 3 m/s at 600 kHz—almost four times the SP speed—while in GHz mode, it peaks at 2.3 m/s at 1000 kHz. When comparing MHz and GHz efficiencies in the BM campaign, MHz mode outperforms GHz mode efficiencies for PRR < 600 kHz across all three speed levels. However, at 1000 kHz and 1.6 m/s MHz mode reaches a no-cut condition, whereas GHz mode fails to cut at 200 kHz at the same speed. This trend highlights the interdependence of the two "duty cycles": shorter intra-pulse intervals for the GHz mode require higher pulse overlap to sustain the full cut compared to MHz mode. Despite this, both efficiencies were lower than those in the SP regime, suggesting greater residual heat accumulation, which increases with intra-burst frequency for this set of process parameters.

For copper, scanning speed climbed up to 3.4 m/s in MHz mode at 200 kHz. Interestingly, the speed increase from SP to BM in MHz mode is consistent at 3.8 times for both materials: from 0.8 to 3 m/s for Al and from 0.9 to 3.4 m/s for Cu. As shown in Fig. 5, unlike other trends, while aluminum MHz max speeds decrease beyond 600 kHz, copper follows this trend across all frequencies. Theoretically, the opposite would be expected, as higher repetition rates would produce a significant rise of temperature and then induce a lower threshold and a higher absorption of the target material [17, 18], facilitating kerf opening. At the same time, an antagonist phenomenon comes into play. Many authors [19, 20] have pointed out both plasma and particle shielding after 200 fs-laser ablation of Al and Cu for high frequencies (1 MHz). Plasma shielding is related to primary material ejection after the laser pulse,

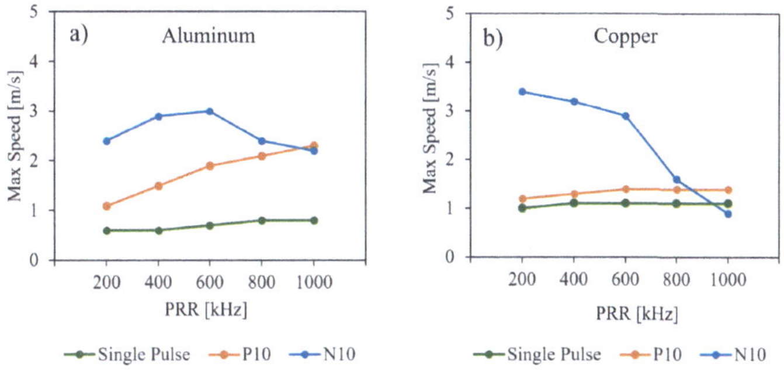

Fig. 5 Maximum scanning speed for each frequency level of **a** aluminum and **b** copper

creating a particle cloud which causes beam defocusing and prevents further heat transfer to the target material, thereby limiting the achievable cutting speeds.

3.2 *Ablation Quality*

In the previous section, it was observed that higher cutting speeds led to greater energy efficiency. However, cuts performed near the maximum speed in the SP mode were found to exhibit burs along the cut edge for pulse overlaps lower than 87%. In this case, heat accumulation at high PRRs positively affects quality, leading to a more consistent kerf. However, exceeding heat accumulation may cause melting along the cut edge, as shown in Fig. 6b.

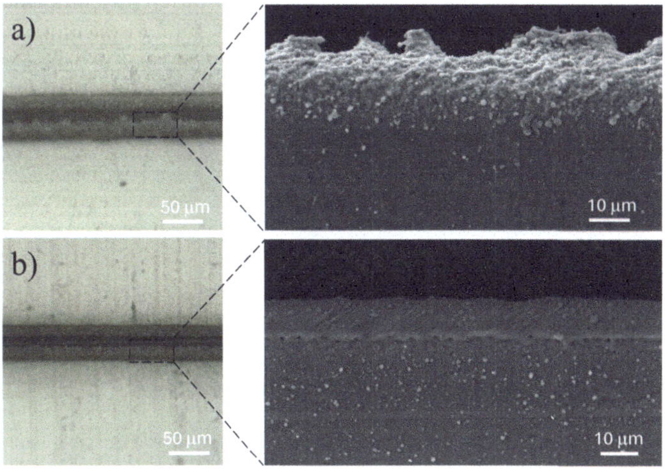

Fig. 6 Aluminum single pulse cuts at maximum speed 800 mm/s carried out at **a** 200 kHz and **b** 1000 kHz

Additionally, SP aluminum processing was accompanied by the occurrence of phase explosion, which caused micrometer liquid particles and gas mixtures to splash out. Per our observation, the greater the pulse overlap and the material removal, the higher the number of particles expelled and then redeposited near the cut edge.

When burst mode is activated, thermal effects exacerbate. Every test performed in the burst mode on aluminum involved *liquid* formation along the cut edge. Even after a fs pulse, a significative quantity of absorbed energy remains within the material, generating heat accumulation if the interval between pulses is too short for a complete heat dissipation in the bulk material. Since sub-pulses occur at intervals (τ_{burst}) shorter than the material relaxation time (τ_{rel}), part of the energy from subsequent sub-pulses accumulates heat. Vaghasiya et al. [21] demonstrated with simulations that the electron temperature increases with rising bursts pulse numbers. If heat accumulation leads to a temperature increase which exceeds the melting point T_m in the non-ablated material next to the interaction zone, the surface remains liquid between two subsequent pulses. Each new pulse then hits a still liquid surface which leads to a completely different, and in particular no longer "cold" process, which will be dominated by hydrodynamic effects and the benefit of ultrafast laser ablation is lost [15]. This concept is illustrated in Fig. 7a. This physical explanation is supported by two facts: first, aluminum is more prone to this phenomenon due to its lower thermal conductivity and melting point; second, increasing the number of pulses P only broadens the molten zone rather than increasing the kerf width for high PRR (explicit in Fig. 7b).

At low frequencies on aluminum, a distinct pattern is observed in both GHz and MHz modes (Fig. 7c). For even pulse numbers, the kerf width increases (reducing efficiency), but this trend is lost when increasing the PRRs or changing material. This phenomenon can be attributed to several factors related to particle plume dynamics and thermal accumulation. Žemaitis et al. [9] reported lower removal efficiencies for the second pulse due to particle shielding, indicating different plume and thermal effects when the process changes (drilling to cutting). As a result, little material is ablated by the third pulse. In copper, this even–odd dependence is absent.

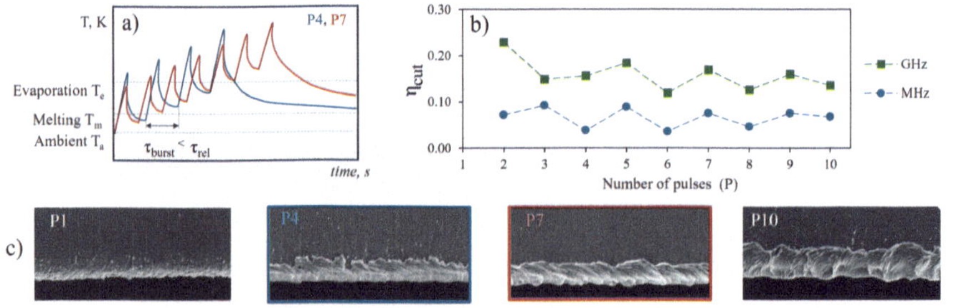

Fig. 7 Number of pulses within a burst influencing melting formation. Possible explanation in: **a** Temperature trend indicating heat accumulation. **b** Cut efficiency for aluminum in GHz and MHz modes when 200 kHz and 0.8 m/s are applied with related SEM images in (**c**)

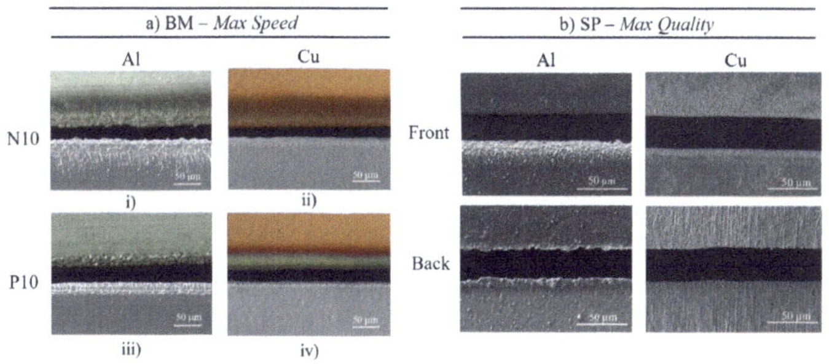

Fig. 8 **a** BM defects formation when max velocities are used for aluminum and copper causing spattering in the MHz mode for lower overlapping and F_{max} values of (i) 60%, 2.8 Jcm^2, (ii) 73%, 1.4 Jcm^2 and dross formation for aluminum at (iii) 82%, 2.8 Jcm^2 and copper along with HAZ at (iv) 97%, 0.57 Jcm^2. **b** High quality aluminum and copper SP cuts performed at 9.4 J/cm^2, O% = 95.6% and 5.7 J/cm^2, O% = 99.8%, respectively

When MHz high speeds are coupled with low PRRs (Overlapping $\leq$76%), hydrodynamic instabilities in molten aluminum cause fingers-like droplets detachment, as shown in Fig. 8a(i). Each image displays an OM view (top) and the corresponding SEM picture (bottom). In GHz cutting, fingers-like spatters are absent, and particle deposition near the molten edge is minimized across all process conditions.

In SP processing of copper, burs form at 200 kHz with high speeds, similar to aluminum but at a higher overlap threshold of 93%. While the front edge maintains good quality across different sets of parameters, the backside's quality improves while decreasing speed, with a perfect cut achieved at 0.1 m/s, as shown in Fig. 8b. Hence, the best balance of energy efficiency and quality is at 0.4 m/s.

Below 400 kHz in SP cuts, the HAZ appears along the cut edges, enlarging as PRRs decrease and is present for all GHz cuts (see Fig. 8a(iv)). Tests confirmed that Argon shielding gas prevents HAZ, confirming it results from heated foil-environment reaction.

4 Conclusions

Aluminum and Copper current collectors for secondary battery manufacturing were ablated using a femtosecond laser at PRRs = 200 kHz – 1 MHz both in the single-pulse and burst mode (MHz and GHz inter pulse frequency). A dual analysis of cut efficiency and quality was conducted, leading to the development of a simplified model that quantifies the fraction of laser energy effectively used for ablation and the portion retained in the material.

The analysis has led to the following findings:

- For aluminum in SP mode, efficiency peaked at intermediate frequencies (400–600 kHz), balancing cutting speed and defect minimization.

- Copper exhibited high-quality front edges in all conditions, though low speeds were required for a consistent kerf across the entire thickness.
- The burst mode allowed for a significant speed increase, though at the cost of increased thermal effects, particularly in aluminum. Specifically, in MHz mode, speeds up to 3.4 m/s were achieved for copper and 3 m/s for aluminum, while GHz mode showed reduced spattering but introduced a heat-affected zone.

While the proposed efficiency model is intentionally simplified, it enables efficient rapid extraction of key insights into SP experimental quality with minimal computational cost. This approach captures the main trends related to intra-burst pulse separation and repetition rate, offering a practical tool for analyzing process conditions and cut edge quality without requiring complex simulations.

References

1. Jeong H, Jang J, Jo C (2022) A review on current collector coating methods for next-generation batteries. Chem Eng J 446:136860. https://doi.org/10.1016/j.cej.2022.136860
2. Pfleging W (2018) A review of laser electrode processing for development and manufacturing of lithium-ion batteries. Nanophotonics 7:549–573. https://doi.org/10.1515/nanoph-2017-0044
3. Yuan Y, Wang H, Han X et al (2024) The local lithium plating caused by anode crack defect in Li-ion battery. Appl Energy 361:122968. https://doi.org/10.1016/j.apenergy.2024.122968
4. Baumann R, Alamri S, Aguilar-Morales AI et al (2022) Advanced remote laser cutting of battery foils using an interference approach. Mater Lett X 14:100138. https://doi.org/10.1016/j.mlblux.2022.100138
5. Lee D, Patwa R, Herfurth H, Mazumder J (2012) Computational and experimental studies of laser cutting of the current collectors for lithium-ion batteries. J Power Sources 210:327–338. https://doi.org/10.1016/j.jpowsour.2012.03.030
6. Angeloni C, Liverani E, Ascari A, Fortunato A (2024) Characterization and process optimization of remote laser cutting of current collectors for battery electrode production. J Mater Process Technol 324:118266. https://doi.org/10.1016/j.jmatprotec.2023.118266
7. Chichkov BN, Momma C, Nolte S et al (1996) Femtosecond, picosecond and nanosecond laser ablation of solids. Appl Phys A 63:109–115. https://doi.org/10.1007/BF01567637
8. Weber R, Graf T (2021) The challenges of productive materials processing with ultrafast lasers. Adv Opt Technol 10:239–245. https://doi.org/10.1515/aot-2021-0038
9. Žemaitis A, Gudauskytė U, Steponavičiūtė S et al (2025) The ultrafast burst laser ablation of metals: Speed and quality come together. Opt Laser Technol 180:111458. https://doi.org/10.1016/j.optlastec.2024.111458
10. Huang J, Shi W, Huang J et al (2021) High speed pulsed laser cutting of anode material for a Li-ion battery in burst mode. Opt Mater Express 11:2300. https://doi.org/10.1364/OME.425816
11. Heidari Orojloo P, Demir AG (2024) Study of burst mode for enhancing the ps-laser cutting performance of lithium-ion battery electrodes. J Laser Appl 36:042013. https://doi.org/10.2351/7.0001417
12. Audouarda E, Pallarèsb D, Romanob J-M, et al. Characterization of batteries materials ablation by femtosecond pulses
13. Žemaitis A, Gaidys M, Brikas M et al (2018) Advanced laser scanning for highly-efficient ablation and ultrafast surface structuring: experiment and model. Sci Rep 8:17376. https://doi.org/10.1038/s41598-018-35604-z
14. Breitling D, Ruf A, Dausinger F (2004) Fundamental aspects in machining of metals with short and ultrashort laser pulses. In: Herman PR, Fieret J, Pique A, et al (eds). San Jose, Ca, p 49

15. Weber R, Graf T, Berger P et al (2014) Heat accumulation during pulsed laser materials processing. Opt Express 22:11312. https://doi.org/10.1364/OE.22.011312
16. Incropera FP, DeWitt DP, Bergman TL, Lavine AS (2007) Fundamentals of heat and mass transfer, 6th edn. Wiley, Hoboken, NJ
17. Joerg S (2012) Characterisation of interaction phenomena in high repetition rate femtosecond laser ablation of metals. https://doi.org/10.13140/2.1.2575.5841
18. Schille J, Schneider L, Loeschner U, et al (2011) Micro processing of metals using a high repetition rate femtosecond laser: from laser process parameter study to machining examples. In: International congress on applications of lasers & electro-optics. Laser Institute of America, Orlando, Florida, USA, pp 773–782
19. König J, Nolte S, Tünnermann A (2005) Plasma evolution during metal ablation with ultrashort laser pulses. Opt Express 13:10597. https://doi.org/10.1364/OPEX.13.010597
20. Ancona A, Döring S, Jauregui C et al (2009) Femtosecond and picosecond laser drilling of metals at high repetition rates and average powers. Opt Lett 34:3304. https://doi.org/10.1364/OL.34.003304
21. Vaghasiya H, Krause S, Miclea P-T (2023) Fundamental study of a femtosecond laser ablation mechanism in gold and the impact of the GHz repetition rate and number of pulses on ablation volume. Opt Mater Express 13:982. https://doi.org/10.1364/OME.474452

An Innovative Mechanistic-Thermal Model for Studying the Effect of MQL in Peck Drilling

Luca Bernini, Paolo Albertelli, and Michele Monno

Abstract Traditional cutting fluids pose sustainability challenges, with high flow rates, limited high-speed effectiveness, and contribute to 7–17% of product cost. Minimum Quantity Lubrication (MQL) emerges as a viable alternative, though its phenomenological effects on peck drilling require further research. In this study, Oxley's orthogonal cutting model is extended to oblique peck drilling, including Johnson–Cook's material characterization. Moving infinitesimal heat sources are adopted to study heat partitioning. Applied to 48 experiments, varying MQL air/oil flow rates, cutting/feed speeds, the model predicts cutting torque, thrust force and workpiece temperature, providing insights into the cutting phenomenology. Correlations were found between cutting parameters and friction angle. Oil flow reduces friction, whereas air flow slightly decreases the heat transfer to the workpiece. This study supports MQL process parameter optimization and enables tool and workpiece temperature prediction.

Keywords MQL peck-drilling · Oblique cutting Oxley's model · Johnson–Cook's material · Analytical mechanistic-thermal model · MQL parameters effect

1 Introduction

Nowadays, environmental impact of machining operations is becoming crucial. Regarding this aspect, involving conventional cutting fluids in the process is critical. Traditional lubricants require high flow rates, and they lose effectiveness at high speeds. Vegetable-based, biodegradable and synthetic fluids represents a viable solution to mitigate the environmental impact of lubrication [1, 2]. Minimum Quantity Lubrication (MQL) is receiving attention in the industrial field, too. In MQL

L. Bernini (✉) · P. Albertelli · M. Monno
Department of Mechanical Engineering, Politecnico Di Milano, Milan, Italy
e-mail: luca1.bernini@polimi.it

MUSP Macchine Utensili Sistemi Di Produzione, Piacenza, Italy

L. Fratini et al. (eds.), *Selected Topics in Manufacturing*, Lecture Notes in Mechanical Engineering, https://doi.org/10.1007/978-3-031-99501-9_6

machining, oils are used as the fluid medium to diminish friction and prevent adhesion between the workpiece, chip and tool, to reduce heat generation and temperature [3].

Drilling is a widely used cutting operation in manufacturing. Despite being a mature technology, it requires specific attention for optimal results. Heat generation during cutting arises from friction and plastic deformation of material. In deep drilling, high depth-to-diameter ratios can cause chip clogging, increasing friction and consequently heat [4]. Excessive temperatures accelerate tool wear, built-up edges, and potential tool plastic deformation [5], while also causing dimensional inaccuracies, oxidation, residual stresses, and microcracks in the workpiece (WP) [6]. Machining temperature is influenced by tool and WP materials, cutting parameters, and lubrication strategy.

Peck drilling is a widely adopted strategy to mitigate the cited problems [4, 5]. It consists in splitting the total drilling length into sequential dives divided by a short period where the drill is returned to the workpiece surface to foster chip evacuation. Indeed, this strategy helps to keep the cutting temperature under control. The study of temperature distribution in metal cutting is hence crucial for determining optimal cutting parameters. Since the primary factors for the temperature increase are the dissipation of the plastic energy and the friction among the chip and the tool, an estimation of the temperature must start from an estimation of the cutting forces.

The prediction of machining forces in drilling has been widely studied using empirical, mechanistic, and finite element methods, all based on Oxley's foundational model [7]. Empirical models, developed to study thrust force, torque, and drilling parameters, were limited by regression fitting and lacked process understanding. Mechanistic models improved accuracy by incorporating process kinematics, assuming cutting forces are proportional to uncut chip area or chip load [8]. This method requires fewer calibration experiments and provides better results across various conditions, by including physical modelling. For example, Chandrasekharan et al. [9] used a mechanistic approach to model forces in drilling Fiber-Reinforced Composite Materials (FRCM). The finite element method (FEM) offers the most detailed predictions, simulating parameters like temperature, friction, and heat transfer [8]. Although computationally intensive, FEM needs only experimental data for validation. Strenkowski et al. [10] exemplified this through a validated FEM model for drilling AISI 1020 steel, confirming its accuracy in predicting thrust force and torque.

In many industrial scenarios, direct temperature measurement is unfeasible due to inaccessibility, thus, temperature is often estimated through models or related parameters. Temperature at the tool-chip interface during metal cutting is a key research focus. Estimation methods include analytical, empirical, and numerical approaches [11]. Patne et al. [12] combined these methods to estimate temperatures during titanium alloy drilling. They used FEM simulations to model the thermal field, validating with infrared (IR) camera and thermocouple measurements. Their findings showed that cutting speed and feed rate affect temperature distribution in the workpiece. Similarly, Segurajauregui et al. [13] found that higher cutting speeds reduced heat transfer to the workpiece, while lower feed rates increased it. Fleischer et al. [14] developed a model for heat flux in FEM-based distortion analysis with MQL. Their experiments revealed significant variation in heat generation with tool wear and cutting parameters. Nevertheless, no research studied temperature and effects of MQL on peck drilling.

With this scope, we developed a mechanistic model by extending Lalwani et al.'s orthogonal theory [15] to oblique cutting. Lalwani et al. modified Oxley's predictive model [7] to include the Johnson and Cook flow stress material model, providing an analytical framework for chip formation and cutting force determination in orthogonal cutting, incorporating assumptions about the shear zones, strain rates, and the effects of temperature on flow stress. Drilling is an oblique cutting process, where the cutting lips are not orthogonal to the cutting speed, generating a third force component. We extend the Lalwani model for peck drilling, drawing on the work of Minukhin [16] and Altintas [17], as described in section *Materials and Methods*. Experimentation is detailed in dedicated subsections. Section *Results and discussion* shows prediction capabilities of the conceived model. Final remarks and future works are drawn in section *Conclusions*.

2 Materials and Methods

This section presents the contribution of the paper in the two fields of mechanistic and thermal modelling of a peck drilling operation.

2.1 Mechanistic Model

The proposed formulation follows Fig. 1 as a reference. The analysis starts from the definition of the web angle ϑ generated between the cutting lip and the radial line to a generic point P of the cutting lip [16, 17]:

$$\sin\vartheta = \frac{t_p}{r} \tag{1}$$

where t_p is the half web thickness, and r the generic radial coordinate of point P.

The helix angle δ_r at point P can then be computed considering the helix angle δ (at $r = R$), where R is the drill-bit radius [16].

$$\tan\delta_r = \frac{r}{R}\tan\delta \tag{2}$$

Two oblique angles can be computed from ϑ and the drill-bit half taper angle κ_t, which are the inclination angle i and the reference angle ε [16, 17]:

$$\sin i = \sin\vartheta \sin\kappa_t \tag{3}$$

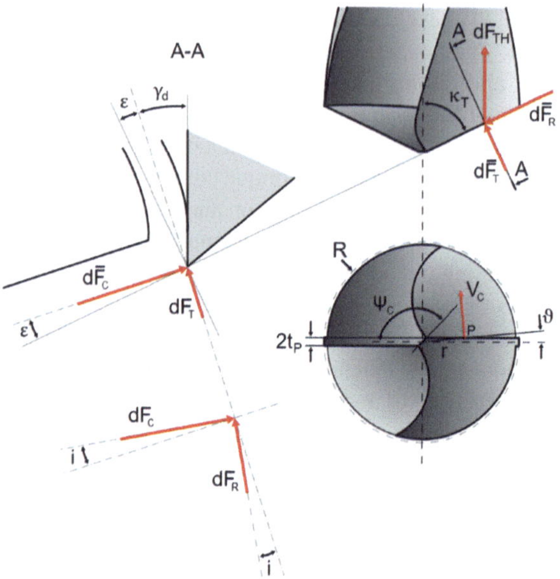

Fig. 1 Representation of angles and forces on drilling oblique cutting

$$\tan \varepsilon = \tan \vartheta \cos \kappa_t \tag{4}$$

From here, it is possible to compute two quantities that are of prominent importance for oblique cutting: the uncut chip thickness t_1, that can be computed through Eq. (5) [16] including the feed per revolution f; the dynamic rake angle, that can be computed through Eq. (6) [16, 17].

$$t_1 = \frac{f}{2} \sin \kappa_t \cos \varepsilon \tag{5}$$

$$\tan \gamma_d = \frac{\tan \delta_r}{\sin \kappa_t} \left(\cos \vartheta + \sin \vartheta \tan \vartheta \cos^2 \kappa_t \right) - \tan \vartheta \cos \kappa_t \tag{6}$$

It is now possible to compose the infinitesimal cutting forces generated in the reference frame of the infinitesimal cutting lip element (forces identified by an overbar) in the tangential, radial and cutting infinitesimal forces in the drill-bit reference frame, according to the reference and inclination angles shown in Fig. 1 [16]:

$$\mathrm{d}\overline{F}_t = \mathrm{d}F_t \cos \varepsilon - \mathrm{d}F_c \sin \varepsilon \tag{7}$$

$$\mathrm{d}\overline{F}_r = \mathrm{d}F_r \cos i - \mathrm{d}F_c \sin i \tag{8}$$

$$\mathrm{d}\overline{F}_c = \mathrm{d}F_c \cos i - \mathrm{d}F_r \sin i \tag{9}$$

Following Lin et al. [18], the cutting dF_c and tangential dF_t forces are independent from the inclination angle. Indeed, the cutting and tangential forces can be computed with the orthogonal cutting formulation (and thus, following the work by Lalwani et al. [15]). The missing force component, dF_r can be derived as a function of dF_c, dF_t, i and the friction angle λ [16]:

$$dF_r = \sqrt{dF_c^2 + dF_t^2}\sin\lambda\cos\gamma_d\tan i \tag{10}$$

In these formulations, it is assumed that chip flow angle η, can be considered equal to the inclination angle i, according to the Stabler's rule [7].

Finally, the thrust force generated on the infinitesimal element of the cutting lip is:

$$dF_{th} = d\overline{F}_t\sin\kappa_t - d\overline{F}_r\cos\kappa_t \tag{11}$$

In order to compute the total cutting torque and thrust force of the drilling operation, the lip is divided in n elements (here, $n = 8$). For thrust force, the indentation component generated by the chisel on the material must be considered [17]. Indeed, differentials were transformed in finite differences:

$$C_t = 2\sum_{j=1}^{n} r_j\Delta\overline{F}_{c,j} \tag{12}$$

$$F_{th} = F_{th,lip} + F_{th,ch} = 2\sum_{j=1}^{n}\Delta F_{th,j} + A_{ch}H_b \tag{13}$$

where, H_b is the Brinell hardness of tool material. The indentation area of chisel edge A_{ch} can be computed from the chisel edge angle ψ_c as [17]:

$$A_{ch} = \frac{2t_p f}{\sin(\pi - \psi_c)\cos\kappa_t} \tag{14}$$

The total drilling power is then defined as the sum of the thrust and cutting powers:

$$P_d = P_{cut} + P_{th} = C_t\omega + F_{th}v_f \tag{15}$$

where ω is the angular velocity of the drill-bit and $v_f = 2\pi\omega f$ is the feed speed. It is now possible to define the thermal energy dissipated during drilling. A partition coefficient (β) is defined to determine the portion of heat flowing either into the WP or in the chip-tool pair:

$$P_d = \left(P_{d,chip} + P_{d,tool}\right) + P_{d,wp} = \beta P_d + (1-\beta)P_d \tag{16}$$

2.2 Thermal Model

A thermal model was conceived starting from the heat equation solution for an instantaneous point source placed in (x_s, y_s, z_s) (Fig. 2 (left)), releasing energy Q at $t = t_s$ [10]. The temperature field generated by the infinitesimal point source, as a function of time is:

$$T(x, y, z, t) = \frac{Q}{(4\pi\alpha(t - t_s))^{3/2}} \exp\left(-\frac{(x - x_s)^2 + (y - y_s)^2 + (z - z_s)^2}{4\alpha(t - t_s)}\right) \quad (17)$$

where α is the diffusivity of the WP material. In order to adapt such model to a peck drilling operation, infinitesimal moving point sources (along the lip and along the trajectory followed by each element of the lip in a revolution) are superimposed. The generated surface is thus approximated through a cone in the space (see Fig. 2 (left)).

The cone must move in space during time, in order to follow the trajectory of the drill-bit. The computation starts from the infinitesimal area of the cone of heat sources, corresponding to the moving infinitesimal point source (Fig. 2 (center)):

$$dA = r d\gamma \frac{d\zeta}{\cos(\kappa_t)} \quad (18)$$

where r is the generic radial position of the infinitesimal area in the cone, $d\gamma$ is the infinitesimal angle, such that $rd\gamma$ is the infinitesimal arc length of the point source, $d\zeta$ is the infinitesimal height of the point source in the drill-bit axial direction. The obtained temperature field at a specific time t, due to a moving conical heat source placed in $\left(x_s(t_s), y_s(t_s), z_p(t_s)\right)$ and releasing power $q(t_s)$ from time 0 to t, can be expressed as:

$$T(x, y, z, t) = \int_0^t \int_0^h \int_0^{2\pi} \frac{q(t_s) r}{(4\pi\alpha(t - t_s))^{\frac{3}{2}} \rho c_p S \cos \kappa_t} \cdot$$

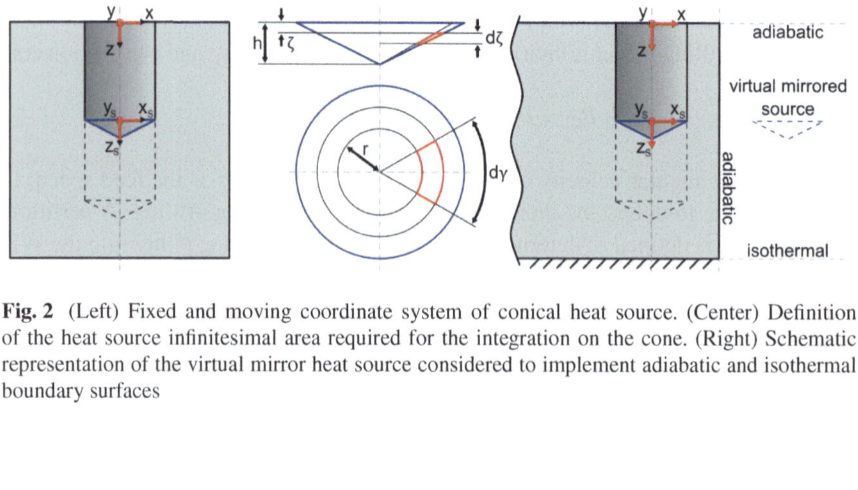

Fig. 2 (Left) Fixed and moving coordinate system of conical heat source. (Center) Definition of the heat source infinitesimal area required for the integration on the cone. (Right) Schematic representation of the virtual mirror heat source considered to implement adiabatic and isothermal boundary surfaces

$$\cdot \exp\left(-\frac{(x - x_s(t_s))^2 + (y - y_s(t_s))^2 + (z - z_s(t_s) - \zeta)^2}{4\alpha(t - t_s)}\right) \mathrm{d}\gamma \mathrm{d}\zeta \mathrm{d}t_s \tag{19}$$

$q(t_s)$ represents the strength of the heat source during peck drilling ($q(t_s) = \beta_1(t_s)\beta$); $\beta_1(t_s)$ is 1 in stationary cutting, 0.3 during dwell times in the WP, 0 elsewhere; ρ is the WP material density; c_p is the specific heat capacity of the WP material; S is the lateral surface of the cone. $z_s(t_s)$ was decomposed into two components: $z_p(t_s)$, the center of the cone with respect to WP surface (following drill-bit trajectory) and ζ, the position of the infinitesimal source within the cone (Fig. 2 (center)). Since Eq. (19) considers an infinite WP, border effects must be introduced as adiabatic and isothermal surfaces. This was done introducing positive ($+q(t_s)$) and negative ($-q(t_s)$) virtual mirrored sources (Fig. 2 (right)). Boundary surfaces are equidistant from the original and virtual mirrored source. Negative heat powers generate negative temperature fields, cancelling temperature increase by the original source on the target surface. Positive heat powers generate opposite thermal flows, cancelling the heat exchanged through the surface.

2.3 Experimental Set-Up

The experimental campaign consisted of 53 drilling tests performed using a Mandelli M5 machine tool on an AISI 316L steel WP (Fig. 3a). Material parameters and thermal properties are reported in Table 1. Each hole was machined following a peck drilling strategy characterized by 5 pecks. The first one used for centering purposes only. The remaining 4 pecks, the focus of our analysis, are 3 mm deep, for a total depth of the hole of 12 mm. Up to 18 holes can be drilled per each workpiece surface. Thus, two WPs coming from the same production batch were used.

A hole is positioned at 6.7 mm from the side of the WP and the distance between two subsequent holes is 13.1 mm (Fig. 3d). The hole has a diameter of 8.8 mm, corresponding to the drill-bit diameter ($2R$).

On the side of the WP, micro-holes were pre-drilled in correspondence of each hole axis. Micro-holes were drilled 8.3 mm far from the surface of the workpiece. Such micro-holes were used as positioning slots for a thermocouple, with the aid of a support comb (Fig. 3c). With a depth of 2.1 mm, micro-holes allowed to position the thermocouple 0.2 mm far from the experimentation hole, in the middle of the third peck. A particular tool holder offered by Gühring was employed to reduce the impact of tool run-out. Two solid carbide drill-bits (DB1 and DB2), Walter A3293TTP-8.8, were used for the experimental tests (Table 2). The trend along the lip was instead derived following Minunkhin et al. [16].

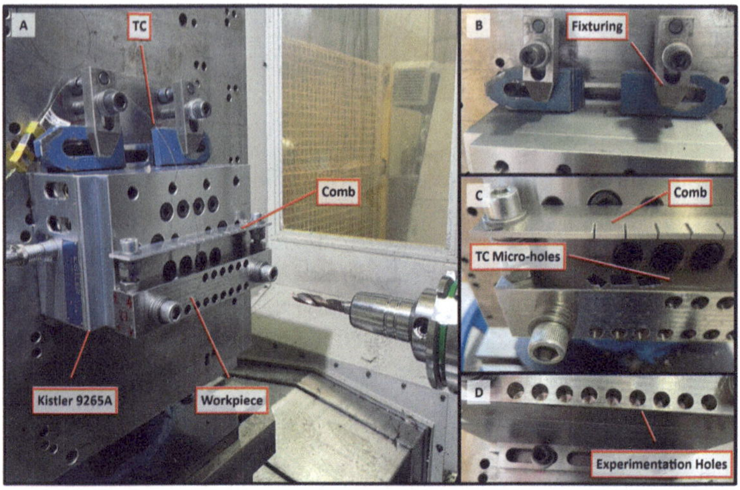

Fig. 3 Details of the experimental setup: **a** overview; **b** fixturing system; **c** comb and micro-holes for the thermocouple; **d** experimentation holes

Table 1 Johson-Cook's model parameters [19] and thermal properties [20] for AISI316L

JC's parameters		Thermal properties	
A	412 MPa	Specific heat capacity, c_p	$4.32\left[\frac{J}{kgK}\right] + 0.13\left[\frac{J}{kgK^2}\right] \cdot T[K]$
B	761 MPa		
C	0.038	Conductivity, k	$9.28\left[\frac{W}{mK}\right] + 0.0157\left[\frac{W}{mK^2}\right] \cdot T[K]$
n	0.510		
m	0.520		

Table 2 Drill-bit parameters

Parameter	Symbol	Value
Cutting edge radius	R	4.4 mm
Cutting edge length	L_c	35 mm
Half taper angle	κ_t	70°

2.4 *Data Acquisition System*

A NI cDAQ9174 was adopted to measure spindle power P_s, , cutting forces (Kistler 9265A dynamometric plate), WP temperature (Tersid MTS-15000 Series Type K ϕ 0.5 mm thermocouple) at 20 kHz, to reduce the effect of electrical noise on power acquisitions. A power meter was used to measure the spindle active power as $P_s(t) = i_1(t)v_1(t) + i_2(t)v_2(t) + i_3(t)v_3(t)$.

Currents were measured through a LEM ring based on Hall's principle. A time interval $\Delta t = 0.05$ s was used for filtering purposes to remove the high frequency noise generally associated to electrical power. The resulting acquisitions are sampled at 20 Hz. Real-time data associated to axis position and velocity of the linear axes and the spindle were acquired through SinuCom NC acquisition system from Siemens. This system was used to access and store quantities from the numerical control of machine tools at a frequency of 250 Hz.

Results in section *Results and discussion* are based on experimental cutting torque and thrust force. Experimental cutting torque is computed based on active spindle power P_s. and spindle efficiey η_s [21], and averaged over pecks.

$$C_{cut}(t) = \frac{P_s(t)}{2\pi\omega\eta_s} \tag{20}$$

$$\overline{C}_{cut} = \frac{1}{4}\sum_{k=1}^{4}\overline{C}_{cut,k} \tag{21}$$

A similar approach was adopted for the thrust force:

$$\overline{F}_{th} = \frac{1}{4}\sum_{k=1}^{4}\overline{F}_{th,k} \tag{22}$$

2.5 Design of Experiments

The experimentation was conceived following a DoE approach, consisting of a $n_e \cdot 2^p + n_c$ design. The chosen parameters were feed per revolution f (with levels 0.048 – 0.112 mm/rev), cutting velocity v_c (with levels 40 – 60 m/ min), MQL air flow rate $\dot{m}_{air}$ (with levels 60 – 100 L/min) and MQL oil flow rate $\dot{m}_{oil}$ (with levels 0 – 10 mL/h). n_e was set to 2 and n_c to 4. The DoE was enriched with additional points (marked with letter E when at lubricating conditions extrema, with PC when at the central point) and validation points (at unseen conditions and marked with V). Moreover, the replicates were distributed on DB1 and DB2, a blocking factor for the randomization of the experiments. For the sake of conciseness, the test were divided in groups based on their parameters (Table 3). The resulting DoE is then listed in Table 4.

Table 3 Parameter groups and associated parameter values

Group	$\dot{m}_{oil}$[mL/h]	$\dot{m}_{air}$[L/ min]	Group	f[mm/rev]	v_c[mm/ min]
LC1	0	60	PP1	0.048	40
LC2	10	60	PP2	0.112	40
LC3	5	80	P	0.080	50
LC4	0	100	PP4	0.048	60
LC5	10	100	PP5	0.112	60
	–	–	PP6	0.048	50
	–	–	PP7	0.112	50
	–	–	PPV	0.096	55

3 Results and Discussion

3.1 Experimental Findings

Firstly, a multivariate linear regression was performed [22] to determine whether or not the cutting conditions and lubricating conditions were influencing thrust force and cutting torque. The regression models ($C_t = 3.385 - 0.040v_c + 44.297f - 0.044\dot{m}_{oil} - 0.003\dot{m}_{air}$, and $F_{th} = 720.29 - 4.26v_c + 12857.00f - 2.39\dot{m}_{oil} - 0.21\dot{m}_{air}$,) properly fit the models with p-values close to 0, and with R^2-adjusted.

indicators of 95% and 90%, respectively. While $\dot{m}_{air}$ is the only parameter not statistically influencing cutting torque, thrust force behaviour is dominated by the feed. This is in accordance with widespread mechanistic models like in [17], where increasing feed-rate increases the chip section, requiring higher effort for cutting materials. For what concerned the relations with MQL parameters, cutting torque is negatively correlated to the oil flowrate, implying less friction forces during cutting.

A second linear regression was performed on the value of temperature measured at the WP in correspondence of the third peck (i.e., the peak temperature except for some rare cases). The model ($T_3 = 295.9 - 1.3v_c - 448.6f - 1.5\dot{m}_{oil} - 0.4\dot{m}_{air}$) showed lower R^2-adj (66%), nevertheless the p-value for the regression is well below 5%,

indicating statistical significance of the model. Residuals associated to PP3 (the central point for cutting conditions) were the more critical, suggesting that the underlying model may not be purely linear. From this ANOVA the effect of each parameter resulted statistically different from zero. Moreover, all the parameters tended to decrease the maximum temperature reached in the workpiece. The negative correlation between temperature, and cutting velocity and feed-rate may be explained with the lower time available for the heat to diffuse in the WP. The negative correlation with oil and air flowrates may instead be affected by friction reduction. In fact, friction generates heat during the cutting action, which is reduced by the presence of oil. The increase of the convective coefficient due to the presence of air, represents

Table 4 Design of experiments

ID	Run order	$\dot{m}_{oil}$ [mL/h]	$\dot{m}_{air}$ [L/min]	f [mm/rev]	v_c [mm/ min]
1	20 (DB1) I 52 (DB2)	0	60	0.048	40
2	2 (DB1) I 51 (DB2)	10	60	0.048	40
3	19 (DB1) I 45 (DB2)	0	100	0.048	40
4	29 (DB1) I 50 (DB2)	10	100	0.048	40
5	1 (DB1) I 31 (DB2)	0	60	0.112	40
6	25 (DB1) I 41 (DB2)	10	60	0.112	40
7	7 (DB1) I 34 (DB2)	0	100	0.112	40
8	22 (DB1) I 42 (DB2)	10	100	0.112	40
9	28 (DB1) I 30 (DB2)	0	60	0.048	60
10	5 (DB1) I 53 (DB2)	10	60	0.048	60
11	10 (DB1) I 37 (DB2)	0	100	0.048	60
12	8 (DB1) I 33 (DB2)	10	100	0.048	60
13	18 (DB1) I 44 (DB2)	0	60	0.112	60
14	21 (DB1) I 35 (DB2)	10	60	0.112	60
15	24 (DB1) I 36 (DB2)	0	100	0.112	60
16	12 (DB1) I 48 (DB2)	10	100	0.112	60
17	11–16 (DB1) I 38–49 (DB2)	5	80	0.080	50
18E	4 (DB1) I 46 (DB2)	0	60	0.080	50
19E	15 (DB1) I 43 (DB2)	10	60	0.080	50
20E	3 (DB1) I 32 (DB2)	0	100	0.080	50
21E	26 (DB1) I 40 (DB2)	10	100	0.080	50
22PC	13 (DB1) I 47 (DB2)	5	80	0.048	50
23PC	6 (DB1) I 39 (DB2)	5	80	0.112	50
24 V	17 (DB1)	0	60	0.096	55
25 V	27 (DB1)	10	60	0.096	55
26 V	14 (DB1)	0	100	0.096	55
27 V	23 (DB1)	10	100	0.096	55
28 V	9 (DB1)	5	80	0.096	55

another possible concurrent cause. From this evidence, MQL effect on heat partitioning among WP, tool and chip, or on the amount of generated heat needed to be assessed.

3.2 Mechanistic Model Predictions

The conceived mechanistic model was applied to all the experimental tests. Model tuning consisted in a non-linear optimization simultaneously involving the cutting torque and the thrust force. For each cutting test separately, the objective function in Eq. (23) was minimized with respect to the friction angle λ, following considerations in section *Experimental findings*. λ is the angle comprised between the resultant of the forces and the force normal to the rake face of the drill-bit (from Merchant's theory [23]). Therefore, as the frictional force decreases, the friction angle λ also decreases.

$$J(\lambda) = \left(\frac{C_{t,model}(\lambda) - C_{t,exp}}{C_{t,exp}}\right)^2 + \left(\frac{F_{th,model}(\lambda) - F_{th,exp}}{F_{th,exp}}\right)^2 \tag{23}$$

The resulting model predictions, in terms of torque and force were reported in Fig. 4, limited to the lubricating conditions LC1 and LC2. The graphs showed the good performances of the conceived model, both in terms of cutting torques and thrust forces. A different behaviour was noticed between the used drill-bit, despite they were of the same type. DB1 generated lower torque and forces than DB2. This may be attributed to the unavoidable slightly different manufacturing geometries. In general, cutting torque predictions were more accurate, whereas thrust forces introduced higher errors, especially for PP2 (low v_c, high f). These results were confirmed by the computation of the relative errors (maximum 14.9% on torque among all the LCs and 21.3% on forces; a peak of 5.4% mean error on torque and -10.7% on forces). Subsequently, a linear regression on the friction angle as a function of cutting and lubricating parameters was conducted (Fig. 5). Graphs were limited to PP3 and LC5 cases. The predicted lambda angle from cutting and lubricating parameters ($\lambda = 49.5 + 0.03v_c + 33.6f - 0.2\dot{m}_{oil} - 0.0\dot{m}_{air}$) was shown together with the simultaneous mean curve prediction interval with 95% confidence. The regression model properly.

described the dependency among the variables (R^2-adj 84% and p-value close to zero). The results coming from this regression model confirmed the considerations in section *Experimental findings*, highlighting the increasing trend of friction coefficient as process parameters become heavier. Moreover, the influence of the oil flow rate on λ was statistically significant, meaning that oil led to a decrement of the friction generated during the process. The friction angle resulted independent from the air flowrate.

3.3 Thermal Model Predictions

Following the preliminary considerations in section *Experimental findings*, an optimization was applied to predict the right WP temperature according to the change

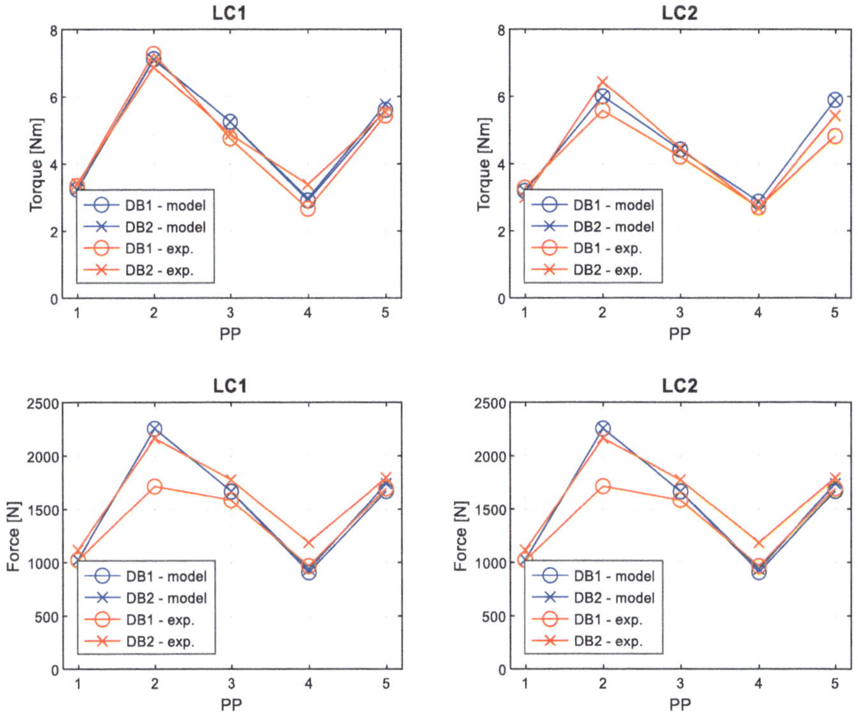

Fig. 4 Cutting torque and force predictions against experimental data for lubricating conditions LC1 and LC2. Both, drill-bit DB1 and DB2 were analyzed

on heat partition, and thus β. The objective function in Eq. (24) was considered:

$$J(\beta) = \left(\frac{T_{3,\text{model}}(\beta) - T_{3,\exp}}{T_{3,\exp}}\right)^2 \quad (24)$$

The cutting power generated during the process was assumed to be entirely converted into heat during cutting, neglecting power losses due to inefficiencies. Nevertheless, since the model was developed for the computation of the temperature field of the WP, only a portion β of the power from Eq. (16) was considered has the input of the thermal model.

As previously done for λ, ANOVA was applied to assess the effect of lubricating and cutting parameters on the β coefficient. The obtained model ($\beta = 0.581 - 0.003v_c - 1.693f - 0.001\dot{m}_{oil} - 0.001\dot{m}_{air}$) was found to properly fit the data, achieving an adjusted R^2-adjusted of 86% and a regression p-value close to zero. The ANOVA also confirmed the preliminary hypotheses regarding the relationships between the heat partition coefficient and the lubricating parameters. In fact, only the p-value related to the oil flow rate resulted not significant. Despite the oil flow rate reducing the amount of generated power (by reducing friction), it

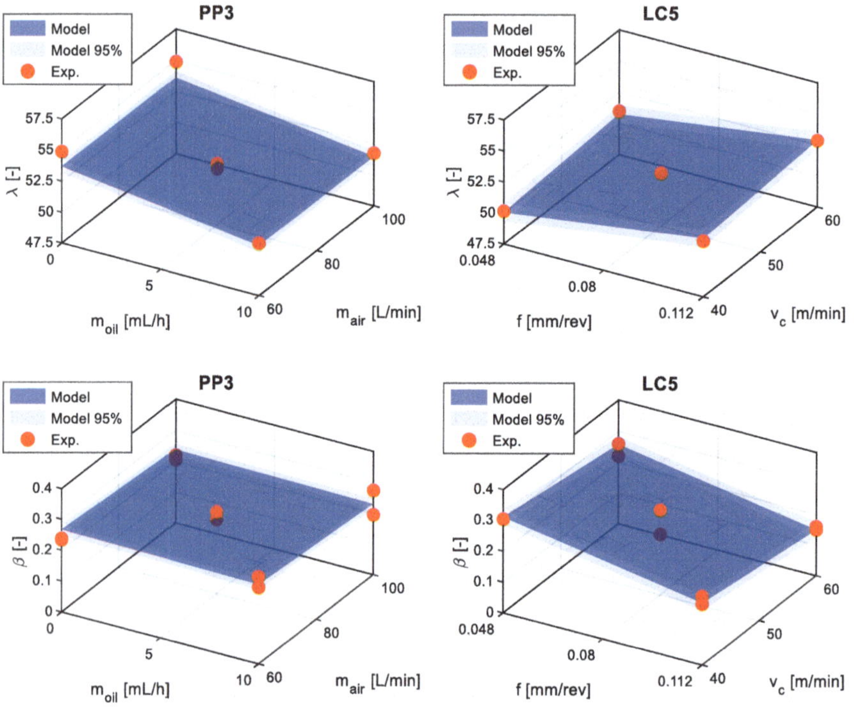

Fig. 5 Regression on the friction angle λ and the heat partition coefficient β. Results are reported for PP3 and LC3 data. 95% simultaneous curve confidence intervals are reported

had no effect on the amount of heat flowing into the WP or the tool. Nevertheless, this effect can be imputed to the air flow rate, which increases the convection effect, bringing part of the heat flow to ambient. Figure 5 graphs showed the trend of β, with respect to LCs and PPs, together with the model predictions and the associated mean curve simultaneous 95% intervals. Both the prediction models (for λ and β) properly represented the set of experimental data.

Two further graphs were reported in Fig. 6. As previously stated, the model was optimised on the third peak temperature, but no optimization was conducted on the thermal trend over the full hole drilling process. These graphs showed the potentiality of the conceived model to properly represent the peck drilling strategy. The only discrepancy, not common to all the cases, between model and experimental measures, relied in the zone after the third temperature peak. In fact, the final peck registered a significant temperature increase in the workpiece (in the experimental measurements). During this peck, the drill-bit is surpassing the thermocouple location, until the final depth of the hole is reached. This means that a cutting temperature increase occurs even if the cutting zone is farther and farther from the measurement point. This suggest that a physical phenomenon is taking place far away from the cutting zone, and in particular it is occurring on the sidewalls of the hole. Such

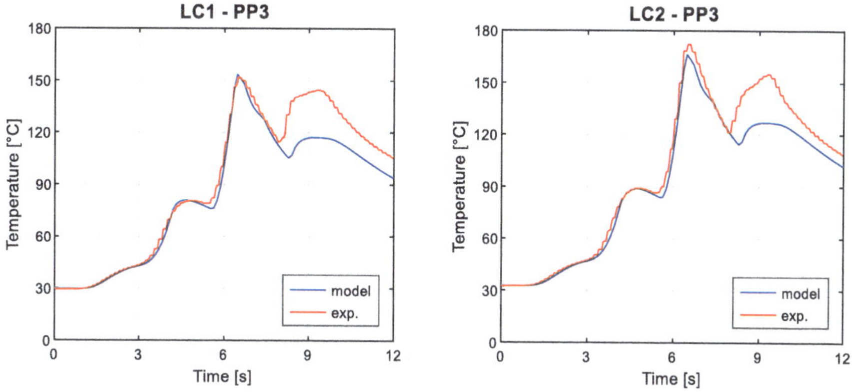

Fig. 6 Comparison between the thermal model predictions and experimental measurements. Data are reported for PP3 process parameters and LC1-2 lubricating conditions

phenomenon was already discussed in literature and is well-known when dealing with deep-hole drilling [24] and involves a thermal interaction between the chip and the drilling sidewall, as well as heat conduction from the tool during chip evacuation. Specifically, the deeper the hole, the more the chip is pressed against the drilled sidewall by the centrifugal force during its evacuation. Indeed, friction is generated, and thus, heat is dissipated into the workpiece.

Additionally, the drill-bit temperature increases as the hole becomes deeper, further increasing heat exchange towards the hole wall. In literature, a depth-dependent force component known as the evacuation force, was introduced to deal with such a phenomenon [24]. However, the experimentation carried out in this paper was not classified as deep drilling, since the depth was only 1.4 times the drill diameter. Consequently, no significant increase in cutting power with depth was observed, but only a thermal effect was visible. Indeed, this behaviour could not be explained by the conceived model, but will be matter of authors' future works.

3.4 *Model Validation*

In order to validate the conceived approach a set of validation experiments was carried out (performed at PP and indicated with letter V in Table 4). Such experiments were performed with different PP than the one used to fit the model, in order to assess regression models robustness on unseen cutting conditions. More specifically, 5 validation tests were obtained scanning all the training LCs (runs 17, 27, 14, 23, and 9 in Table 4). The validation procedure involved multiple quantities: λ, C_t and F_{th} for the mechanistic part, β and T_3 for the thermal part.

For the mechanics of cutting, the true values of λ were obtained by optimization on the validation tests, using Eq. (23), whereas C_t and F_{th} were measured experimentally; the validation values were obtained by predicting λ through linear regression and using the conceived mechanistic model to predict C_t and F_{th}. For the thermal part, the true values of β were obtained following the optimization process given in (24) on the validation tests, whereas T_3 was measured experimentally; the validation values were obtained by predicting β through the regression model and thus partitioning the output power predicted by the mechanistic part. Once the input power fraction was known the conceived thermal model was applied and T_3 was extracted. The comparison between the true and validation friction angle λ was shown in Fig. 7 (left). The mean relative error of λ predictions on the different lubricating conditions was 1.1%. All the prediction intervals included the true λ obtained by optimization, suggesting excellent prediction capabilities for the conceived model. Secondly, the results on the cutting torque and thrust force predicted from the mechanistic model were shown in Fig. 8 (left). The mean relative prediction error under different lubricating conditions, with respect to the measured torque values, was 9.2%, whereas the mean error for the thrust force was 2.0%. In this case, higher discrepancies were found on LC1 and LC4. The same procedure was adopted for the validation of the heat partition coefficient. The comparison between the optimized β and the one computed with the regression model was shown in the Fig. 7 (right). The mean relative error of β predictions on the different lubricating conditions was 17.9%. Moreover, the mean relative error of T_3 predictions on the different lubricating conditions is 13.9% (Fig. 8 (right)). The higher errors in β and T_3 may be linked to the imperfect linearity of the heat partition process and third peak temperature model, as identified in section *Experimental findings*. In fact, PPV cutting parameters were closer to the central point of the DoE than to extrema, where the highest discrepancy between regression and experimental data was found.

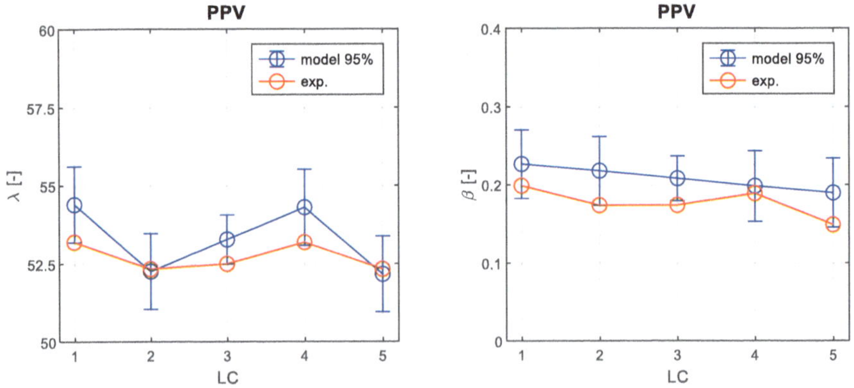

Fig. 7 Comparison between optimized (true) λ (left) and β (right), and the validation counterparts

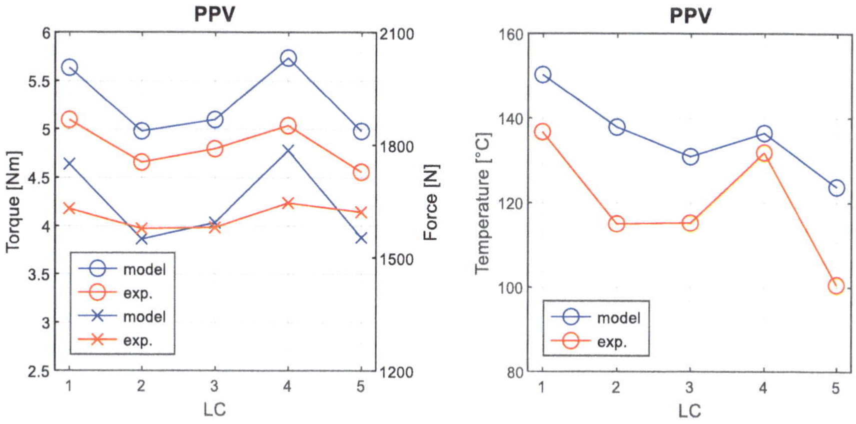

Fig. 8 Comparison between experimental C_t, F_{th} (left), T_3 (right), and the validation counterparts

4 Conclusions

In this paper an innovative mechanistic and thermal model was conceived to understand the phenomenology behind MQL lubrication effect in peck drilling. The comparison of measured and predicted cutting torque and thrust force validated the mechanistic model under different lubrication conditions. A multivariate linear regression model for the friction angle revealed positive correlation with feed and cutting speed, negative correlation with oil flow rate and absence of correlation with air flow rate.

The thermal model successfully predicted heat transfer, and a regression model for the heat partition coefficient showed negative correlation with cutting speed and feed rate (with feed rate having a stronger effect), absence of correlation with oil flow rate, slightly negative correlation with air flow rate.

It was found that MQL oil flow rate does not reduce temperature by altering the fraction of heat entering the material. Instead, it decreases the total power (heat) generated by reducing friction, leading to lower temperatures in the workpiece. The increase in convection cooling due to air flow rate reduces the heat partition coefficient, since only conduction was modelled. Indeed, oil primarily contributes to chip formation, whereas air facilitates oil delivery and enhances the heat exchange coefficient between the machined surface and the environment. The inclusion of friction generation due to chip in deep holes drilling will be matter of future works by the authors.

References

1. Debnath S, Reddy MM, Yi QS (2014) Environmental friendly cutting fluids and cooling techniques in machining: a review. J Clean Prod 83:33–47. https://doi.org/10.1016/j.jclepro.2014.07.071
2. Kundrak J, Varga G (2013) Use of coolants and lubricants in hard machining, Tehnicki Vjesnik-Technical. Gazette 20:1081–1086
3. Sayuti M, Sarhan AAD, Salem F (2014) Novel uses of SiO_2 nano-lubrication system in hard turning process of hardened steel AISI4140 for less tool wear, surface roughness and oil consumption. J Clean Prod 67:265–276. https://doi.org/10.1016/j.jclepro.2013.12.052
4. Pizzi M, De Gaetano F, Ferroni M, Boschetti F, Annoni M (2023) A deep-hole microdrilling study of pure magnesium for biomedical applications. Micromachines 14:132. https://doi.org/10.3390/mi14010132
5. Hiremath SS (2022) Mechanical peck drilling of Inconel 625: parametric analysis and grey fuzzy logic based multi-response optimization. J Micromanuf 25165984221135713. https://doi.org/10.1177/25165984221135713
6. Soori M, Arezoo B (2022) A review in machining-induced residual stress. J New Technol Mater 12:64–83
7. Oxley PLB (1989) The mechanics of machining: an analytical approach to assessing machinability. Ellis Horwood Publisher
8. Flachs JR, Salahshoor M, Melkote SN (2014) Mechanistic models of thrust force and torque in step-drilling of Al7075-T651. Prod Eng Res Devel 8:319–333. https://doi.org/10.1007/s11740-014-0531-5
9. Chandrasekharan V, Kapoor SG, DeVor RE (1995) A mechanistic approach to predicting the cutting forces in drilling: with application to fiber-reinforced composite materials. J Eng Ind 117:559–570. https://doi.org/10.1115/1.2803534
10. Strenkowski JS, Hsieh CC, Shih AJ (2004) An analytical finite element technique for predicting thrust force and torque in drilling. Int J Mach Tools Manuf 44:1413–1421. https://doi.org/10.1016/j.ijmachtools.2004.01.005
11. Pereira Guimarães BM, Da Silva Fernandes CM, Amaral De Figueiredo D, Correia Pereira Da Silva FS, Macedo Miranda MG (2022) Cutting temperature measurement and prediction in machining processes: comprehensive review and future perspectives. Int J Adv Manuf Technol 120:2849–2878. https://doi.org/10.1007/s00170-022-08957-z
12. Patne HS, Kumar A, Karagadde S, Joshi SS (2017) Modeling of temperature distribution in drilling of titanium. Int J Mech Sci 133:598–610. https://doi.org/10.1016/j.ijmecsci.2017.09.024
13. Segurajauregui U, Arrazola PJ (2015) Heat-flow determination through inverse identification in drilling of aluminium workpieces with MQL. Prod Eng Res Devel 9:517–526. https://doi.org/10.1007/s11740-015-0631-x
14. Fleischer J, Pabst R, Kelemen S (2007) Heat flow simulation for dry machining of power train castings. CIRP Ann 56:117–122. https://doi.org/10.1016/j.cirp.2007.05.030
15. Lalwani DI, Mehta NK, Jain PK (2009) Extension of Oxley's predictive machining theory for Johnson and Cook flow stress model. J Mater Process Technol 209:5305–5312. https://doi.org/10.1016/j.jmatprotec.2009.03.020
16. Minukhin I (2013) An improved method of cutting forces prediction for the primary cutting edges of twist drills, masters. Concordia University. https://spectrum.library.concordia.ca/id/eprint/977065/. Accessed 6 March 2025
17. Altintas Y (2012) Manufacturing automation: metal cutting mechanics, machine tool vibrations, and CNC design. Cambridge University Press, Cambridge. https://doi.org/10.1017/CBO9780511843723
18. Lin GCI, Mathew P, Oxley PLB, Watson AR (1982) Predicting cutting forces for oblique machining conditions. Proc Instit Mech Eng 196:141–148. https://doi.org/10.1243/PIME_PROC_1982_196_015_02

19. Benmeddour A (2021) Experimental investigation and numerical prediction of the effects of cutting tool geometry during turning of AISI 316L steel, period. polytech. Mech Eng 65:293–301. https://doi.org/10.3311/PPme.16844
20. Kim CS (1975). Thermophysical properties of stainless steels. https://doi.org/10.2172/4152287
21. Albertelli P (2017) Energy saving opportunities in direct drive machine tool spindles. J Clean Prod 165:855–873. https://doi.org/10.1016/j.jclepro.2017.07.175
22. Montgomery DC (2012) Design and analysis of experiments. John Wiley & Sons, Inc. https://doi.org/10.1198/tech.2006.s372
23. Merchant ME (1944) Basic mechanics of the metal-cutting process. J Appl Mech 11:A168–A175. https://doi.org/10.1115/1.4009380
24. Han C, Zhang D, Luo M, Wu B (2018) Chip evacuation force modelling for deep hole drilling with twist drills. Int J Adv Manuf Technol 98:3091–3103. https://doi.org/10.1007/s00170-018-2488-6

Novel Approaches for Layer-Level Functionally Graded Materials via PBF-LB/M

Vito Errico, Paolo Posa, and Sabina Luisa Campanelli

Abstract Powder Bed Fusion-Laser Beam (PBF-LB/M) enables the production of Functionally Graded Materials (FGMs) with tailored properties for specific applications. This study explores a novel method for creating FGMs within layers using a custom powder separation system that integrates seamlessly with existing PBF-LB plants employing blade/roller spreading. Two distinct approaches were implemented: the ex-situ method, where powders were pre-mixed before deposition in transition zones; and the other, which employed an innovative setup for in-situ real-time powder mixing during deposition. AISI 316L stainless steel and CoCrMo alloy were chosen for their excellent mechanical properties, biocompatibility and corrosion resistance, making them optimal for aerospace and biomedical applications. Results from microstructural analysis, mechanical testing and chemical characterization confirm the reliability and cost-effectiveness of this approach, highlighting its potential to advance multi-material AM.

Keywords Additive manufacturing · Powder bed fusion · Functionally graded materials

1 Introduction and Background

The advent of Functionally Graded Materials (FGMs) has signified a seminal advancement in the domain of materials science, with the potential to customize transitions in properties such as strength, thermal, toughness, electrical, and corrosion resistance, within a single component [1]. This tunability is especially valuable for industries with extreme performance demands, like aerospace, automotive, biomedical, molding and energy, where parts must endure harsh service conditions [2]. By gradually altering composition, microstructure or density, FGMs avoid stress concentrations and brittle intermetallic phases that often plague abrupt material interfaces [3, 4]. Additive Manufacturing (AM) has further unlocked the potential of

V. Errico (✉) · P. Posa · S. L. Campanelli
Department of Mechanics, Mathematics and Management, Politecnico Di Bari, Bari, Italy
e-mail: vito.errico@poliba.it

L. Fratini et al. (eds.), *Selected Topics in Manufacturing*, Lecture Notes in Mechanical Engineering, https://doi.org/10.1007/978-3-031-99501-9_7

FGMs [5] by offering unparalleled precision in spatial material distribution, a level of control that traditional techniques cannot match [6]. In particular, Powder Bed Fusion–Laser Beam (PBF-LB/M) has proven to deliver high resolution and excellent surface quality for graded components [7]. Nevertheless, realizing FGMs with PBF-LB/M at the layer level still faces hurdles, namely powder cross-contamination before and after building, limited throughput, the challenges of economically integrating FGM capability into existing systems, and the difficulty of scaling production to industrial volumes [1, 8].

To overcome these challenges, innovative strategies are required to adapt commercial PBF-LB/M machines to these advanced applications [9, 10]. Recent studies have attempted to address these limitations by developing novel setups and methodologies. For instance, Nadimpalli et al. [11] utilized an open-architecture PBF-LB/M system, known as the Aurora Labs system, equipped with individual powder hoppers (hopper powder feeding) with extrusion screws at the outlet that control powder feed, coupled with an inert gas flow that facilitates powder transport and mixing. Other studies have explored external modifications to powder delivery systems. Wang et al. [12] installed powder feeders outside the building chamber to produce multi-material FGMs (316L-CuSn10-18Ni300-CoCr alloys) using a custom-designed Dimetal-100 SLM system equipped with four powder hoppers (hopper powder feeding). This system incorporates a flow regulation mechanism that precisely controls and monitors the powder feeding rate for each hopper. Other researchers have modified the SLM Solutions 280HL system (hopper powder feeding) to enable multi-material fabrication. Wits et al. [13] have modified the recoater of this system (partitioning it internally) to allow the deposition of two materials in a single build, successfully printing SS316L-IN718, though interface cracking remained a challenge. More recently, Daram et al. [14] equipped the same system with dual powder feeders, enabling in-situ mixing and deposition of distinct metal powders. This approach was used to fabricate Ni–Ti FGMs with smooth compositional gradients, but delamination issues persisted.

However, it should be noted that the aforementioned solutions enable the fabrication of FGM components exclusively in the building direction (between layers). Innovative powder delivery mechanisms have also been developed to advance FGMs processing at layer-level. For instance, Wen et al. [9] introduced the Compositionally Graded Alloys (CGA) system (hopper powder feeding), which features a specially designed partition within the powder hopper to separate different materials into distinct sections, enabling composition variation layer-level instead of between layers. In addition, Lesko et al. [10] proposed the Graded Alloy Processing (GAP) system (hopper powder feeding), which integrates a traditional recoater with a multi-material hopper capable of depositing powder at specific points on the layer. Additionally, this system is equipped with laser or vacuum techniques to remove previously deposited powder. Furthermore, more recently patented systems have demonstrated the ability to fabricate FGMs with three-dimensional compositional gradients, both between and within layers. These systems signify a substantial advancement in overcoming the constraints of traditional methodologies. Wei et al. [6, 15] presented a six-channel multi-material PBF-LB/M system (vibrating nozzle) (patent WO2020002951A1), equipped with programmed ultrasonic powder dispensers. These powder dispensers are equipped with a vibrating nozzle and, when combined with a vacuum sucker for powder removal, are capable of creating smooth gradients and complex geometries. In a similar vein, Aerosint SA unveiled the 'Selective Powder Deposition' system (patterning drums) (patent US20220234290A1) [16]. This system uses an electrophotographic mechanism to precisely control powder placement within a single layer. In this approach, each rotating patterning drum is dedicated to depositing one specific powder, which allows for highly accurate formation of FGMs within the layer.

However, the majority of existing solutions are custom-built, which increases costs and complexity and limits their scalability in industrial settings. Consequently, such solutions remain inaccessible for broader industrial adoption. Furthermore, the integration of intralayer FGM capabilities into existing conventional PBF-LB/M systems remains an unresolved challenge, highlighting the need for more cost-effective and adaptable approaches to FGM AM. One strategy that has been identified to address this problem is to improve the functionality of existing PBF-LB/M systems, particularly those employing blade- or roller-based powder deposition mechanisms and the supply of the new powder layer through a chamber or tank located next to the building platform. Rather than introducing new PBF-LB/M machines [15, 16] or implementing overhead hoppers for powder delivery [10], a group of researchers has recently patented a novel equipment for layer-level multi-material or FGM fabrication (IT202200027369A1), with a view to modifying the aforementioned traditional systems. This design modification, which is inexpensive and easy to implement, has the potential to simplify the integration of FGM intralayer capabilities and expand the range of applications that can be realized with commercial PBF-LB/M systems, thus making them more versatile and cost-effective for industrial use [2, 17]. Unlike previous methodologies, this new approach facilitates the fabrication of intralayer FGMs without requiring substantial modifications to commercial systems. This innovation addresses the critical challenges of powder contamination and material compatibility while maintaining the economic feasibility essential for industrial adoption [18].

In this study, two approaches were employed to produce discontinuous functionally graded materials (dFGMs) within a single layer [1, 19]. These approaches involved the use of the customized powder separation system (mentioned above) that seamlessly integrates with existing PBF-LB/M setups. The ex-situ approach entails the pre-mixing of the powders prior to the loading stage, followed by their placement within a transition region between the materials under investigation. Conversely, the in-situ method incorporates a real-time mixing mechanism within the PBF-LB/M machine (in-situ real-time mixing deposition), allowing precise control over material gradients during the spreading powder process. In-situ approach not only enhances process flexibility but also mitigates issues such as powder oxidation and mixing inhomogeneity caused by differing material properties. To validate this methodology, AISI 316L stainless steel and CoCrMo alloy were selected as model materials due to their complementary properties, including high corrosion resistance, biocompatibility, and mechanical performance. These materials are of particular relevance for aerospace and biomedical applications. To the best of the authors' knowledge, this printing strategy is the first to exhibit the manufacturing of layer-level dFGM CoCrMo-AISI 316L parts fabricated using PBF-LB/M process. The results from microstructural analysis, mechanical testing, and chemical characterization underscore the potential of this approach to advance multi-material AM technologies. This research fills a significant gap between experimental and industrial applications of intralayer FGMs in AM, offering a scalable and economically viable solution. In contrast to earlier methods that relied on costly custom-built systems, the proposed approach leverages existing PBF-LB/M infrastructure, aligning with industry trends towards sustainable and resource-efficient manufacturing. The ability to produce FGMs with tailored properties has profound implications for various industries, as it enables the creation of lightweight, high-performance components and biocompatible implants with enhanced mechanical integrity.

2 Materials and Methods

2.1 Innovative Equipment for Layer-Level Multi-Material Fabrication of dFGMs via PBF-LB/M

In this study, a removable powder separation system has been devised to enable the fabrication of dFGMs at the layer-level using conventional PBF-LB/M machines. As illustrated in Fig. 1, which exemplifies the two case studies, two custom dividers have been designed and manufactured to partition the powder bed into distinct volumes, each containing a different material.

The first divider (Fig. 1a) consists of two flat walls, positioned parallel to each other at a nominal distance of 10 mm, thereby creating an interstice in which the layer-level dFGM is produced. In contrast, the second divider (Fig. 1b) presents a complex geometric wall that separates the powder bed into two volumes, establishing a homogeneous mixing zone at the interface (10 mm thick) for fabricating the dFGM. It is important to note that both dividers are designed to have identical external dimensions, ensuring seamless compatibility with the existing powder bed layout. However,

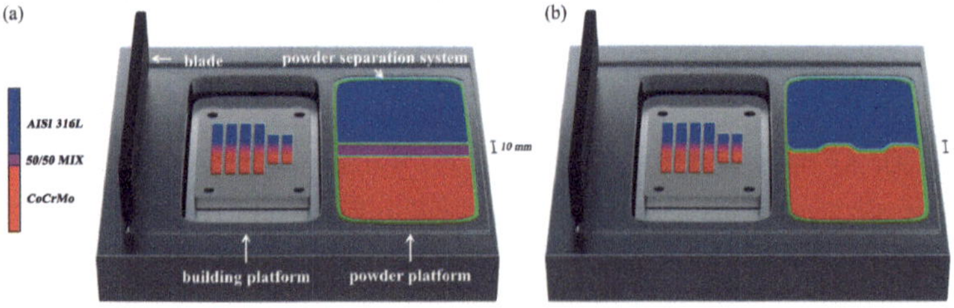

Fig. 1 Schematic diagram of the experimental set-up for fabricating intralayer dFGMs using a conventional blade-based PBF-LB/M system. **a** Ex-situ approach. **b** In-situ approach

to ensure the reproducibility of this fabrication process in other conventional PBF-LB/M systems, it is essential to emphasize several key precautions. Prior to loading the powder, the divider is inserted into the powder platform, thereby creating separate chambers for the respective powders. Following the loading of the powders, the divider is then removed, thereby enabling the powders to interact at the interface without initiating premature and undesired mixing [2, 17, 20]. During the subsequent spreading phase, the blade (or roller) traverses the whole powder bed, mixing the two powders within the designed transition zone and establishing a gradual compositional change within the same layer. In the first approach (Fig. 1a), the powders are pre-mixed and subsequently loaded into a 10 mm-wide gap, transitioning in the single layer from 100% Material A to a 50%/50% mixed zone (by volume) and finally to 100% Material B. The second approach (Fig. 1b) embodies the "*in-situ real-time mixing deposition*" strategy, wherein the powders are mixed during the real-time recoating process through the ad-hoc design of an interface divider. The entire procedure requires no changes to the commercial PBF-LB/M system, thus preserving cost-effectiveness and scalability. This innovative strategy represents a further advancement beyond the patented methodology (IT202200027369A1), by substantially broadening the industrial applicability of FGMs.

2.2 Materials and Specimens' Fabrication

In this study, CoCrMo and AISI 316L powders with particle sizes between 15 and 45 μm were utilized. The powders, which were produced by the gas atomization process (argon for CoCrMo and nitrogen for AISI 316L), exhibited a spherical shape and were found to be defect-free, thus rendering them suitable for the PBF-LB/M process. The rationale behind the selection of this combination of materials stemmed from their complementary properties. CoCrMo alloys are renowned for their elevated mechanical strength, exceptional wear resistance, and superior corrosion resistance, attributable to the formation of a stable passive film, along with their noteworthy hardness, rendering them especially well-suited for high-stress applications and biomedical devices [21]. Conversely, AISI 316L exhibits greater ductility and ease of processing, enabling the fabrication of complex geometries while still ensuring good corrosion resistance (particularly in marine and biomedical environments), lower costs, and improved thermal stability during cyclic heating and cooling [22]. Table 1 presents the certified chemical composition of each powder as provided by the manufacturer (Mimete Metal Powders). The elemental composition was determined through Inductively Coupled Plasma Optical Emission Spectrometry (ICP-OES) following the ASTM E1479-16 and ASTM E2594-20 standards. Additionally, the content of carbon (C), nitrogen (N), and sulfur (S) was assessed using the Combustion and Fusion Analytical Technique in accordance with the ASTM E1019-18.

The fabrication of all specimens was performed using an M1 Cusing machine (Concept Laser), setting the same processing parameters, which were optimized

Table 1 Certified chemical compositions (wt.%) of CoCrMo and AISI 316L powders

$CoCrMo_{wt.\%}$					AISI $316L_{wt.\%}$				
Fe	**Cr**	**Ni**	**Si**	**Mn**	**Fe**	**Cr**	**Ni**	**Si**	**Mn**
0.17	28.30	0.10	0.70	0.70	Bal	17.90	13.90	0.44	1.11
Co	**Mo**	**S**	**N**	**C**	**Co**	**Mo**	**S**	**N**	**C**
Bal	5.80	–	0.10	0.12	–	2.70	–	–	0.02

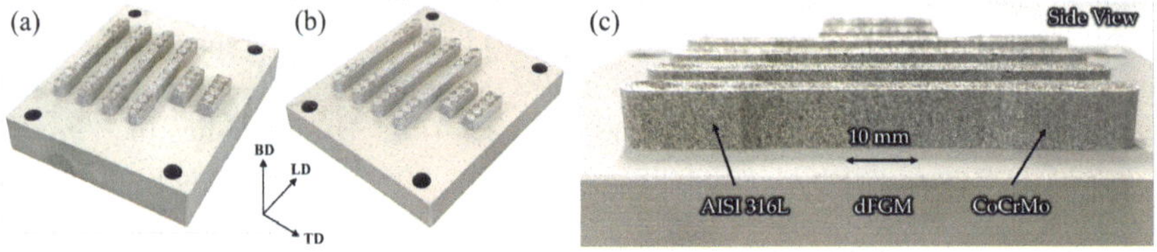

Fig. 2 Layer-level dFGMs fabricated by PBF-LB/M. **a** Ex-situ deposition approach. **b** In-situ deposition approach. **c** In-situ printing with highlighted FGM interface zone

through preliminary studies, for both materials. Specifically, the jobs were carried out with a laser power of 100 W, a scanning speed equal to 180 mm/s, a layer thickness of 30 μm, a fixed laser beam diameter equal to 200 μm, and a hatch spacing of 140 μm. The selection of these parameters was made with the objective of achieving optimal material consolidation, thereby ensuring that the fabricated components consistently attained a relative density that exceeded 99%. This was confirmed through optical microscopy analysis of cross-sections obtained from metallographic specimens. Figure 2 illustrates the printed dFGM specimens obtained using both deposition approaches.

In total, three print jobs were performed: one dedicated to the deposition of single material (CoCrMo) and two focused on the fabrication of dFGMs. Each job included the production of four dogbone-shaped tensile specimens (designed in accordance with ASTM E8/E8M-15a for powder metallurgy), one prismatic specimen for metallographic and chemical analysis, and one prismatic specimen for hardness testing. In both dFGMs cases, the material interface was precisely positioned at mid-length to facilitate a thorough mechanical and microstructural characterization.

2.3 *Metallographic Analysis and Mechanical Characterization of Specimens*

A comprehensive metallurgical and mechanical characterization was carried out to evaluate the effectiveness of the proposed approach for the fabrication of intralayer dFGMs and to assess the quality of the fabricated components. A preliminary metallographic examination was conducted to analyze the interface integrity and to determine the density of the processed materials. Specimens were sectioned, embedded in resin, and polished in accordance with standard metallographic preparation procedures outlined in ASTM E3–11 (2017). The microstructural observations were made using a Nikon Eclipse MA200 inverted optical microscope to inspect the microstructural features and identify any potential defects. For a more detailed investigation, separate chemical etching treatments were performed to characterize the microstructures of individual materials. Specifically, CoCrMo was subjected to immersion etching with HCl for 8 h, while AISI 316L was subjected to immersion etching with a glycerol reagent for 15 s, prepared with the following proportions: 5 mL of HNO_3, 10 mL of HCl and 15 mL of glycerol. To evaluate possible cross-contamination between the two materials and investigate the interface area in more detail, a chemical analysis was conducted using a Zeiss Sigma 300 VP SEM (Scanning Electron Microscope) and an Energy Dispersive X-ray Spectroscopy (EDS). This analysis was particularly useful for differentiating the two printing approaches used, allowing for a clearer understanding of their distinct microstructural characteristics and the effects on the interface.

The mechanical characterization focused primarily on evaluating the material transition zone. Vickers microhardness measurements were performed using a Shimadzu HMV-G hardness tester in accordance with the UNI EN ISO 6507–1:2023 International Standard, applying a 2 kg load for a 10-s dwell time. Additionally, Brinell hardness tests were carried out according to UNI EN ISO 6506–1:2014, using a 2.5 mm diameter ball and a load of 187.5 kg. This dual hardness testing approach allowed the characterization of the dFGM region both at micro and macro-scale. Finally, uniaxial tensile tests were conducted utilizing a Instron 3369 Universal Testing machine equipped with a 50 kN load cell, following the ASTM standard E8/E8M-15a to assess the mechanical properties of the intralayer-graded region in terms of ultimate tensile strength (UTS) and elongation at failure ($E_{\%}$). The results obtained from these tests provided valuable insights into the mechanical behavior and structural reliability of the fabricated functionally graded materials.

3 Results and Discussion

3.1 Microstructural and Chemical Analysis of dFGMs Specimens

The microstructural investigation of the layer-level dFGMs provided valuable insights into the effects of different deposition approaches on the interface characteristics. Initial observations, at low magnifications, using optical microscopy (OM) revealed the presence of numerous semicircular melt pool morphologies, resembling typical fish-scale patterns (see Fig. 3a and b) [23, 24]. At higher magnifications (see Fig. 3c and d), the OM micrographs exhibited key characteristics such as fine cellular

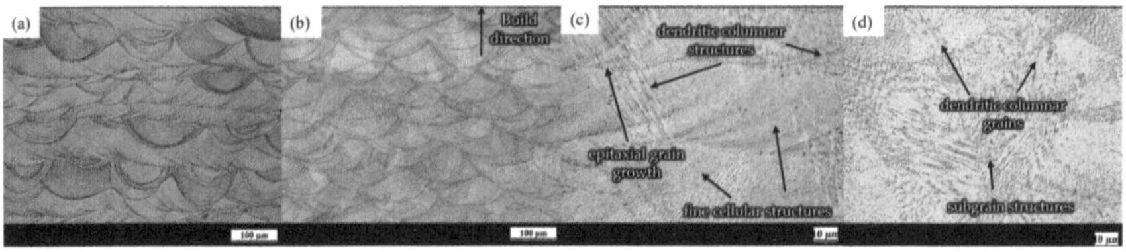

Fig. 3 Results of OM micrograph of single material areas at different magnifications: **a** OM of CoCrMo at 200×; **b** OM of AISI 316L at 200×; **c** OM of CoCrMo at 1000×; **d** OM of AISI 316L at 1000×

structures, indicating rapid cooling rates and significant undercooling during solidification [21, 24]. Specifically, the microstructure for both materials displayed cellular structures, with elongated dendritic columnar grains following the primary heat flow direction due to the sequential layer deposition [21, 25]. The CoCrMo alloy samples (Fig. 3a and c) exhibited a prominent abundance of parallel columnar grains, while the microstructure of AISI 316L (Fig. 3b and d) exhibited similar columnar grain formation but with additional complexity, as subgrain structures within the grains showed different orientations. These subgrains have been shown to highlight the dynamic thermal interactions occurring during deposition, where grains grow across multiple melt pool boundaries due to repeated remelting [25]. Further OM analysis of individual materials revealed epitaxial grain growth across layers and melt pool boundaries [21, 24].

In addition, despite the setting of identical process parameters for the printing of both materials under investigation, the relative density of the dFGM and single material metallographic samples was >99.8%, which can be attributed to the similarity of the properties of both alloys, including density$_{[gcm\text{-}3]}$ ($8_{AISI\ 316L}$—8.3_{CoCrMo}), melting temperature$_{[°C]}$ ($1370\text{-}1410_{AISI\ 316L}$ – $1350\text{-}1430_{CoCrMo}$), thermal conductivity$_{[Wm\text{-}1\ K\text{-}1]}$ ($16\text{-}21_{AISI\ 316L}$—$13_{CoCrMo}$), and thermal expansion$_{[K\text{-}1x10\text{-}6]}$ ($16_{AISI\ 316L}$—$11\text{-}13_{CoCrMo}$) [26]. Adequate overlap between the individual tracks and layers, coupled with good flowability of the virgin powder, also helped prevent the various defects usual in the PBF-LB/M process. This confirmed the high density achieved with the process parameters used.

Furthermore, a more in-depth investigation was conducted on the 10 mm interface region. A careful selective etching of the interface region revealed distinct banding patterns, with alternating zones dominated by CoCrMo (light region) or AISI 316L (dark region), in the in-situ approach (see Fig. 4a); whereas a homogeneous region (dark and light) is observed in the ex-situ approach (see Fig. 4b). Despite the evident distinction between the materials in these areas, they collectively contribute to an overall mixed area of 50%/50% by volume. However, in comparison with the ex-situ method (see Fig. 4b), the in-situ approach resulted in a more pronounced interface differentiation (see Fig. 4a). Furthermore, the detailed microstructural analysis confirmed that the interface in the samples fabricated with both techniques, in-situ and ex-situ, is free of severe defects, such as lack of melting, critical porosity, delamination or cracks, highlighting the excellent metallurgical bonding between the materials

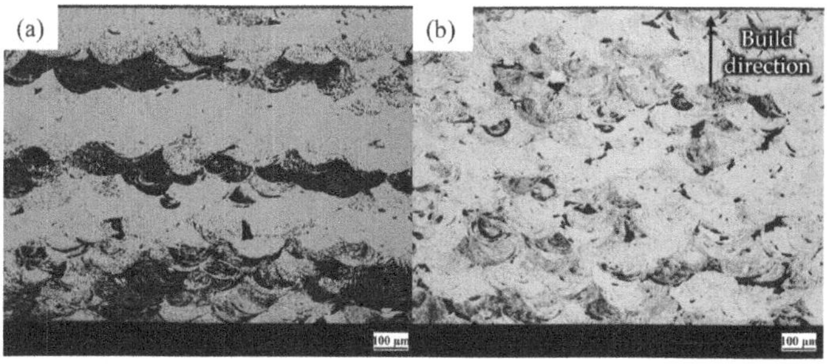

Fig. 4 Interface microstructure at 100× magnification: **a** in-situ deposition approach; **b** ex-situ deposition approach

and ensuring the structural integrity of the interface with this new printing approach. Consequently, it can be concluded that the in-situ deposition method with real-time mixing offers significant advantages due to its ability to produce a dFGM interface by reducing preparation time, as it is not necessary to pre-mix powders (as in the case of the ex-situ approach), avoiding the potential oxidation of powders during the pre-mixing phase, and reducing the complexity of the divider, which consists of a single wall (albeit non-linear), compared to the alternative system.

As regards chemical investigation, energy-dispersive X-ray spectroscopy (EDS) analysis, including elemental mapping, line scans, and quantitative assessments, was performed to assess how the chemical composition changes along the interface of the dFGM specimens. The results obtained corroborated the interfacial microstructural characteristics observed for both deposition approaches. As illustrated in Figs. 5 and 6, the EDS maps and line scans revealed the dominant elements in each alloy: Fe for AISI 316L and Co for CoCrMo, confirming the compositional transitions at the interface. Specifically, as demonstrated in Fig. 5a, which presents EDS mapping for the in-situ approach over two consecutive layers, there is clear visualization of Fe concentration in the upper layer (AISI 316L) and Co in the lower layer (CoCrMo). Conversely, Fig. 5b, which illustrates the ex-situ approach under analogous conditions, demonstrates a more blended distribution of Fe and Co, suggesting enhanced mixing between the two alloys.

A similar trend was observed in the EDS line scans (see Fig. 6), conducted over an extended area covering more than 50 overlapping layers. Figure 6a, corresponding to the in-situ deposition, reveals distinct peaks and valleys in the elemental profiles, where Co peaks correspond to Fe valleys and vice versa. This pattern indicates well-defined compositional layering. Conversely, Fig. 6b, representing the ex-situ approach, shows overlapping Fe and Co profiles, confirming a more intermixed interface region.

In addition, quantitative EDS analysis was performed to obtain insights into the distribution trends of the characteristic elements in both alloys (Fe, Ni, and Cr for AISI 316L; Co, Cr, and Mo for CoCrMo). The results, displayed in Fig. 7, highlight the variations in elemental concentrations across the interface. To facilitate

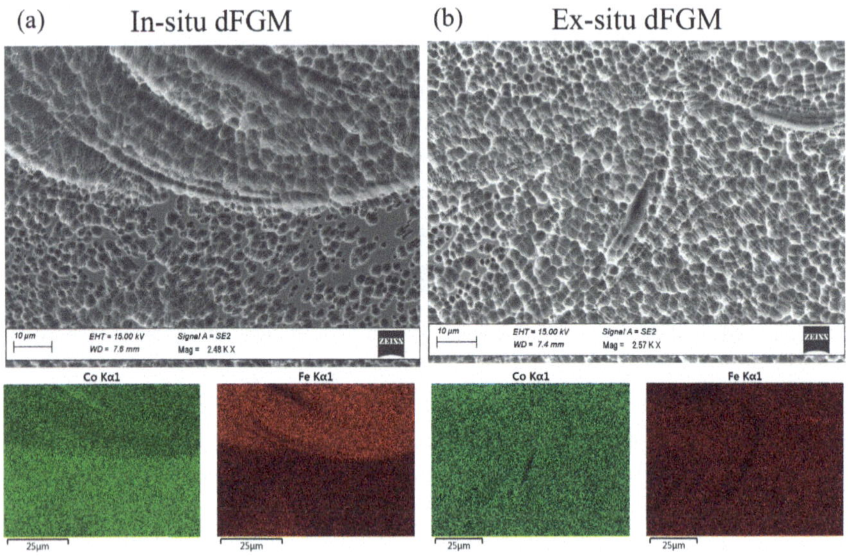

Fig. 5 EDS elemental maps for two consecutive layers in the interface region: **a** in-situ deposition approach; **b** ex-situ deposition approach

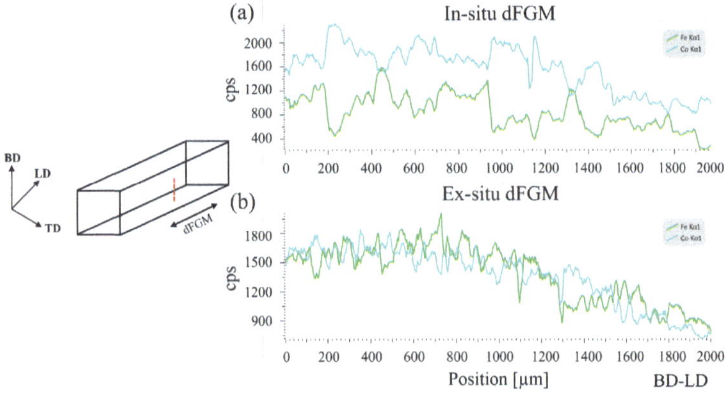

Fig. 6 EDS line scan profiles in the interface region: **a** in-situ deposition approach; **b** ex-situ deposition approach

direct comparison, the measured values for each element were normalized to their maximum concentration within the analyzed region. This normalization, which transforms the data into a dimensionless scale ranging from 0 to 1, enhances clarity in visualizing concentration gradients. The distinction between the two deposition approaches is once again evident. In the ex-situ method, the interface exhibits a relatively constant 50%/50% distribution of the characteristic elements of both alloys (see Fig. 7a). Conversely, the in-situ method (Fig. 7b) demonstrates a more gradual

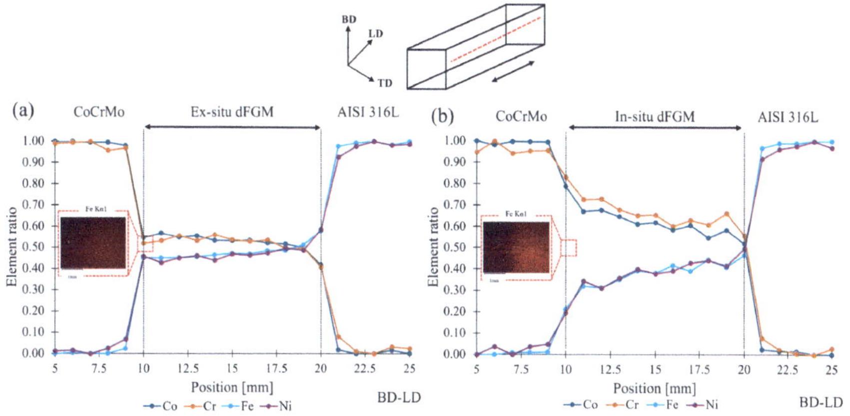

Fig. 7 Normalized EDS quantitative analysis of characteristic elements across the interface: **a** ex-situ deposition approach; **b** in-situ deposition approach

change in the characteristic elements when transitioning from one alloy to another. This result is likely influenced by the 200 × 200 μm^2 survey area used for EDS quantitative analysis, which may have sampled regions with a predominance of one material over the other.

However, the results obtained confirm the absence of cross-contamination beyond the designated 10 mm interface region. It is noteworthy that the concentrations of Fe and Co approach zero in areas where these elements are not present, thereby ensuring a sharp transition. This finding is of particular significance, as it demonstrates the capability of the custom-built deposition system implemented in this study to achieve dFGM fabrication at the layer-level without contamination in individual material regions. The issue of cross-contamination poses a significant challenge in the fabrication of layer-level FGM or multi-material components and is well-documented in the literature, particularly in traditional PBF-LB/M systems that rely on blade-based powder distribution spreading [7, 8, 27]. Nevertheless, for the low percentage of powder arising from the dFGM-single material interface zone and thus experiencing slight cross-contamination, various post-processing separation methods are being investigated. These include magnetic separation for materials with unique magnetic characteristics and sieving for powders with different particle size distributions. Additionally, other techniques like electrostatic separation and dense medium separation are being explored, though they are currently less developed than magnetic and sieving separation [28]. These processes are pivotal in fostering a circular economy by facilitating material recovery and reuse, reducing waste, and decreasing reliance on virgin powders. It is important to note that successful recycling does not necessarily require total separation. The reuse of partially separated powders while maintaining material integrity is a viable option in scenarios where a regulated level of cross-contamination is deemed acceptable and achieving extremely high purity is not a prerequisite. This flexibility is further enhanced in AM multi-material or FGMs, leading to improved sustainability and resource efficiency.

3.2 *Mechanical Characterization of Layer-Level dFGMs: Hardness and Tensile Test Analysis*

The change in mechanical properties across the interface was first evaluated using hardness tests and the results are presented in Fig. 8 and Table 2.

In particular, for the Vickers microhardness (HV) tests, three replicate measurements were taken, and the results are shown in the graph as averages with standard deviations. For the Brinell hardness (HB) test, a single value was calculated for each position on the sample (see Table 2). The results, shown in Fig. 8, confirm the findings obtained from the previous microstructural and chemical analyses. Specifically, a stepwise trend was observed, starting at around 220–230 HV in the AISI 316L region and gradually increasing to approximately 400 HV in the CoCrMo region for both deposition methods. These results are in agreement with the literature on the same materials [23, 29]. However, slightly different values of hardness were found for the two deposition approaches. The ex-situ method exhibited hardness values that

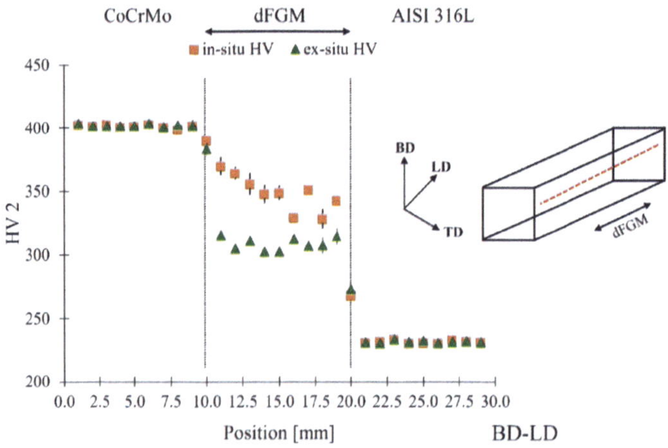

Fig. 8 Low-force Vickers hardness trend along the BD-LD section of dFGM specimens

Table 2 Results of Brinell hardness test along the BD-LD section of dFGM specimens

Approach	Position [mm]										
	2.5	5	7.5	10	12.5	15	17.5	20	22.5	25	27.5
In-situ	229	229	229	260	321	329	329	363	393	393	393
Ex-situ	229	229	234	285	285	298	298	373	393	393	393

were approximately 10% lower than those of the in-situ approach for both HV and HB. Additionally, as confirmed by the chemical analysis, the ex-situ process results in a hardness profile with relatively constant values in the interface region, whereas the in-situ method shows a less step-like and more ramp-like trend between the two materials. It is important to note that this observation is more pronounced in the Vickers microhardness results (Fig. 8), which are more pointed due to the smaller indentation area (with a diagonal size of 200 μm). This smaller area could potentially sample regions where one material predominates over the other, thus influencing the data. Conversely, the Brinell hardness (Table 2) exhibited a constant trend in the interface region for the in-situ method. Finally, to further explore the innovative concept of using a traditional PBF-LB/M machine for fabricating dFGMs, a comparative study was performed to assess the final tensile properties of the two dFGM configurations examined in this work. These were benchmarked against single-material specimens, specifically, CoCrMo produced in the current study and AISI 316L obtained from a previous investigation by the authors [2, 17]. Prior to testing, all samples were meticulously polished to ensure the absence of any surface imperfections. The experimental outcomes for the three batches, in conjunction with the AISI 316L reference, are comprehensively outlined in Table 3.

For each batch, the ultimate tensile strength (UTS) and percentage elongation at failure ($E_{\%}$) were calculated as mean values over four samples, with the corresponding standard deviations provided. It is noteworthy that the CoCrMo specimens exhibited a brittle fracture behavior ($E_{\%}$ ~ 5%) and reached an average UTS of 1150 ± 12.28 MPa. These properties are in line with the expected performance for this alloy and exceed those typically achieved by conventional processing routes, as referenced by the ASTM F75 for cast CoCrMo alloys [29]. Conversely, the AISI 316L samples exhibited a classical metallic response under tensile loading, with an average UTS of 706 MPa and an elongation of approximately 40%, accompanied by significant necking prior to fracture [2, 17]. This ductile behavior is consistent with literature reports, confirming that the mechanical performance of AISI 316L is comparable to, or even surpasses, the standard expectations [30]. The dFGM specimens, meanwhile, displayed UTS values similar and sometimes softly higher than those of AISI 316L. Specifically, the ex-situ dFGM samples reached an average UTS of 717.80 ± 10.68 MPa, while the in-situ dFGM specimens recorded an average UTS of 717.50 ± 4.69 MPa. However, both approaches exhibited a significantly reduced elongation, but higher than the single CoCrMo material, which was about 12.30% for in-situ and 11.40% for ex-situ. Of particular significance is the observation that

Table 3 Comparative results of tensile properties for dFGMs (ex-situ and in-situ), CoCrMo, and AISI 316L

Mechanical properties	dFGMs approach		CoCrMo	AISI 316L [17]
	In-situ	Ex-situ		
UTS [MPa]	717.50 ± 4.69	717.80 ± 10.68	1150 ± 12.28	706.29 ± 14.93
$E_{\%}$	12.30 ± 2.40	11.40 ± 2.35	5 ± 2.80	41.24 ± 2.36

deformation occurred within the more ductile AISI 316L region, with fracture initiation occurring beyond the interface. This observation indicates that the interfacial strength in the dFGMs exceeds the UTS of the AISI 316L, ensuring that failure does not occur at the transition. In summary, the results suggest that the transition zone, formed by either deposition approach, effectively combines a high-strength alloy (CoCrMo) with a highly ductile austenitic steel (AISI 316L) to create a graded material possessing an optimal balance of strength and ductility. The mechanical interlocking achieved within this region plays a key role in preventing premature fracture at the interface [31]. This performance is comparable to literature results on FGM components produced via DED-LB with similar material combinations, which showed similar properties (lower UTS around 480 MPa and slightly higher elongation at failure between 17 and 19%) [25].

3.3 Comparison of Multi-Material Powder Deposition Systems for Layer-Level FGM in PBF-LB/M: Current Approaches and Emerging Solutions

Although significant progress has been made in the field of FGMs using the PBF-LB/M process, widely documented in recently published review articles [1, 3, 7, 8, 26, 27], this technology still needs refinement to meet practical requirements. A major gap in current research is the integration of newly developed systems for FGM fabrication into conventional PBF-LB/M machines; this remains a challenging task due to their complex architecture, size, and functional demands, especially since many commercial systems are proprietary and closed-source. These challenges limit the adoption and scalability of FGMs solutions in industry. It is important to note that current efforts in FGMs processing have focused on the development of deposition systems, but the process itself also needs attention [8]. The current efficiency and accuracy in depositing gradients and multiple powders need to be improved to meet the basic requirements of applications, enabling the printing of useful multi-material components. A key issue in multi-material printing is cross-contamination between materials, which must be addressed. Powder recycling in multi-material processing is also a challenge, as powders cannot be easily separated from one another. Table 4 compares the various layer-level FGMs deposition systems for PBF-LB/M, highlighting that each system has its own specific advantages and disadvantages.

Based on the results presented in Table 4, the powder delivery mechanism developed in this study, particularly the *in-situ real-time mixing deposition approach*, exhibits advantages over current state-of-the-art systems. Its seamless integration with conventional PBF-LB/M machines, reduced cross-contamination, and cost efficiency underscore its significant technological impact. This approach, which bears similarities to laser welding, facilitates the fabrication of components made of diverse materials in a single print job. This is achieved by strategically dividing the powder

Table 4 Comparison of benefits and drawbacks of different layer-level FGMs deposition systems for PBF-LB/M

Powder delivery mechanisms	Cross contamination level	Layer-level discontinuous FGMs, dFGMs	Layer-level continuous FGMs, cFGMs	Level of productivity	Scalability	Costs
Hopper powder feeding	High	✓	✓	Slow	Medium	Low
Vibrating nozzle	Medium	✓	✓	Very slow	Low	High
Patterning drums	High	✓		Slow	Low	Very high
In-situ approach	Medium	✓	✓	Tolerable	High	Very low
Ex-situ approach	Medium	✓		Tolerable	High	Very low

*Some results were extracted from Musatto's work[8]

platform and tailoring the print preparation process. While this innovative method shows considerable promise, it is not without limitations. In particular, the pre-set deposition sectors restrict the range of producible components, thus constraining the full potential of multi-material 3D fabrication. Consequently, there is a necessity for ongoing efforts to be dedicated to overcoming these challenges and enhancing the versatility and applicability of the system.

4 Conclusions

In this work, a novel powder separation system was successfully integrated into a conventional PBF-LB/M setup with the objective of enabling the fabrication of layer-level discontinuous functionally graded materials (dFGMs) CoCrMo-AISI 316L. This was achieved using two distinct deposition strategies: an ex-situ approach with pre-mixed powders and an in-situ real-time mixing method.

The primary conclusions of the study are outlined below.

- Interface integrity: microstructural investigations (OM and SEM) revealed that both approaches yield an interface free of cracks, porosity, and significant defects, demonstrating excellent metallurgical bonding between AISI 316L and CoCrMo. In addition, the in-situ method produced a more pronounced interface differentiation, with distinct banding patterns, than a homogenous biphasic region obtained with the ex-situ approach.
- Chemical and elemental distribution: EDS maps and quantitative analysis of characteristic elements confirmed that the interface exhibited a sharp transition with

minimal cross-contamination in the dFGM-single material transition region; in fact, the elemental signatures of each alloy remain distinctly preserved, with Fe concentrated in the AISI 316L zones and Co dominant in the CoCrMo regions, demonstrating the effective segregation achieved by the deposition strategy. Furthermore, EDS maps and line scan profiles showed that the in-situ approach yielded clearly defined and alternating elemental peaks, indicative of compositional layering among the different layers, whereas the ex-situ method resulted in a more blended, near-equal distribution of the characteristic elements.
- Mechanical properties: hardness measurements (Vickers and Brinell) exhibited a stepwise increase in hardness from the ductile AISI 316L region to the high-strength CoCrMo zone. The in-situ method demonstrated a smoother, ramp-like hardness transition and exhibited, in the interface region, hardness values approximately 10% higher than the ex-situ method. Furthermore, tensile tests indicated that fracture consistently occurs within the AISI 316L region, confirming that the interfacial bond strength exceeds that of the weaker material.
- Process efficiency and scalability: a comparative analysis with existing multi-material deposition systems highlights that the proposed in-situ real-time mixing strategy is not only cost-effective and easy to implement without increasing printing time but also offers reduced cross-contamination and enhanced scalability. This method leverages conventional blade-based PBF-LB/M systems, thereby broadening the range of applications for FGMs without requiring significant modifications to existing equipment.

Overall, the results of this study demonstrate that traditional PBF-LB/M systems can be modified to fabricate high-quality layer-level FGMs. The innovative approaches presented here successfully overcame key limitations associated with multi-material additive manufacturing and open new opportunities for producing components that require a combination of strength and ductility. Future work will focus on further process optimization, addressing deposition constraints and exploring advanced powder recycling strategies, to fully realize the industrial potential of this versatile technology.

References

1. Kumar P, Sharma SK, Singh RKR (2023) Recent trends and future outlooks in manufacturing methods and applications of FGM: a comprehensive review. Mater Manuf Processes 38:1033–1067
2. Posa P, Errico V, Angelastro A, Campanelli SL (2024) Layer-level fabrication of continuous functionally graded materials (cFGMs) via powder bed fusion—laser Beam technology. CIRP J Manuf Sci Technol 55:247–260
3. Reichardt A, Shapiro AA, Otis R et al (2021) Advances in additive manufacturing of metal-based functionally graded materials. Int Mater Rev 66:1–29
4. Wang D, Liu L, Deng G et al (2022) Recent progress on additive manufacturing of multi-material structures with laser powder bed fusion. Virtual Phys Prototyp 17:329–365

5. Tay YWD, Lim JH, Li M, Tan MJ (2022) Creating functionally graded concrete materials with varying 3D printing parameters. Virtual Phys Prototyp 17:662–681
6. Wei C, Sun Z, Chen Q, et al (2019) Additive manufacturing of horizontal and 3D functionally graded 316L/Cu10Sn components via multiple material selective laser melting. J Manuf Sci Eng Trans ASME 141
7. Kavousi Sisi A, Ozherelkov D, Chernyshikhin S, et al (2024) Functionally graded multi-materials by laser powder bed fusion: a review on experimental studies. Progress in Additive Manufacturing
8. Mussatto A (2022) Research progress in multi-material laser-powder bed fusion additive manufacturing: a review of the state-of-the-art techniques for depositing multiple powders with spatial selectivity in a single layer. Results Eng 16:100769
9. Wen Y, Zhang B, Narayan RL et al (2021) Laser powder bed fusion of compositionally graded CoCrMo-Inconel 718. Addit Manuf 40:101926
10. Lesko C, Walker J, Middendorf J, Gockel J (2021) Functionally graded titanium-tantalum in the horizontal direction using laser powder bed fusion additive manufacturing. JOM 73:2878–2884
11. Nadimpalli VK, Dahmen T, Valente EH, et al (2019) Multi-material additive manufacturing of steels using laser powder bed fusion. In: Proceedings of the 19th international conference & exhibition, Bilbao, ES, June 2019
12. Wang D, Deng G, Yang Y et al (2021) Interface microstructure and mechanical properties of selective laser melted multilayer functionally graded materials. J Cent South Univ 28:1155–1169
13. Wits WW, Amsterdam E (2021) Graded structures by multi-material mixing in laser powder bed fusion. CIRP Ann 70:159–162
14. Daram P, Hiroto T, Watanabe M (2023) Microstructure and phase evolution of functionally graded multi-materials of Ni–Ti alloy fabricated by laser powder bed fusion process. J Market Res 23:5559–5572
15. Wei C, Li L, Zhang X, Chueh YH (2018) 3D printing of multiple metallic materials via modified selective laser melting. CIRP Ann 67:245–248
16. Rafiee M, Farahani RD, Therriault D (2020) Multi-material 3D and 4D printing: a survey. Adv Sci 7:1902307
17. Errico V, Posa P, Liang L et al (2023) Layer-level AISI 316L–18Ni (300) maraging multi-material fabrication via laser-powder bed fusion. Mater Sci Eng, A 886:145731
18. Feenstra DR, Banerjee R, Fraser HL et al (2021) Critical review of the state of the art in multi-material fabrication via directed energy deposition. Curr Opin Solid State Mater Sci 25:100924
19. Yadav S, Liu S, Singh RK et al (2024) A state-of-art review on functionally graded materials (FGMs) manufactured by 3D printing techniques: advantages, existing challenges, and future scope. J Manuf Process 131:2051–2072
20. Errico V, Posa P, Fusco A et al (2023) Intralayer multi-material structure stainless-steel/nickel-superalloy fabricated via laser-powder bed fusion process. Manuf Lett 35:11–15
21. Coutinho Saraiva BR, Novotný L, Carpentieri B et al (2023) Effect of cyclic loading on microstructure and crack propagation in additively manufactured biomaterial Co–Cr–Mo alloy. J Market Res 26:3905–3916
22. Narasimharaju SR, Zeng W, See TL et al (2022) A comprehensive review on laser powder bed fusion of steels: processing, microstructure, defects and control methods, mechanical properties, current challenges and future trends. J Manuf Process 75:375–414
23. Errico V, Fusco A, Campanelli SL (2022) Effect of DED coating and DED + Laser scanning on surface performance of L-PBF stainless steel parts. Surf Coat Technol 429:127965
24. Li H, Wang M, Lou D et al (2020) Microstructural features of biomedical cobalt–chromium–molybdenum (CoCrMo) alloy from powder bed fusion to aging heat treatment. J Mater Sci Technol 45:146–156
25. Sommer N, Kluge P, Stredak F et al (2021) Additive manufacturing of compositionally-graded AISI 316L to CoCrMo structures by directed energy deposition. Crystals (Basel) 11:1043

26. ZainElabdeen IH, Ismail L, Mohamed OF et al (2024) Recent advancements in hybrid additive manufacturing of similar and dissimilar metals via laser powder bed fusion. Mater Sci Eng, A 909:146833
27. Schneck M, Horn M, Schmitt M et al (2021) Review on additive hybrid- and multi-material-manufacturing of metals by powder bed fusion: state of technology and development potential. Progress Addit Manuf 6:881–894
28. Horn M, Prudzilko P, Anstaett C, et al (2018) Powder separation strategies for recycling in multi-material additive manufacturing
29. Sing SL, Huang S, Yeong WY (2020) Effect of solution heat treatment on microstructure and mechanical properties of laser powder bed fusion produced cobalt-28chromium-6molybdenum. Mater Sci Eng, A 769:138511
30. Liverani E, Toschi S, Ceschini L, Fortunato A (2017) Effect of selective laser melting (SLM) process parameters on microstructure and mechanical properties of 316L austenitic stainless steel. J Mater Process Technol 249:255–263
31. Bartolomeu F, Carvalho O, Gasik M, Silva FS (2023) Multi-functional Ti6Al4V-CoCrMo implants fabricated by multi-material laser powder bed fusion technology: a disruptive material's design and manufacturing philosophy. J Mech Behav Biomed Mater 138:105583

Towards Digital Twin in Additive Manufacturing: A Multi-fidelity Approach for Enhancing LPBF Process Modeling with In-situ Data

Matteo Bugatti, Giulio Valenti, and Bianca Maria Colosimo

Abstract Process modeling for additive manufacturing is essential to enable fully digital optimization workflows and reduce dependence on experimental studies to identify design flaws such as overheating, excessive thermal stress, and deformation. However, accurate track-level predictions in Laser Powder Bed Fusion (LPBF) demand significant computational resources and detailed material parameters, which are difficult to measure across the broad temperature ranges involved in the process. While post hoc calibration has been widely used to enhance model accuracy by aligning predictions with measurable outcomes, it still relies on manufacturing physical parts for validation. This study bridges the gap between physics-based simulation models and real-world processes by developing a multi-fidelity approach that incorporates in-situ data to enhance model accuracy. By employing machine learning to integrate simulations with real-time observations of temperature evolution, the multi-fidelity model improves predictive capabilities while reducing the experimental effort for first-time right production. This represents the first step towards the development of an AM digital twin for dynamically adjusting process parameters and controlling the printing process.

Keywords Additive manufacturing · Artificial intelligence · Virtual modeling

1 Introduction

Additive manufacturing (AM), particularly laser powder bed fusion (LPBF), has become a prominent technique for metal component fabrication due to its capacity to produce fully functional parts with high geometric complexity. Despite these advantages, LPBF is associated with significant challenges, including high defect rates, limited process repeatability, and difficulties in part qualification. A major limitation arises from the use of fixed process parameter sets—for bulk, down-skin, and

M. Bugatti (✉) · G. Valenti · B. M. Colosimo
Dipartimento di Meccanica, Politecnico di Milano, Milan, Italy
e-mail: matteo.bugatti@polimi.it

L. Fratini et al. (eds.), *Selected Topics in Manufacturing*, Lecture Notes in Mechanical Engineering, https://doi.org/10.1007/978-3-031-99501-9_8

up-skin—assigned based on the relative position between each layer section and the outer surface of the part, without consideration of local geometry or thermal boundary conditions. This generalized approach can result in substantial spatial variations in thermal histories within a single layer, especially in regions with differing thicknesses or proximity to free surfaces. Since thermal stability is critical for ensuring dimensional accuracy, mechanical integrity, and surface quality, such variations can lead to undesirable outcomes. High thermal gradients may induce residual stresses, deformation, or cracking, while excessive heat accumulation can cause internal and surface defects such as keyhole porosity, over-melting, and swelling. Conversely, insufficient energy input may result in lack-of-fusion and incomplete melting.

In-situ monitoring can be used to detect thermal inhomogeneities and heat accumulation in the part [1]. Both high-speed video imaging in the visible and infrared range have been used to monitor the print and detect temperature inhomogeneities in the part [2–4]. While in-situ monitoring can be used to detect unexpected anomalous conditions during the print, the correction of systematic errors in the print, e.g., wrong process parameters, or lack of support structures, requires some degree of process modelling. In fact, when paired with an appropriate control strategy [5], in-situ monitoring can be effectively used to correct for systematic errors. As a result, while in-situ monitoring provides valuable insights into the process, it remains a reactive (feed-back) approach, correcting for anomalies only after their occurrence.

However, many instances of heat accumulation follow systematic patterns and could, in principle, be anticipated by modeling the effects of process parameters and varying boundary conditions on the thermal history of the part. Process modelling through simulation represents a more effective strategy to prevent systematic issues through feed-forward control.

The literature presents various LPBF process simulation models aimed at predicting temperature fields with varying fidelity. Part-scale simulations are the most common but offer only qualitative thermal history predictions due to the simplifications needed for full-scale feasibility. These models neglect critical factors such as laser scan strategy and track sequence, which significantly influence local thermal behavior. Meso-scale simulations provide more accurate predictions of localized phenomena by incorporating detailed process physics, but they face key limitations: (i) they are computationally infeasible for full-scale parts, and (ii) they require extensive calibration of temperature-dependent material properties. While simulations can yield accurate, physics-based insights, they cannot capture unexpected process deviations arising from real-world variability, unlike in-situ monitoring approaches.

In literature, models have been proposed to combine the strengths of both simulation and process monitoring [6–8]. These models typically integrate real-time data from thermal imaging with physics-based thermal simulations to improve predictions or classification performances. Gaikwad et al. [6] demonstrate how combining experimentally measured thermal data (with a pyrometer) with simulation predictions obtained using a surrogate model based on graph theory enhances the capability of a classification model based on support vector machine to predict flaw occurrence, compared to using the same model fed only with data from the pyrometer or graph theory alone. The integration of real-time data and simulation has been explored not only for active process control but also to enhance prediction accuracy and computational efficiency. Li et al. [7] developed a surrogate model based on Gaussian Processes (GP) for a FEM thermal simulation of a fused deposition modeling (FDM) process. They applied Bayesian calibration using thermal data from an IR camera to tune unknown simulation parameters, such as the heat convection coefficient and latent heat of fusion, while also correcting model discrepancies. This resulted in a faster model capable of estimating the thermal field 40 times faster than a standalone FEM. Similar methodology have been employed for modeling the Laser Powder Bed Fusion (LPBF) process. Ye et al. [9] replaced a FEM-based thermal model and a Phase Field model for microstructure prediction with GP-based surrogates combined with Bayesian calibration. The calibration was performed using experimental data on melt pool width, enabling a faster and more adaptive predictive model. Bevans et al. [10] developed a k-Nearest Neighbors (kNN) model that merges thermal data from an experimentally validated physics-based computational model with experimental data from optical tomography and thermal imaging. Their approach enables the prediction of porosity onset, melt pool depth, grain size, and microhardness, further demonstrating the potential of data-driven models in refining LPBF process monitoring and control.

All the models presented above combine simulation and in-situ data for either model calibration, or for improving predictions or classification accuracy. However, through the continuous exchange of information and model updates between the simulation model and the process data, a digital twin of the process can be effectively implemented to control the output of the print. This approach could enable more precise and adaptive control, addressing the limitations of relying solely on one method or the other.

Malik et al. [11] created a surrogate model based on k-Nearest Neighbors (kNN), combining a FEM simulation with ex-situ metallographic data to capture melt pool depth and predict lack-of-fusion defects. They then integrated a control method based on in-situ data that leverages a classification model to adjust process parameters in response to defect predictions. Li et al. [12] developed a parameterized physics-based digital twin (PPB-DT) that predicts melt pool geometries and defects through a high-fidelity computational model calibrated with experimental data ex-situ data. They further extended this approach with a machine learning model, which enables both monitoring and control of the melt pool by adjusting process parameters.

Digital twin approaches have been proposed for direct energy deposition (DED) processes as well. Liao [13] developed a physics-informed neural network (PINN) surrogate model by combining FEM-based simulation data with real-time thermal measurements from an infrared camera, replacing the final layer of the FEM model with experimentally obtained temperature data. This model was coupled with a proportional-integral (PI) controller, which compares predicted and measured temperatures to dynamically regulate laser process parameters.

Similarly, Karkaria et al. [8] developed a surrogate model utilizing Long Short-Term Memory (LSTM) based machine learning, incorporating Bayesian inference to predict temperature distributions across different spatial locations. Starting from a FEM simulation enriched with experimental thermal data, they further integrated a Bayesian optimization framework for time-series process optimization. This setup allows them to determine the optimal laser power profile, aiming to achieve specific mechanical properties in DED-manufactured components.

In preparation of a digital twin framework for LPBF processes, a data fusion approach is proposed to combine the strengths of simulation and in-situ monitoring to predict the temperature history of the part for new unexplored process conditions. This approach is based on the concept of multi-fidelity, which refers to the integration of data from multiple sources with different levels of accuracy and resolution for developing one synthetic surrogate model [14]. Fast part-scale thermal simulation can be used for temperature prediction, but it is based on several simplifying hypotheses, making it a "low-fidelity" source of data. On the other hand, in-situ monitoring with high-speed thermal imaging provides "high-fidelity" information on the print but lacks prediction capabilities.

The proposed approach is based on Gaussian Process (GP) regression, which is a non-parametric machine learning technique that can be used to model complex functions [15]. The GP model combines simulation results and in-situ data in a single model to improve the prediction accuracy obtained using simulation only. The proposed approach is developed and tested using simulated and observed data from a test campaign based on simple geometries.

The results of the model are compared with the in-situ measurements and with a competitor "black-box" (BB) model trained only on in-situ data. The rest of the paper is organized as follows:

- Section 2 describes the experimental setup used to collect the in-situ data and the details of the finite element simulation model. The proposed data fusion approaches are also discussed in this section.
- Section 3 presents the results of the data fusion approach and compares it with the in-situ measurements and the competitor "black-box" model.
- Section 4 concludes the paper and discusses future work perspectives on the topic.

2 Materials and Methods

Part of the experimental campaign and corresponding in-situ thermal data reported in [4] were used to develop this work. The main characteristics of the experiment and the monitoring setup are reported in the following sections.

2.1 Experimental Campaign

The first goal is to generate a dataset that comprises high and low-fidelity data that could be used to train a multi-fidelity model. To study process conditions characterized by different heating and cooling cycles while still being realistic, an experiment was set up to print 4 simple cube-like samples with 5×5 mm base and varying the

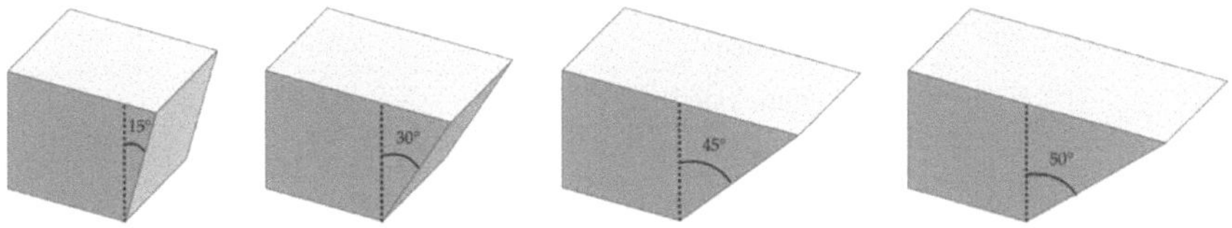

Fig. 1 Inclination angles of overhang samples

Table 1 Process parameters used for the experiments

Parameter	Value
Material	AISI 316L
Particle size distribution (D10-D90)	15–45 μm
Laser power (W)	200
Laser mode	Continuous wave (CW)
Scan speed (mm/s)	900
Hatch distance (mm)	0.05
Layer thickness (mm)	0.03
Hatch strategy	Meandering

overhang angle (OH [°]) between 15° and 50° (Fig. 1). The main process parameters are reported in Table 1.

Inclined surfaces are known to be a challenging feature in LPBF because of the lack of support and the worse heat dissipation of the loose metal powder bed, but it is an almost inevitable process condition when printing complex shapes. The monitoring setup used to acquire the high-fidelity temperature data is described in Sect. 2.2. The model developed to simulate the printing process and generate the low-fidelity data is described in Sect. 2.3.

2.2 *In-situ Thermography*

A high-speed medium wave infrared (MWIR) camera, the FLIR X6901sc, was used to monitor the process. The camera is mounted off-axis (30° inclination from the vertical) and placed outside of the build chamber. The approximate working distance between the camera and the target is 500 mm. The build chamber window used by the camera is made of synthetic sapphire, which is characterized by a very high transmissivity for the wavelengths of interest, (transmissivity > 85% in the 2000–5000 nm range).

The main camera settings are reported in Table 2.

Because of the wide range of temperatures covered during the process (from melting temperature to close to ambient temperatures) the super-framing option was used to increase the dynamic range of the camera. This allowed to monitor the whole

Table 2 MWIR camera settings

Parameter	Value
Frame rate (fps)	667
Frame size (px)	224 × 72
Resolution (μm/px)	≈ 300

range of temperatures covered during the process (300–1600 K) while keeping a reasonably high frame rate (667 fps).

A total of 4 consecutive layers—one for each hatch pattern—were monitored for all the available samples at 1/2 of the total height of the print. This allowed to capture all the different combinations of process conditions (i.e., hatch patterns and overhang angle). To estimate the true temperature of the printed part from the IR videos, the global equivalent emissivity, i.e., a combination of transmittance of the optical chain and emissivity of the observed materials, was measured. The transmittance was experimentally estimated following the procedure described in the ASTM E1897-97 and then multiplied by a constant emissivity for AISI 316L that found in literature ($\varepsilon = 0.22$, [16]).

The temperature estimated from the in-situ measurements served as the ground truth for model fitting. However, uncertainties in the temperature estimation—arising from factors such as temperature and material-dependent emissivity, camera setup (e.g., viewing angle, distance to target, build chamber temperature), and sample characteristics (e.g., surface roughness)—were not explicitly accounted for at this stage. This is because all models were trained using the same ground truth data and are therefore subject to the same potential bias.

2.3 *Simulation Model*

A single-layer transient implicit thermal simulation was developed to replicate in a virtual environment the experimental conditions observed in-situ.

The finite element method (FEM) model was implemented using the commercial software ABAQUS. Custom user subroutines were developed to add functionalities to the standard transient FEM model, specifically:

- DFLUX to simulate the laser heat flux (DFLUX).
- SDVINI and USDFLD to track the change of phase during the process and assign the corresponding material properties to each element.

A larger domain was modeled compared to the actual layer size of the part (Fig. 2).

This allows to simulate realistic heat conduction between the part and the surrounding powder bed. The size of the excess volume was determined using the thermal length $\sqrt{4 \cdot \alpha \cdot t}$, where α is the thermal diffusivity and t is the total simulation time [17]. This extra length was added in all three directions to the original

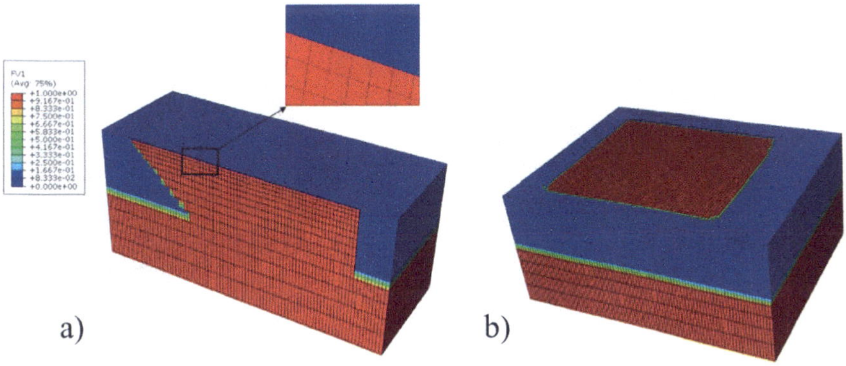

Fig. 2 Simulation domain. The FV1 field is used to track the phase of the material from powder (FV1 = 0, blue) to solid (FV1 = 1, red). The cross section of the simulation domain after recoating (left) shows the thin continuous powder layer deposited on top. After laser exposure (right) the powder layer is selectively melted and solidified in the desired shape

volume of the part to make sure the boundary conditions did not influence the heat transfer during the simulation.

Temperature-dependent thermal properties were used for the simulation, as shown in Fig. 3.

In the powder state, the conductivity was scaled down to 1% of conductivity of the solid metal. The density was also scaled down using the measured apparent density as a reference. To simulate the laser heat flux accurately, either as a 2D or a 3D heat flux, a fine mesh is required in the heated area to discretize the melt pool. Additionally, a small time increment should be used not to incur any numerical instabilities when simulating the large thermal gradient induced by the laser. This approach is not applicable to large part-scale models, as it would require excessive computational time to simulate every layer. A different approach proposed in the literature disregards the laser beam shaped heat flux and applies an equivalent heat input by exposing the part slice at the same time or activates the slice at melting temperature [18]. However, this approach lowers the peak temperatures, therefore

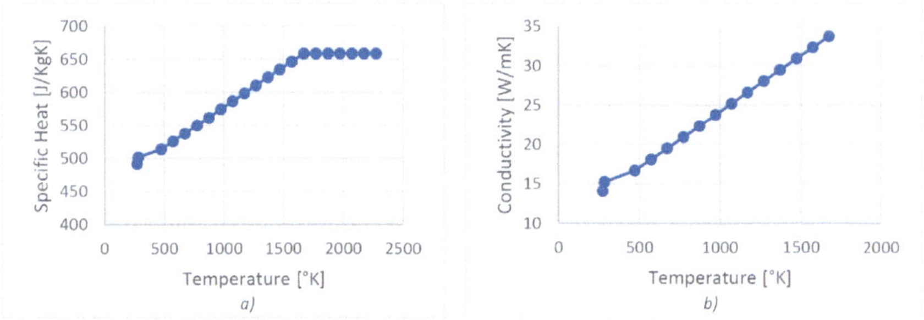

Fig. 3 AISI 316L specific heat (**a**) and conductivity (**b**)

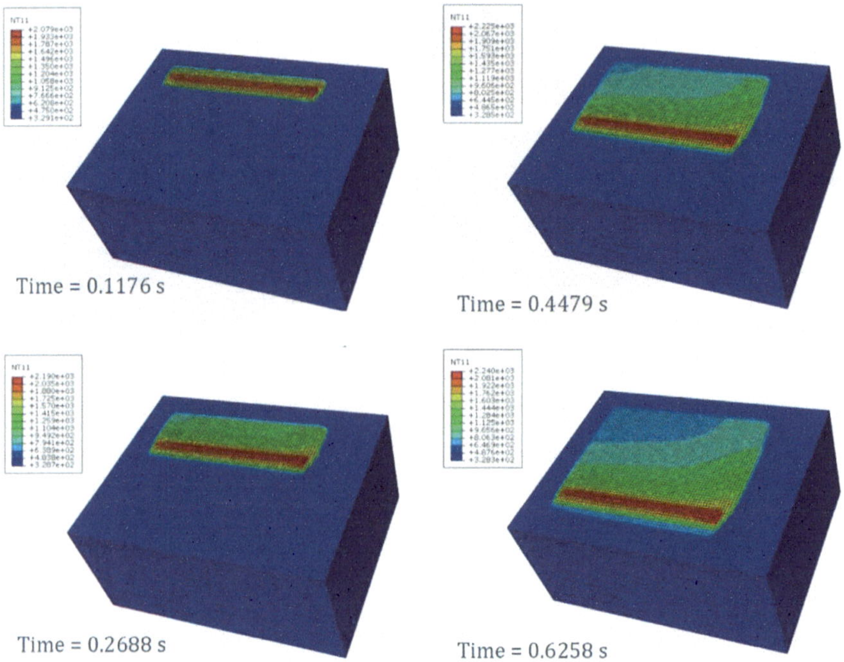

Fig. 4 Example of the equivalent laser track heat flux for the 30° overhang sample

phase change cannot be modeled properly and is usually neglected. To overcome the limitations of these two approaches, a different method was used for modelling the heat source [19, 20]. The heat input relative to each individual laser track was applied as a moving heat source, effectively simulating the laser exposure over the layer during the scanning time (Fig. 4). This way, most local heating effects can be captured by the transient simulation, while requiring a fraction of the computational power to run.

2.4 Data Analysis

With surface temperature data available from both sources of information, i.e., in-situ thermography (high-fidelity) and simplified heat transfer simulation (low-fidelity), the following sequence of steps were performed:

1. Data registration: A registration operation between the nodes and the pixels of the thermal video is performed.

2. Data averaging: Temporal and spatial resolution is different between the two datasets. Therefore, nodal values that fall in the area corresponding to each individual pixel were averaged to have a 1-to-1 correspondence of extracted cooling profiles from the two sources.
3. Data extraction and model training/testing: Extraction of individual cooling profiles for training the models, i.e.:
    a. Data fusion model: Showing the strategy that was explored to combine the two sources of information.
    b. Competitor model: An alternative, completely data-driven model, trained only on in-situ thermography data to make predictions.

At this preliminary stage, only the cooling profile near the overhang region were extracted for training the models. To reduce noise and border effects, the cooling profiles from the mid-point of the overhang edge were extracted for a "super-pixel" made of 3×2 grid of regular pixels (Fig. 5).

After data extraction, two models were developed, i.e.:

- Data fusion (DF) model: this model is trained to combine the output thermal simulation with in-situ thermography temperature measurements to provide more accurate temperature predictions using just a single input regressor, i.e., the time after laser pass, t.
- Competitor model: a completely data driven model (also called "black-box", BB) that provides temperature predictions for unobserved overhang geometries by modeling the overhang angle. In fact, compared to the DF model, the BB model is using two regressors, t and the overhang surface angle (OH).

For both models, Gaussian process regression (GPR) was employed. Gaussian processes are a powerful non-parametric supervised learning method that has been used extensively for regression tasks. GPR is an extremely versatile tool with some peculiar properties, such as (i) interpolation between observations, (ii) kernels to adapt to different datasets, and (iii) including the mean and standard deviation as output, allowing it to give a probabilistic approach to the prediction.

Further details on Gaussian processes can be found in [15]. Details on the implementation of GPR for the two models are provided in the following sections.

Data fusion model. To combine thermal simulation data (low-fidelity) with in-situ thermography (high-fidelity) measurements, the ratio between the temperatures of

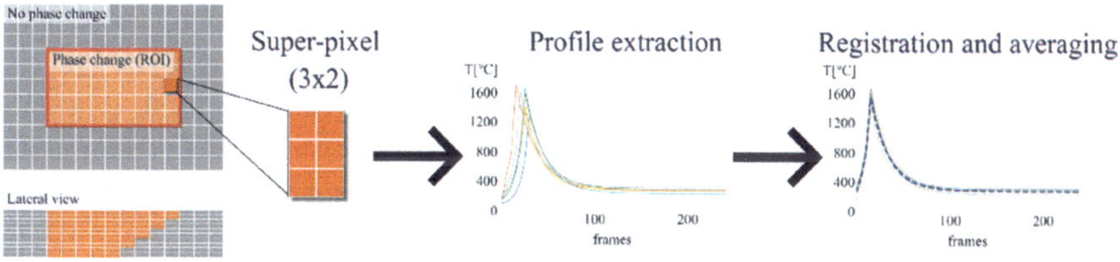

Fig. 5 Profile extraction procedure

the two cooling profiles were estimated.

$$r(t) = T_{Hi-Fi}(t)/T_{Lo-Fi}(t)$$

The correction term, $r(t)$, captures the ratio between high and low fidelity temperatures as a function of time after laser pass, and it was modelled with a simple Gaussian process:

$$r(t) = f(t) + \varepsilon, \quad \varepsilon \sim N\left(0, \sigma^2\right)$$

where $f(t) \sim GP\left(\mu(t), k\left(t, t'\right)\right)$ represents a Gaussian process with mean function $\mu(t)$ and a squared exponential (radial basis function, RBF) kernel for the covariance function:

$$k\left(t, t'\right) = \sigma_f^2 \exp\left(-\frac{\left(t - t'\right)^2}{2l_t^2}\right)$$

where σ_f^2 is the signal variance, and l_t is the length-scale parameters controlling smoothness in the input dimension. By modeling the correction term, we could adjust the temperatures extracted from the simulation to match more closely the IR camera measurements, thus compensating for any modeling error in the simulation. The new "multi-fidelity" temperature profile will combine the modelled correction term, $r(t)$, with low-fidelity temperature readings to obtain a more accurate estimate of the true temperature:

$$T_{Mu-Fi}(t) = r(t) \cdot T_{Lo-Fi}(t)$$

Competitor model. A competitor model trained solely with in-situ thermography (high-fidelity) data was also developed to compare the performance of the data-fusion models with a completely data-driven model.

GP regression was used to fit the competitor, black-box (BB) model on the monitoring data. The model included the overhang angle as regression term, and it was trained to predict the cooling profile of the top surface after laser pass.

A Gaussian Process Regression (GPR) model was developed to fit the competitor, black-box (BB) model to the monitoring data. The model aims to capture complex thermal behavior influenced by both temporal and geometric factors. Specifically, the two input regressors are the time after the laser pass (t) and the inclination angle of the overhang surface (OH), which together determine the local cooling dynamics. A similar procedure to the one described for the DF model was used, with the main differences being (i) the direct modelling of the temperature, T, and (ii) the additional OH in the input regressors. In fact, the cooling temperature, T, is modeled as a Gaussian process:

$$T(t, OH) = f(t, OH) + \varepsilon, \quad \varepsilon \sim N\left(0, \sigma^2\right)$$

where $f(t, OH) \sim GP\left(\mu(t, OH), k\left((t, OH), (t', OH')\right)\right)$ represents a Gaussian process with mean function $\mu(t, OH)$ and a squared exponential kernel (the same type used for the DF model, radial basis function RBF) for the covariance function:

$$k\left((t, OH), (t', OH')\right) = \sigma_f^2 \exp\left(-\frac{(t-t')^2}{2l_t^2} - \frac{(OH-OH')^2}{2l_{OH}^2}\right)$$

where σ_f^2 is the signal variance, and l_t and l_{OH} are the length-scale parameters controlling smoothness in each input dimension.

3 Results and Discussion

An example of temperatures extracted from the super-pixel in correspondence of the overhang region is shown in Fig. 6 for both in-situ thermography and thermal simulation.

The cooling profiles of the in-situ thermography (Fig. 6a) and the thermal simulation (Fig. 6b) show a good agreement in terms of shape and heat accumulation for the different samples. As expected, the cooling rate decreases as the inclination of the overhang surface increases in both datasets. However, there is a stark difference in terms of temperature values. While in-situ thermography readings peak at approximately 1200 K, the simulated temperatures peak at approximately 2100 K. This probably stems from the differences in resolution between the two tools. The approximate resolution of the camera is 3 times larger than the element size (100 μm) used for the simulation. Therefore, in-situ thermography is not capable of capturing the phase change and the corresponding temperature peak with the same resolution of

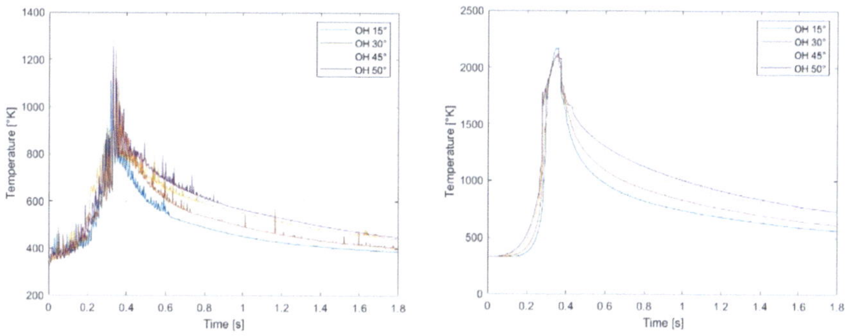

a) In-situ thermography *b) Thermal simulation*

Fig. 6 Temperature profiles of the overhangs

the thermal simulation. Other effects might be present, e.g., inaccurate thermal properties or heat flux modeling, however, these are the unknown sources of discrepancy that motivate data fusion and this work specifically.

All models, data fusion (DF) and competitor black-box (BB), were trained on data from overhangs at 15° and 45° to evaluate the accuracy in predicting cooling profiles for both interpolated angles (30°) and extrapolated angles (50°). To avoid overfitting, a fivefold cross-validation strategy was employed. The dataset was randomly partitioned into training and validation sets while ensuring uniform coverage of the target OH levels and time points. The final hyperparameters were selected based on their average performance across folds in terms of root mean square error (RMSE). The cooling profiles predicted with the two data fusion model and the competitor model are shown in Fig. 7.

From a qualitative comparison between the predictions of the two models, we see a good prediction ability from both models. The 30° overhang predictions show a very good overlap for both DF and BB models. However, when the models are used to extrapolate the cooling behavior for the 50° overhang, the observed profile falls outside of the BB model 95% confidence interval more frequently compared to the DF model (Fig. 7b). The cooling profile predicted by the DF model seems to follow more closely the real cooling profile, whereas the BB predicts a faster cooling rate compared to the real one. For a more quantitative comparison between the performance of the two models, the prediction error was computed for the predicted cooling profiles.

The plot of Fig. 8 clearly illustrates the behavior that the qualitative comparison was suggesting. While the average prediction error of the BB model is more stable than the DF model when interpolating (Fig. 8a), DF predictions stay stable even when extrapolating outside of the training region (Fig. 8b). For both inclinations, the observed cooling profiles falls within the confidence interval of the DF prediction. This is not true for the BB model, which, despite showing a better performance at predicting the 30° cooling profile, it struggles to keep the same stability for the 50°

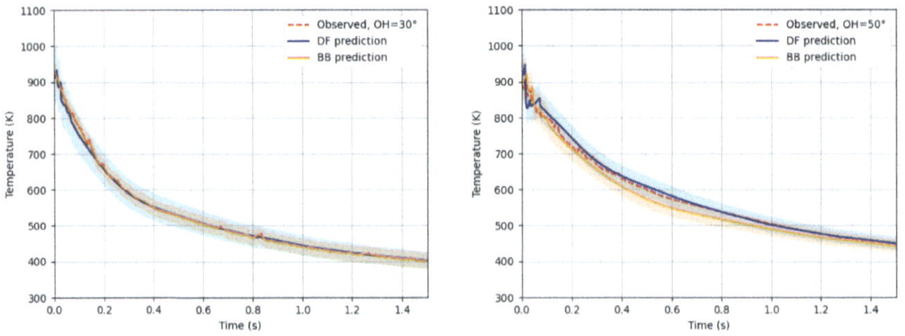

a) Predictions for 30° overhang *b) Predictions for 50° overhang*

Fig. 7 Predicted cooling profiles for 30° and 50° overhang. The 95% confidence interval around the prediction is also reported

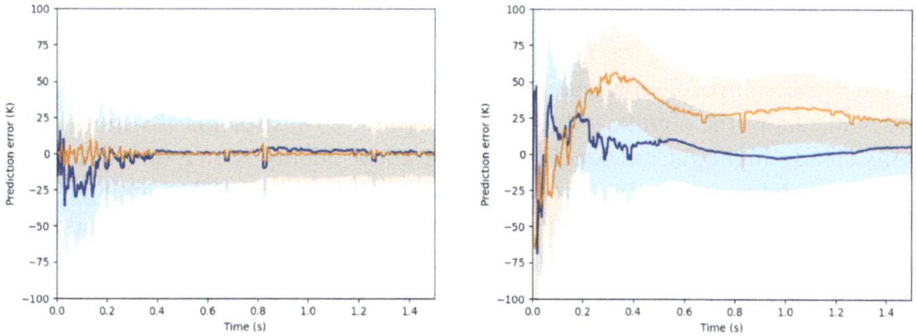

a) Predictions for 30° overhang *b) Predictions for 50° overhang*

Fig. 8 Prediction error of data fusion and black-box model

overhang condition. This confirms that the information captured by the simulation is effective at guiding the prediction even outside of the training range. Some systematic deviation is still present at the very beginning of the cooling (up to $t = 0.2s$ from the laser pass), which indicates that temperature peak is not correctly reconstructed by the DF model. This could be due to the noisy temperature reading when the laser scans across the overhang, generating the corresponding temperature peak. Because of the data preparation procedure, the actual peak gets averaged together with the neighboring pixels, resulting in a lower peak and different temperature reading. This effect needs to be investigated further to improve prediction at peak. However, it does not seem to affect the overall prediction of the cooling profile, which correctly identifies the systematic heat accumulation phenomena related to the geometrical features of the part. These preliminary results demonstrate that the DF approach allows to compensate for all sources of uncertainty that can significantly influence the simulation accuracy, e.g., temperature-dependent material properties, boundary conditions, including equivalent heat flux, while still offering the capabilities of a "white box" model based on physics to make predictions outside of the training window.

4 Conclusions

The results show that the data fusion approach presented in this work outperforms the model based solely on in-situ data, producing more accurate temperature predictions. This proves that, by properly modelling the systematic bias that there is between in-situ thermography and thermal simulation, it is possible to predict heat accumulation and other temperature-related information by correcting the low-fidelity simulation data and reconstructing the corresponding high-fidelity temperature measurements. This approach shows the potential of correcting for potential inaccuracies in simplified simulation by modifying directly the output instead of spending extensive computational power and hundreds of iterations to calibrate for the tens of inputs the finite element simulation relies on. Overall, the data fusion predictions show a smaller average prediction error when considering all the tested configurations. This is thanks to the significantly higher stability of DF over the BB model. The results clearly highlight the benefits of using a low-fidelity source of information, rather than relying only on high-fidelity data to predict outside of the training region. While the approach shows promising results at the sample level, its generalization to different geometries, materials, scan strategies, and process parameters requires further investigation. Variations in these factors may lead to thermal behaviors not captured by the current model, limiting its predictive accuracy outside the training domain. Additionally, scaling the method to larger parts will increase the computational cost of thermal simulations, which may limit its applicability for real-time, feed-forward control. Future work will aim to evaluate the data fusion approach on new geometries and scan strategies, with a focus on computational efficiency to assess its suitability for integration into a digital twin framework that enables continuous information exchange between model predictions and in-situ measurements for enhanced process control.

5 Competing Interest

The authors declare that they have no known competing financial interests or personal relationships that could have appeared to influence the work reported in this paper.

Funding The research was supported by MICS (Made in Italy—Circular and Sustainable) Extended Partnership and received funding from the European Union Next-GenerationEU (PIANO NAZIONALE DI RIPRESA E RESILIENZA (PNRR)—MISSIONE 4 COMPONENTE 2, INVESTIMENTO 1.3—D.D. 1551.11-10-2022, PE00000004).

References

1. Grasso M, Remani A, Dickins A et al (2021) In-situ measurement and monitoring methods for metal powder bed fusion: an updated review. Meas Sci Technol 32:112001. https://doi.org/10.1088/1361-6501/ac0b6b
2. Grasso M, Repossini G, Colosimo B (2017) In-situ monitoring of metal additive manufacturing processes via spatter signature. In: Proceedings of the XIII AITeM Conference. pp 1–10
3. Bugatti M, Colosimo BM (2022) Towards real-time in-situ monitoring of hot-spot defects in L-PBF: a new classification-based method for fast video-imaging data analysis. J Intell Manuf 33:293–309. https://doi.org/10.1007/s10845-021-01787-y
4. Bugatti M, Colosimo BM (2024) Predicting the roughness of overhanging surfaces in laser powder bed fusion via in-situ thermal imaging. J Manuf Process 124:1340–1348. https://doi.org/10.1016/j.jmapro.2024.06.050
5. Vasileska E, Demir AG, Colosimo BM, Previtali B (2020) Layer-wise control of selective laser melting by means of inline melt pool area measurements. J Laser Appl 32:022057. https://doi.org/10.2351/7.0000108

6. Gaikwad A, Imani F, Yang H et al (2019) In situ monitoring of thin-wall build quality in laser powder bed fusion using deep learning. Smart Sustain Manuf Syst 3:98–121. https://doi.org/10.1520/SSMS20190027
7. Li J, Jin R, Yu HZ (2018) Integration of physically-based and data-driven approaches for thermal field prediction in additive manufacturing. Mater Des 139:473–485. https://doi.org/10.1016/j.matdes.2017.11.028
8. Karkaria V, Goeckner A, Zha R et al (2024) Towards a digital twin framework in additive manufacturing: machine learning and Bayesian optimization for time series process optimization. J Manuf Syst 75:322–332. https://doi.org/10.1016/j.jmsy.2024.04.023
9. Ye J, Mahmoudi M, Karayagiz K et al (2022) Bayesian calibration of multiple coupled simulation models for metal additive manufacturing: a Bayesian network approach. ASCE-ASME J Risk Uncertain Eng Syst Part B Mech Eng 8:011111. https://doi.org/10.1115/1.4052270
10. Bevans BD, Carrington A, Riensche A et al (2024) Digital twins for rapid in-situ qualification of part quality in laser powder bed fusion additive manufacturing. Addit Manuf 93:104415. https://doi.org/10.1016/j.addma.2024.104415
11. Malik AW, Mahmood MA, Liou F (2024) Digital twin–driven optimization of laser powder bed fusion processes: a focus on lack-of-fusion defects. Rapid Prototyp J 30:1977–1988. https://doi.org/10.1108/RPJ-02-2024-0091
12. Li Y, Mojumder S, Lu Y et al (2023) Statistical parameterized physics-based machine learning digital twin models for laser powder bed fusion process
13. Liao S, Toward a digital twin of metal additive manufacturing: process optimization
14. Jiang P, Zhou Q, Shao X (2020) Multi-fidelity surrogate models. In: Jiang P, Zhou Q, Shao X (eds) Surrogate model-based engineering design and optimization. Springer, Singapore, pp 55–87
15. Rasmussen CE, Williams CKI (2008) Gaussian processes for machine learning, 3. print. MIT Press, Cambridge, Mass
16. Experimental determination of the emissivity of powder layers and bulk material in laser powder bed fusion using infrared thermography and thermocouples. https://www.mdpi.com/2075-4701/10/11/1546. Accessed 10 Mar 2025
17. Carslaw HS, Jaeger JC, Carslaw HS, Jaeger JC (1986) Conduction of heat in solids, 2nd edn. Oxford University Press, Oxford, New York, Second Edition
18. Zaeh MF, Branner G (2010) Investigations on residual stresses and deformations in selective laser melting. Prod Eng 4:35–45. https://doi.org/10.1007/s11740-009-0192-y
19. Chen Q, Liang X, Hayduke D et al (2019) An inherent strain based multiscale modeling framework for simulating part-scale residual deformation for direct metal laser sintering. Addit Manuf 28:406–418. https://doi.org/10.1016/j.addma.2019.05.021
20. Irwin J, Michaleris P (2016) A line heat input model for additive manufacturing. J Manuf Sci Eng. https://doi.org/10.1115/1.4033662

Combining FEM Simulations and Machine Learning for Skin Contamination Prediction in Aluminum Alloy Extrusion Processes

Riccardo Pelaccia, Marco Negozio, and Adrian Hugh Alexander Lutey

Abstract Skin contamination is a critical defect in aluminum alloy extrusion, leading to rejection of contaminated profiles due to poor mechanical properties. Experimental, empirical and numerical modeling approaches to identifying skin contamination in industry are either time-consuming, inaccurate or expensive. The present study integrates finite element method (FEM) simulations and machine learning (ML) to predict skin formation and evolution, offering a powerful tool for optimizing extrusion and advancing smart manufacturing practices. FEM simulations are calibrated and validated against several industrial case studies, after which artificial neural networks (ANNs) are trained with the simulation dataset, significantly reducing the computational cost of future predictions. The proposed approach not only enables accurate prediction of skin contamination but also provides insight into the die and process parameters most critical to skin defect formation.

Keywords Machine learning · Extrusion · Virtual modeling

1 Introduction

The extrusion of aluminum alloys is a highly efficient manufacturing process for producing profiles with complex geometries and excellent mechanical properties [1, 2]. However, a critical quality issue is the skin defect, also known as billet skin contamination, or back-end defect [3, 4]. This defect develops due to differences between the inner billet material and its outer "skin", which is altered during the direct chill (DC) casting process and further affected by surface impurities such as oxides,

---

R. Pelaccia (✉)
Dipartimento di Scienze e Metodi dell'Ingegneria, Università degli Studi di Modena e Reggio Emilia, Reggio Emilia, Italy
e-mail: riccardo_p@unimore.it

M. Negozio · A. H. A. Lutey
Dipartimento di Ingegneria dei Sistemi e delle Tecnologie Industriali (DISTI), Università degli Studi di Parma, Parma, Italy

L. Fratini et al. (eds.), *Selected Topics in Manufacturing*, Lecture Notes in Mechanical Engineering, https://doi.org/10.1007/978-3-031-99501-9_9

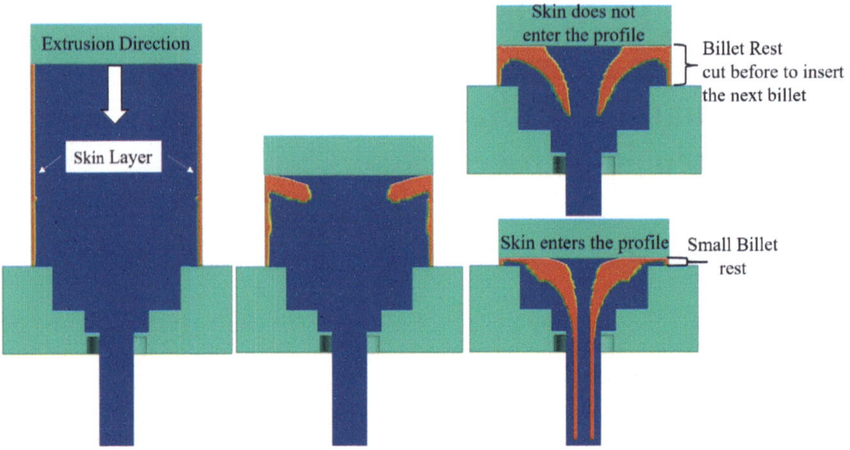

Fig. 1 Schematic of billet skin flow during the extrusion process

residual lubricant and dust. Under high friction and deformation conditions during extrusion, the contaminated skin gradually mixes with the billet material, forming a layer that degrades both the quality and mechanical performance of the final product [5, 6]. Figure 1 shows the evolution of skin flow during extrusion. High friction levels at the billet-container and billet-ram interfaces prevents the skin from immediately flowing towards the die, instead promoting flow towards the moving ram. Only after a certain ram stroke, the billet skin starts to flow into the die, thus leading to profile contamination if the billet rest, or remaining material in the container that is cut and scrapped, is too small.

The billet rest length is therefore one of the primary factors influencing the severity of the skin defect [7]. At the end of an extrusion run, the remaining billet material near the die exit cools, promoting adhesion to the die wall. During the subsequent extrusion run, the material at the interface is torn off, leaving a mark on the surface called the stopmark (Fig. 2). If the billet skin flows within the die but does not become part of the profile, it is extruded after the stopmark and discarded along with the inevitable scrap from charge welds [7, 8]. Conversely, if the billet rest is too short, the defect may extend beyond the designated stopmark, necessitating additional scrapping and thus increasing production costs. Moreover, an excessively long billet rest can result in unnecessary scrap of uncontaminated material, reducing overall process efficiency.

Optimization of the billet rest length is therefore important for maintaining both product quality and cost-effectiveness. In addition to billet rest length, evolution of the skin defect is highly sensitive to several other parameters such as billet geometry, die design, extrusion ratio and billet quality. The geometry of the billet, particularly its length-to-diameter ratio, significantly affects material flow during extrusion. Longer billets can enhance productivity by reducing the frequency of billet changes, but this can lead to formation of the skin defect at an early ram stroke [7, 9]. Design of

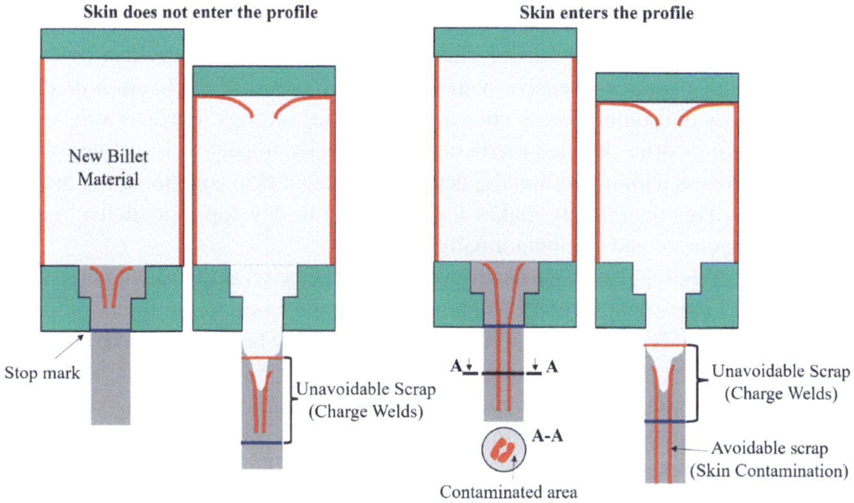

Fig. 2 Assessment of industrial scrap

the extrusion die also plays a critical role in determining material flow patterns and behavior of the billet skin. Factors such as die angle, bearing length and the distance between die exits influence the formation of shear zones between dead metal zones and flow zones during extrusion. Adjustments in die design, including optimizing the distance between die exits, can reduce the extent of billet skin contamination, thereby improving the quality of the extruded product [10, 11]. The extrusion ratio, defined as the ratio of the cross-sectional area of the billet to that of the extruded profile, directly impacts the extrusion load, the manner in which the material flows and consequently defect formation [4, 7, 9]. Finally, the initial thickness of the billet skin, influenced by factors such as DC-casting conditions and surface treatments, affects the extent of contamination in the final product. Thicker oxide layers or surface impurities can lead to earlier formation of the skin defect within extruded profiles, leading to internal defects that compromise mechanical properties [12, 13]. Therefore, controlling skin thickness through proper billet preparation and handling is crucial for ensuring product integrity.

Numerical investigations utilizing the finite element method (FEM) have significantly contributed to understanding the mechanisms of billet skin contamination. These studies have demonstrated that its extent is highly sensitive to friction conditions at the billet-container and billet-ram interfaces, as well as the initial surface quality of the billet [4, 12, 13]. Elevated temperatures and rapid changes in ram speed further worsen frictional heating, which not only alters the flow stress of the material but can also accelerate oxidation on the billet surface. Such thermomechanical interactions have been investigated through both experimental and numerical approaches, leading to improved predictive models for optimizing die design and process parameters to minimize skin defect formation [4, 14–16]. Despite the valuable insights provided by FEM studies, a significant drawback of this approach is the high computational cost associated with performing detailed simulations. Running a comprehensive FEM analysis for each set of process parameters can be time-consuming and resource-intensive, which limits the practical application of these simulations for real-time process control in industrial settings. Furthermore, while FEM simulations offer detailed predictions, the precise impact and relative weight of various process parameters and die design features on skin contamination are not always clear. This uncertainty makes it challenging to develop a predictive model that is both accurate and computationally efficient.

To address these challenges, a novel approach has been developed within this study that leverages the strengths of FEM simulations while overcoming their computational limitations. Two hundred and fifty simulations have been performed, covering a wide range of extrusion conditions and parameter variations affecting skin flow behavior. This dataset has been employed to train an artificial neural network (ANN) for predicting evolution of the skin defect based on key input parameters such as billet and die designs and the extrusion ratio, amongst others. The reliability of the numerical model to create an accurate training dataset for the ANN has been demonstrated through experimental validation of outcomes obtained for five industrial case studies. The ultimate goal of this work is to develop a methodology that offers a comprehensive framework for analyzing and optimizing the extrusion process. This approach enhances predictive capabilities, reduces computational effort and supports process sustainability by improving resource efficiency.

2 FEM Simulation

2.1 Generation of Dataset for ANN Training and Testing

The initial dataset for training and testing the ANN was generated via numerical simulations using FEM software Qform Extrusion UK. Figure 3 illustrates the geometric parameters selected as ANN input variables, including billet length (BL), billet diameter (BD), die entrance diameter (D1), skin layer thickness (ST) and die length (DL). Additional input parameters included the extrusion ratio (ER), the ratio (R1) between the billet area and the feeding area (for solid profiles) or the circular area inscribing the feeding area (for hollow profiles), and the ratio (R2) between the latter and the feeding area. R2 is equal to one for solid profiles and greater than one for hollow profiles due to the presence of the mandrel.

The skin layer thickness and billet geometric parameters primarily influence the amount of defect entering the die, while the die geometric parameters affect material flow behavior during extrusion, thereby determining how the skin layer deforms and flows. To reduce the number of variables and required simulations, the effects of varying the process parameters (i.e. ram speed and billet temperature) and the material were neglected, as these mainly affect material resistance under deformation, whereas geometric parameters predominantly govern material flow behavior. Process

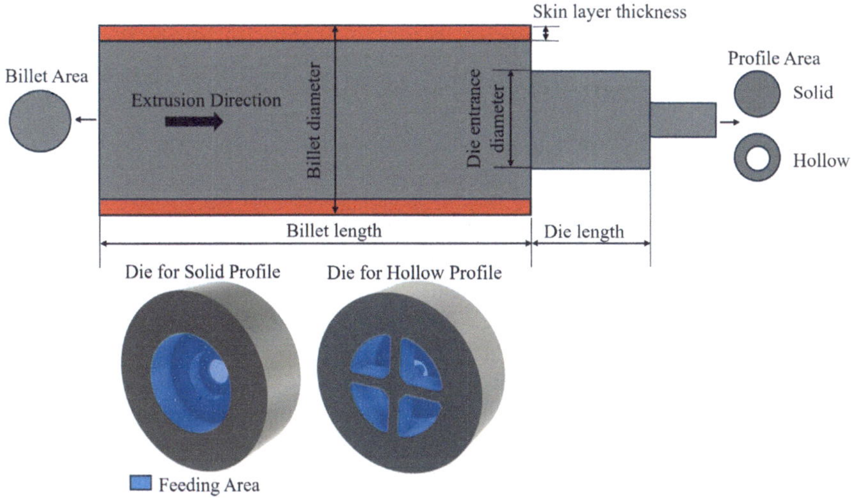

Fig. 3 Geometric parameters selected as input variables for ANN training and testing

parameters were therefore held constant at realistic industrial values, while AA6060 aluminum alloy was selected as the material to be extruded. Future works will focus on the effects of process parameters and the extruded material.

A total of 250 simulations were performed by selecting at least four levels of each input variable within industrially relevant ranges. Variation in the die's geometric parameters resulted in fourteen distinct die designs, seven for producing solid bar profiles and seven for producing hollow tubes. Table 1 summarizes the tested ranges of each parameter. Table 2 presents constant process parameters used in the simulations.

Qform Extrusion UK employs an arbitrary Lagrangian–Eulerian (ALE) approach in which the die and profile mesh are fixed while the billet mesh follows the deformation path imposed by the ram stroke. Figure 4 shows an example of the solid and hollow profiles meshed within the software. Double symmetry was exploited,

Table 1 Parameter ranges employed for generation of dataset for ANN training and testing

Variable	Min. value	Max. value
Billet length, BL (mm)	250	1000
Billet diameter, BD (mm)	50	250
Die entrance diameter, D1 (mm)	35	180
Skin layer thickness, ST (mm)	0.125	0.5
Extrusion ratio, ER	14	150
R1	1.03	5.22
R2	1	2.42
Die length, DL (mm)	50	150

Table 2 Constant process parameters employed for generation of dataset for ANN training and testing

Parameter	Value
Extruded material	AA6060
Ram speed (mm/s)	5
Billet temperature (°C)	480
Die temperature (°C)	480
Container temperature (°C)	450
Ram temperature (°C)	400
Billet rest (mm)	10

thereby modeling one quarter of the entire geometry. For simulation of skin flow, a post-processing subroutine was used, where point tracking of nodes within the skin layer was performed throughout the entire process. A refined mesh was therefore employed within the billet, with a maximum element size of 0.1 mm for the skin layer.

Within Qform Extrusion UK, complete adhesion is typically applied at the tool-aluminum interface, except within bearing regions where the profile attains its final shape and the high velocities involved promote a slip condition [17–19]. A friction factor of one was therefore assigned to all aluminum-tool interfaces except for the bearings, where a slip condition was applied with the friction factor automatically calculated by the software. Material properties for AA6060 aluminum alloy were defined using the corresponding material card from the Qform UK library.

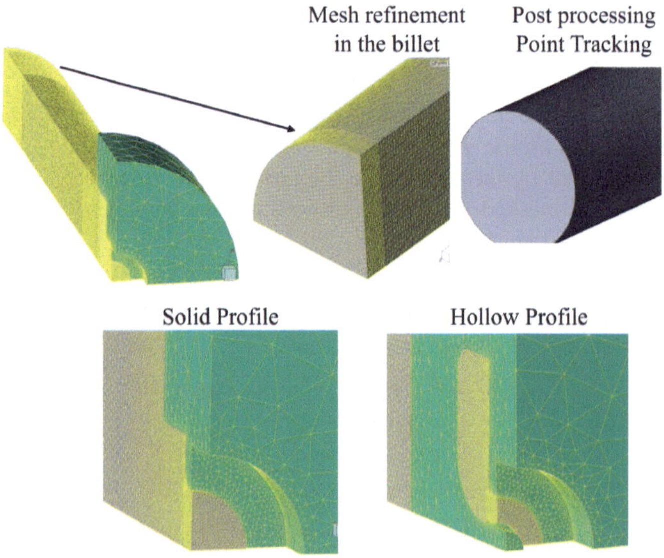

Fig. 4 Example of mesh for solid and hollow profiles employed for generation of dataset for ANN training and testing

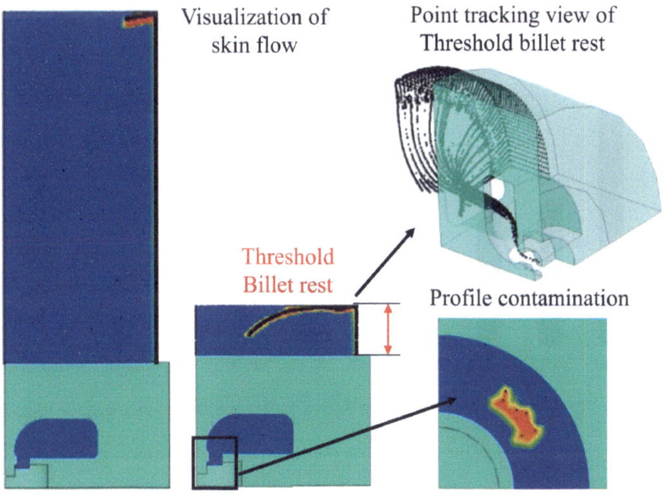

Fig. 5 Example of numerical assessment of skin contamination within Qform UK

Simulation outcomes provided the billet rest lengths at which the skin defect entered the extruded profile (Fig. 5). This length will be referred to as the "threshold billet rest" from hereon in for simplicity. If the skin defect remains within the die, it is discarded along with the unavoidable charge weld defect (Fig. 2). A billet rest length of 10 mm was selected as the stop condition for all simulations, as smaller values are neither industrially significant nor practically realistic. Consequently, when the skin defect did not enter the profile, the threshold billet rest value was set to 0, even though the actual value was potentially in the range 0–10 mm.

Figure 5 shows an example of simulated skin flow evolution and assessment of the threshold billet rest for a tube profile. Figure 6 shows the relative frequency distribution of threshold billet rest values obtained from simulations. These outcomes indicate that the relative frequency decreased from the 1–10 mm range to values exceeding 100 mm over the tested parameter range.

2.2 Validation of Simulation and Generation of Independent Test Dataset for ANN

Five experimental case studies comprising three hollow and two solid profiles were selected to validate the numerical model and generate independent test datasets for the ANN model. Figure 7 reports the geometry of all profiles imported into Qform Extrusion. Experimental data for Profile A was taken from the literature [3], while those for profiles B to E were taken from industrial case studies analyzed in previous works [20, 21]. Table 3 summarizes the process parameters employed within simulations, while Table 4 reports specific variables calculated as inputs required for the

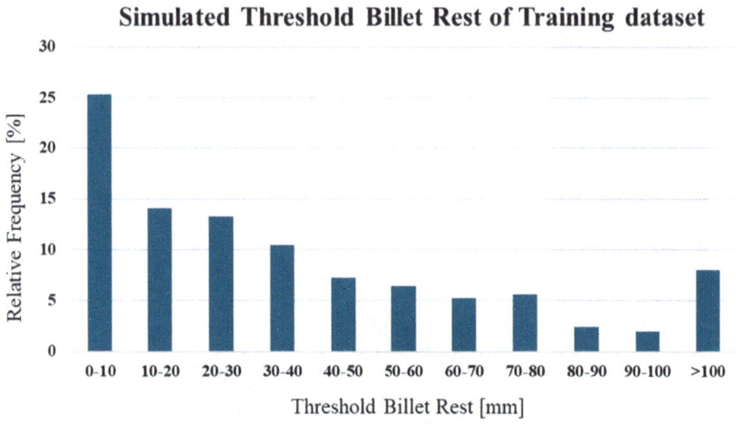

Fig. 6 Relative frequency distribution of threshold billet rest values calculated with FEM simulations employed for generation of dataset for ANN training and testing

ANN. For non-circular entrances (i.e., profiles B, C and D), the die entrance diameter was calculated as the equivalent diameter of a circle having the same area as the entrance.

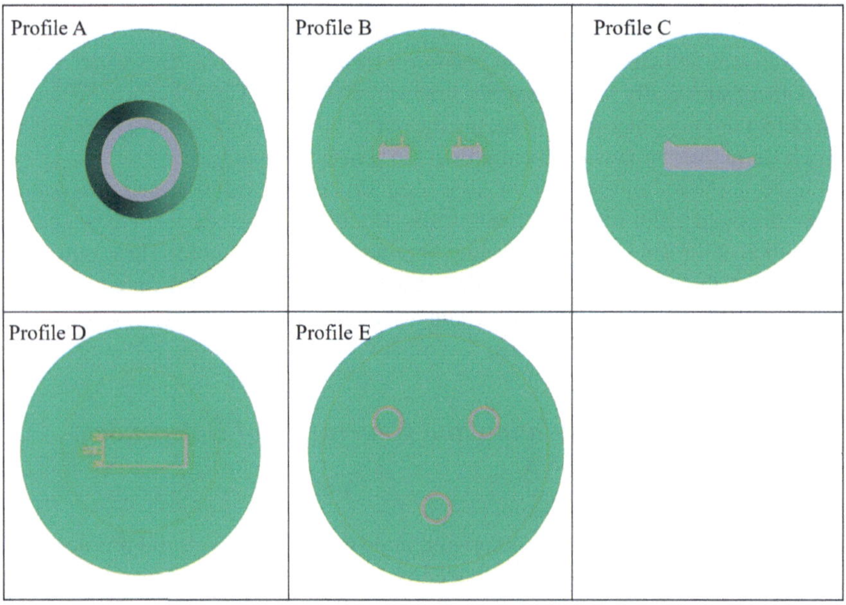

Fig. 7 Experimental case studies employed for validation of FEM simulation and generation of independent test dataset for ANN

Table 3 Process parameters employed in experimental case studies and simulations

Parameter	Profile A	Profile B	Profile C	Profile D	Profile E
Extruded material	AA1100	AA6063	AA6082	AA6060	AA6082
Ram speed (mm/s)	1	6.44	7.6	8	4.4
Billet temperature (°C)	400	480	480	480	440
Die temperature (°C)	400	450	450	510	480
Container temperature(°C)	400	430	440	430	420
Ram temperature (°C)	400	380	380	420	400
Friction factor (except for bearings)	1	1	1	1	1
Billet rest (mm)	4	15	15	44	30

Table 4 Calculated input variables for ANN for all experimental case studies

Variable	Profile A	Profile B	Profile C	Profile D	Profile E
Skin layer thickness (mm)	0.5	0.25	0.25	0.23	0.3
Billet length (mm)	50	670	916	880	647
Billet diameter (mm)	49.5	264	264	211	286
Die entrance diameter (mm)	44	89.1	88.7	163.5	200.9
R1	1.27	8.8	8.9	1.66	1.19
R2	1.65	1	1	1.9	3
Die length (mm)	25	72	45	145	117
Extrusion ratio	20.9	44.8	19.7	27.3	46.4

Experimental analysis consisted of evaluating skin contamination within the profile area at various distances from the stop mark to evaluate both defect evolution and billet rest lengths at which the defect would enter the profile (the minimum billet rest to avoid profile contamination). Profiles were cut into 100 mm samples, with each specimen marked, ground using P220 and P500 abrasive papers and etched with a caustic soda solution (30% NaOH in water at 60 °C with an etching time of 45–90 s). Once all samples were etched, images of each cross section were acquired. Using CAD software, the percentage skin contamination was calculated by dividing the defect area by the total cross-sectional area for each specimen. For Profile A [3], the analysis was conducted by cutting the material flow at different ram strokes to evaluate the evolution of skin flow from the container to the die (Fig. 8).

Figure 8 presents a comparison of experimental and FEM simulation outcomes for Profile A at two different ram strokes in two different cutting planes. It is immediately clear that the numerical model accurately replicates skin flow within the die. In this specific case, the skin remained in the die without contaminating the profile at the end of the extrusion process (billet rest: 4 mm). Figures 9 and 10 report the evolution of profile contamination as a function of distance from the stop-mark for Profiles B and C. The latter is negative in the graph by convention, because the defect enters

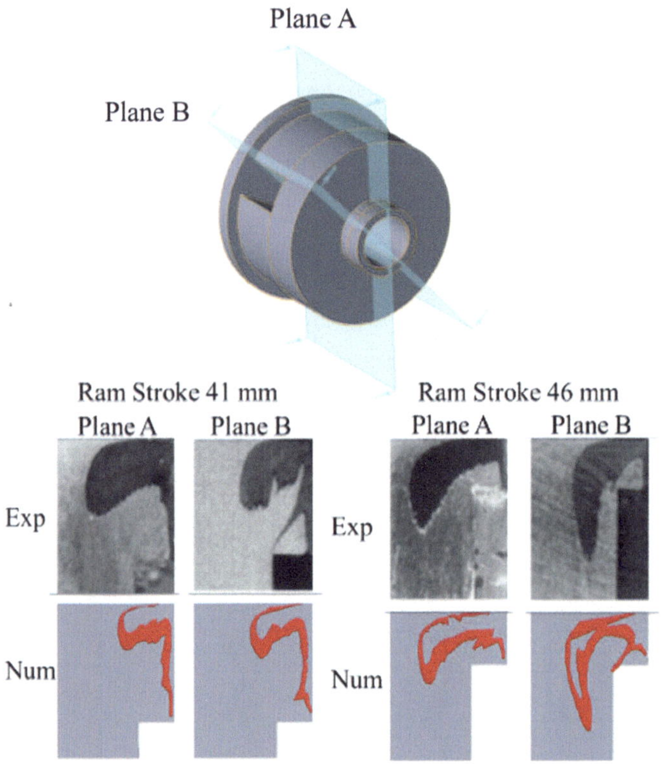

Fig. 8 Comparison of experimental and simulated skin flow for Profile A

the profile at the end of one extrusion run, and consequently before extrusion of the stopmark, which occurs at the beginning of the subsequent run.

Experimental data for Profile B were acquired up to − 2150 mm from the stopmark coinciding with the industrial scrap, clearly not enough to avoid a contaminated product. Therefore, it was not possible to verify whether the numerical prediction of defect onset was correct in this case. However, moving towards the stopmark, both the shape and amount of surface contamination were predicted well by the FEM model. Numerical results obtained for Profile B were therefore considered as being accurate in forecasting appearance of the defect. Industrial scrap of 4500 mm was excessive for Profile C, with the FEM model predicting defect onset at − 3600 mm compared to the experimental value of − 3700 mm. Overestimation of the percentage skin contamination was instead observed from − 1500 mm up to the stopmark. This discrepancy, which was nevertheless considered acceptable, was caused by the presence of charge weld defect near the stopmark. When the charge occurred, it was mixed with the skin contamination, making the two defects difficult to distinguish experimentally. Nonetheless, the numerical prediction was considered reliable as the outcome of industrial interest is proper prediction of defect onset in order to select the

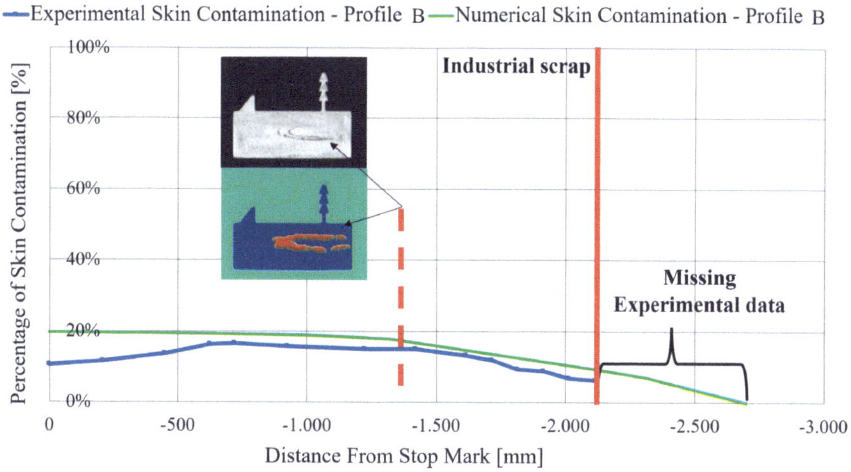

Fig. 9 Comparison of experimental and simulated profile contamination for Profile B

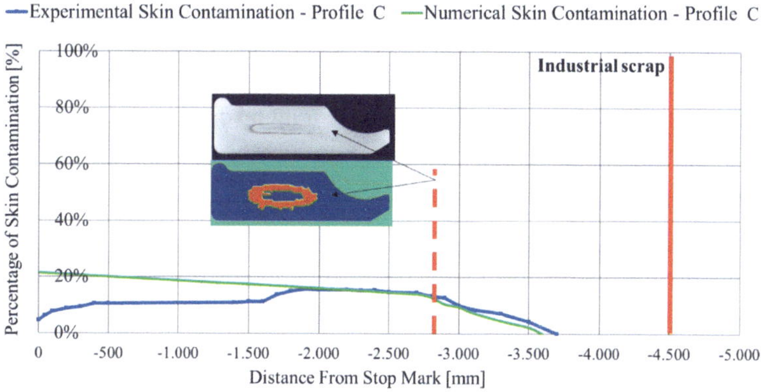

Fig. 10 Comparison of experimental and simulated profile contamination for Profile C

optimal billet rest. For Profiles D and E, the billet rest was sufficient to avoid contamination, with experimental campaigns confirming that both profiles remained uncontaminated. Simulations using the actual billet rest values agreed with experimental outcomes, further validating the model's ability to predict the threshold billet rest to prevent contamination. As such, despite gaps in some experimental measurements, the numerical predictions demonstrated high accuracy across diverse industrial case studies with various shape complexities, process parameters and extruded materials.

3 Artificial Neural Network

3.1 ANN training

As outlined previously, input parameters for training and testing ANNs were chosen based on known factors influencing evolution of skin contamination [3, 4, 7, 9, 13, 14] including the billet length (BL), billet diameter (BD), die entrance diameter (D1), skin thickness (ST), extrusion ratio (ER), R1, R2 and die length (DL). The output parameter was the threshold billet rest at which the skin defect entered the extruded profile, with predictions based on data deriving from FEM simulations. The full dataset was randomly split into 80% training data and 20% test data, with the latter utilized to evaluate the accuracy of the trained ANN. Fully connected neural networks for regression were trained using the *fitrnet* function in MATLAB, employing a Broyden-Fletcher-Goldfarb-Shanno quasi-Newton algorithm (LBFGS) to minimize the mean squared error (MSE), or loss. Grid ANN hyperparameter optimization was performed over the ranges given in Table 5 to determine the influence of the activation function, number of hidden neuron layers and number of neurons per hidden layer on the accuracy of ANN predictions. Due to the limited size of the dataset, relatively simple ANN architectures were considered, involving a maximum of two hidden layers with 20 neurons each. This configuration minimized the risk of overfitting by preventing the network from simply memorizing training data instead of generalizing effectively to unseen datasets. Tested hyperparameter ranges during optimization and the final ANN configuration employed for predicting skin contamination are shown in Table 5.

The results of hyperparameter optimization are presented in Fig. 11, where the test set loss (MSE) is displayed as a function of the activation function and number of neurons per layer for ANNs with 1 and 2 hidden layers. The activation function clearly has a strong influence on the accuracy of predictions and the variability of error over the considered hyperparameter range. The MSE was generally lowest with a ReLU activation function, achieving an MSE below 10 mm^2 over a relatively large range. While ANNs with a Sigmoid activation function and 2 hidden layers yielded comparable MSE to ReLU in some cases, values were consistently lower with ReLU for both 1 and 2 layers. A tanh activation function yielded highest MSE across the entire range of hyperparameters, implying that this was not appropriate

Table 5 ANN hyperparameter ranges employed for optimization, together with final ANN configuration employed for predicting skin contamination

	Optimization	Final
Activation function, fully connected layers	Rectified linear unit (ReLU) Sigmoid Hyperbolic tangent (tanh)	ReLU
Number of fully connected layers	1–2	1
Number of neurons per layer	1–20	15

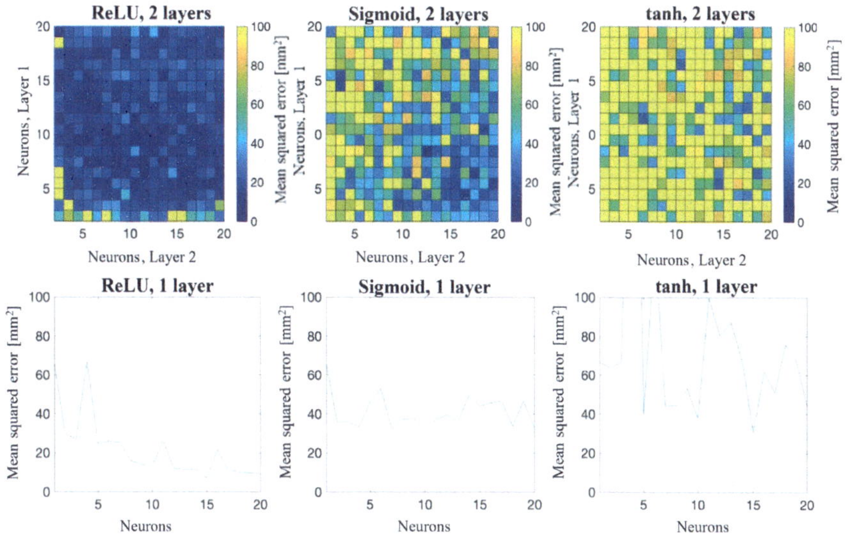

Fig. 11 Test set loss (MSE) of ANNs trained with hyperparameters over the ranges given in Table 5

for predicting skin contamination. With a ReLU activation function and 1 hidden layer, the MSE decreased rapidly with increasing neurons up to 15 neurons (MSE = 7.2 mm^2), after which values oscillated but did not improve significantly. Introducing an additional layer of neurons did not lead to further benefits in terms of reducing the MSE but offered a wide range of combinations achieving good performance. Due to the limited size of the dataset and the requirement for generalization to the experimental case studies, the configuration chosen for the final ANN for prediction of skin contamination comprised a single hidden layer with 15 neurons and a ReLU activation function.

3.2 *ANN Testing*

A comparison between trained ANN and FEM outputs for the threshold billet rest in the test dataset is provided in Fig. 12a, where ANN outputs are denoted 'predicted' and FEM outputs 'true'. The red line represents exact correspondence between the predicted and true values, while the blue dashed lines represent 10 mm deviations to either side, which were considered industrially acceptable. The vast majority of ANN predictions accurately correspond to values of contaminated billet rest calculated by FEM simulations, with only one point exceeding a deviation of ± 10 mm. Interestingly, while threshold billet rest values below 10 mm could not be determined with FEM simulations (outcomes in this case were assigned values of 0 mm), the

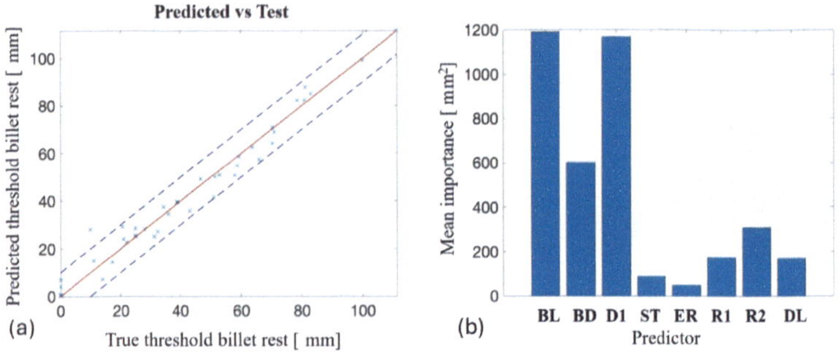

Fig. 12 **a** Comparison between trained ANN (predicted) and FEM (true) outputs for threshold billet rest in the test set. **b** Permutation importance of input parameters employed for training the same ANN. Red line represents exact correspondence between predicted and true values. Blue dashed lines represent 10 mm deviations between predicted and true values

ANN provided predictions in the range 0–10 mm, suggesting that higher resolution of calculated values could be achieved with the trained ANN than with FEM simulations for low values of threshold billet rest based on data from outside this range.

Figure 12b presents the mean permutation importance of each input variable for the same trained ANN. This explainable artificial intelligence (XAI) metric evaluates the increase in MSE upon randomly shuffling a given input variable, providing an indication of the importance of each variable for predicting correct outcomes with a specific model. The parameters most strongly affecting model accuracy were the billet length, billet diameter and die entrance diameter, while R1, R2 and the die length were of moderate importance. Skin thickness and extrusion ratio were of lesser overall influence; however, their inclusion was considered necessary for correct physical representation of the process.

3.3 *ANN Testing with Independent Data*

Table 6 compares the outcomes of experiments, FEM models and ANN predictions of the threshold billet rest for the industrial case studies outlined previously. Comparison between experiments and the FEM model again confirmed the reliability of the FEM model, thus justifying the use of numerical simulations to create the training dataset for the ANN. Direct prediction of the billet rest length with the ANN was good for Profiles A and C, while the accuracy was more limited for the other profiles despite the relatively low MSE achieved with the test dataset during hyperparameter optimization. This outcome is likely due to differences in the characteristics and geometry of profiles employed in the industrial case studies compared to those employed to generate training data. In addition, some of the ANN variables calculated for the industrial case studies were out of the ranges considered in the training

Table 6 Comparison of experimental, FEM simulation and ANN predictions of threshold billet rest for industrial case studies

Threshold billet rest (mm)	Experimental	FEM	ANN
Profile A	< 4	< 4	4.4
Profile B	> 63	75.3	137
Profile C	203	198	195
Profile D	< 44	< 44	51.5
Profile E	< 30	< 30	46

dataset (e.g. R1 for Profile B and R2 for Profile E), thus leading to extrapolated results. This outcome highlights the importance of generating high quality training data for correct and complete representation of the process over the parameter range required for predicting outcomes. Based on these outcomes, the dataset employed for training and testing the ANN can be considered representative of round solid and hollow profiles but less so of those with marked asymmetry.

4 Conclusion

This study has integrated FEM simulations and machine learning (ML) to predict skin contamination in aluminum alloy extrusions. A numerical dataset was generated within Qform Extrusion UK to train an artificial neural network (ANN). The accuracy of FEM simulations for generating reliable data was demonstrated by validating the numerical code with different industrial case studies The final ANN configuration, featuring a single hidden layer with 15 neurons and a ReLU activation function, was able to predict FEM outcomes in the test set within $\pm$ 10 mm for all but one data entry. ANN model predictions were closely aligned with experimental measurements for five additional case studies, confirming the ability of the ANN to generalize correctly.

These results underscore the potential of combining FEM-based datasets with ML to significantly reduce computational cost while preserving high predictive accuracy. Moreover, the approach offers valuable insight into the geometric parameters that most critically affect skin defect formation. Analysis of the ANN permutation importance, an explainable artificial intelligence metric, highlighted the importance of the billet length, billet diameter and die entrance diameter on the threshold billet rest, with R1, R2, die length, skin thickness and extrusion ratio exhibiting lower impact.

Overall, this methodology provides a powerful tool for optimizing extrusion processes, reducing material waste and advancing smart manufacturing practices. Future work will focus on expanding the dataset, refining the ANN architecture and extending the model to include additional alloys and more complex extrusion geometries.

Funding PNRR–M4C2INV1.5, NextGenerationEU-Avviso 3277/2021—ECS_00000033 ECOSISTER-spk 3.

References

1. Hattori CS, Almeida GFC et al (2021) Microstructure and fatigue properties of extruded aluminum alloys 7046 and 7108 for automotive applications. J Market Res 14:2970–2981. https://doi.org/10.1016/j.jmrt.2021.08.085
2. Negozio M, Pelaccia R, Donati L (2023) Reggiani, B (2023) Numerical investigation of the surface recrystallization during the extrusion of a AA6082 aluminum alloy under different process conditions. Int J Adv Manuf Technol 129:1585–1599. https://doi.org/10.1007/s00170-023-12397-8
3. Kim Y-T, Ikeda K (2000) Flow behavior of the billet surface layer in porthole die extrusion of aluminum. Metall Mater Trans A 31:1635–1643. https://doi.org/10.1007/s11661-000-0173-4
4. Valberg HS, Lefstad M, de Moraes Costa AL (2020) On the mechanism of formation of back-end defects in the extrusion process. Proced Manuf 47:245–252. https://doi.org/10.1016/j.promfg.2020.04.207
5. Yu J, Zhao G et al (2019) Microstructures of longitudinal/transverse welds and back-end defects and their influences on the corrosion resistance and mechanical properties of aluminum alloy extrusion profiles. J Mater Process Technol 267:1–16. https://doi.org/10.1016/j.jmatprotec.2018.12.006
6. Liu Y, Wang X et al (2022) Formation mechanism and optimization strategy of surface back-end defects in miniature complex hollow extruded profile. J Mater Process Technol 308:117726. https://doi.org/10.1016/j.jmatprotec.2022.117726
7. Saha P (2018) Extrusion and drawing of aluminum alloys. In: Anderson K, Weritz J, Kaufman JG (eds) Aluminum science and technology. ASM International, p 0. https://doi.org/10.31399/asm.hb.v02a.a0006534
8. Reggiani B, Pinter T, Donati L (2020) Scrap assessment in direct extrusion. Int J Adv Manuf Technol 107:2635–2647. https://doi.org/10.1007/s00170-020-05127-x
9. Chien K, Robbins P, Jowett C, Wang Y (2018) Extrusion productivity, part I—Billet geometry. Light Metal Age 76(2)
10. Qamar SZ, Pervez T, Chekotu JC (2018) Die defects and die corrections in metal extrusion. Metals. https://doi.org/10.3390/met8060380
11. Giarmas E, Tzetzis D (2022) Optimization of die design for extrusion of 6xxx series aluminum alloys through finite element analysis: a critical review. Int J Adv Manuf Technol 119:5529–5551. https://doi.org/10.1007/s00170-022-08694-3
12. Lou S, Wang Y et al (2017) Analysis and prediction of the billet butt and transverse weld in the continuous extrusion process of a hollow aluminum profile. J Mater Eng Perform 26:4121–4130. https://doi.org/10.1007/s11665-017-2771-y
13. Bayat N, Carlberg T (2016) Surface formation in direct chill (DC) casting of 6082 aluminium alloys. IOP Conf Series Mater Sci Eng 117:012038. https://doi.org/10.1088/1757-899X/117/1/012038
14. Hatzenbichler T, Buchmayr B, Umgeher A (2007) A numerical sensitivity study to determine the main influence parameters on the back-end defect. J Mater Process Technol 182:73–78. https://doi.org/10.1016/j.jmatprotec.2006.07.013
15. Park N, Song et al (2021) Numerical and experimental investigation on the surface defect generation during the hot extrusion of Al6063 alloy. Materials 14.https://doi.org/10.3390/ma14226768
16. Kniazkin I., Pelaccia R., Negozio M., et al. (2023) Investigation of the skin contamination predictability by means of QForm UK extrusion code. Mater Res Proceed 28:543–552. https://doi.org/10.21741/9781644902479-59
17. Bandini C, Reggiani B, Donati L, Tomesani L (2015) Code Validation and development of user routines for microstructural prediction with Qform. Mater Today Proceed 2:4904–4914. https://doi.org/10.1016/j.matpr.2015.10.052
18. Stebunov S, Biba N et al (2011) Influence of contact friction conditions on thin profile simulation accuracy in extrusion. In: Key engineering materials, vol 491. Trans Tech Publications, Ltd., pp 35–42. https://doi.org/10.4028/www.scientific.net/kem.491.35

19. Biba N, Rezvykh R, Kniazkin I (2017) Coupled modelling of aluminium profiles extrusion and product quality improvement by means of simulation. AIP Conf Proc 1896:140010. https://doi.org/10.1063/1.5008166
20. Negozio M, Pelaccia R (2024) Numerical, experimental, and analytical investigation of the skin contamination evolution in the extrusion of different industrial profiles. Int J Adv Manuf Technol 131:4727–4738. https://doi.org/10.1007/s00170-024-13329-w
21. Pelaccia R, Negozio M et al (2024) Extrusion benchmark 2023: effect of die design on profile speed, seam weld quality and microstructure of hollow tubes. Key Eng Mater 988:47–62. https://doi.org/10.4028/p-gNXRC5

A Two-Step Simulation Approach to Predict Surface Micro-textures Replication in Injection-Molded Polymeric Parts

Keltoum Oubellaouch, Marco Sorgato, and Barbara Reggiani

Abstract Predicting the microstructures' replication in injection-molded parts is challenging due to the large scale difference between macro- and microscale features, requiring significant computational time. To address this, a two-step simulation strategy is proposed: first, simulating the macro-cavity using Moldflow, and then applying the results as boundary conditions for the micro-cavity simulation in COMSOL Multiphysics. The behavior of the polymer melt at the microscale differs from that at the macroscale due to effects like melt-mold surface tension, polymer no flow temperature and melt-mold heat transfer coefficients. These factors are incorporated into the COMSOL model to better predict flow dynamics at the microscale. Experimental tests were conducted to evaluate these effects, comparing numerical results with experimental data under various conditions. This work demonstrates the potential of the model to improve the accuracy of microstructure replication in injection molding.

Keywords Multi-physics simulation · Micro injection · Polymer flow · Moldflow · COMSOL

K. Oubellaouch (✉) · B. Reggiani
University of Modena and Reggio Emilia, Reggio Emilia, Italy
e-mail: keltoum.oubellaouch@unimore.it

K. Oubellaouch
En&Tech, University of Modena and Reggio Emilia, Reggio Emilia, Italy

M. Sorgato
University of Padova, Padova, Italy

B. Reggiani
INTERMECH, University of Modena and Reggio Emilia, Reggio Emilia, Italy

L. Fratini et al. (eds.), *Selected Topics in Manufacturing*, Lecture Notes in Mechanical Engineering, https://doi.org/10.1007/978-3-031-99501-9_10

1 Introduction

Computer-aided engineering (CAE) software tools, such as Moldflow®, Moldex3D®, and Sigmasoft®, have proven to be powerful tools for simulating and optimizing the cavity filling, packing, and cooling phases in injection molding of macroscopic parts. However, when it comes to micro-injection molding, where parts or features are on the micrometer scale, the flow behavior of the polymer melt can differ significantly from that in conventional injection molding. As the system scale decreases, the high surface to volume ratio amplifies interfacial effects, such as wall slip, surface tension, and micro-scale viscosity, which become more pronounced and influence the flow dynamics in ways that traditional macroscopic models may not adequately capture [1]. Wall slip, where polymer melt moves along the mold wall instead of sticking to it, becomes crucial in micro- and nanostructures due to intense shear forces, rapid cooling, and wall-induced rigidity [1]. This effect arises when shear stress surpasses a critical threshold; below that, adhesion prevails. In micro-injection molding, in addition to the high surface-to-volume ratio, the flow resistance push shear stress higher, making wall-slip more likely. However, its impact varies with flow conditions, shear rates, and material properties. For example, Yao and Kim [2] showed that for features under 1 μm, wall slip can significantly alter flow velocity, capturing this with a modified power-law slip model. Similarly, surface tension plays a critical role in micro-injection molding, as capillary forces can dominate over inertial and viscous forces at small scales. Surface tension effects are particularly important in the filling of micro-cavities, where they can influence the flow front shape and lead to incomplete filling or air entrapment [3].

Despite these challenges, many studies have attempted to use macroscopic CAE software to predict the filling of micro-cavities. For example, Brulez et al. [4] used Moldflow to simulate the filling of a microgroove by employing a multi-scale meshing strategy. However, due to the limitations of Moldflow in generating highly precise meshes, they first created an advanced mesh with multiple mesh sizes in COMSOL Multiphysics and then imported it into Moldflow. Similarly, Rytka et al. [5] utilized this type of mesh and confirmed that only by using a high-density mesh in the micro-cavity could the filling of the micro-groove be accurately captured. Lin and Young [6] simulated the filling of a micro-pillar on the surface of a plate using Moldflow. They found that the simulation predicted the micro-pillar to be entirely filled during the filling phase, whereas experimental results showed that complete filling was only achieved after the packing phase [6]. Similarly, Shen et al. [7] simulated the filling of low-aspect-ratio microfeatures using Moldflow. Their simulation predicted complete filling of the microfeatures, which was attributed to the use of a coarse mesh that could not accurately capture the flow behavior at the micro-scale. Another parameter that is widely discussed in the literature and significantly influences simulation predictions is the heat transfer coefficient H_{tc} between the mold and the melt polymer [8]. At the microscale, the presence of microfeatures alters contact area and surface roughness, while trapped air and potential chemical changes induced during the generation of the microfeatures further complicate the determination of H_{tc}. These factors make it challenging to understand how H_{tc} changes for surfaces with microfeatures, necessitating experimental calibration to select an appropriate value for accurate simulations. Liu and Gehde [9] findings indicate that the heat transfer coefficient rises as the contact area expands, a phenomenon driven by increased surface roughness. Based on their calculations, the average values varied between 20,000 and 35,000 W m^{-2} K^{-1}, depending on the surface roughness and specific injection parameters. Kim and Turng [10] conducted a work on an optical part featuring a micro-groove, interestingly they found that the Heat transfer coefficient varies between 500 and 1000 W m^{-2} K^{-1}, depending on the filling speed. Similarly, Kuhn et al. [11] used a low value of this coefficient in their study (850 W m^{-2} K^{-1}). Whereas, Tofteberg and Andreassen [12], Rytka et al. [5] and Brulez et al. used a higher value of H_{tc} (30,000 W m^{-2} K^{-1}) and they found good correlations. Another critical parameter that significantly influences the reliability of simulations is the temperature threshold at which the polymer stops flowing. As the polymer temperature decreases and approaches the transition temperature, its flow ceases. Some studies consider this threshold to be exactly the glass transition temperature (T_g) for amorphous polymers and the crystallization temperature (T_c) for semicrystalline polymers and they determined it using Differential Scanning Calorimetry (DSC) [13]. However, other studies have introduced the concept of the no flow Temperature (T_{no}), which is generally higher than the transition temperature. For amorphous polymers, T_{no} typically ranges between T_g + 20 K and T_g + 70 K, while for semicrystalline polymers, it is estimated to be 10 to 80 K below the melting temperature (T_m) [14].

Given the limitations of macroscopic CAE tools in accurately capturing micro-scale flow behavior, this study proposes a hybrid modeling approach that combines the strengths of Moldflow for macroscopic simulations with COMSOL Multiphysics for micro-scale simulations. By leveraging Moldflow for the overall cavity filling and COMSOL for detailed micro-feature analysis, this approach aims to overcome the challenges associated with micro-injection molding simulations and provide more accurate predictions of replication ratios and flow behavior at the micro-scale.

2 Materials and Methods

2.1 Polymer Properties

The polymer used in this study is TOPAS 5013L-10, a Cyclic Olefin Copolymer (COC) from TOPAS Advanced Polymers, chosen for its excellent microstructure replication and low shrinkage, ensuring dimensional accuracy in injection-molded parts. Its thermal properties, including heat capacity and thermal conductivity (Fig. 1a, b), critical for flow behavior, were obtained from the Moldflow database (2024). Additional properties are summarized in Table 1.

The T_{no} of Topas 5013L-10 was determined using the Melt Flow Index (MFI) method by Rytka et al. [5]. The polymer was dried at 100 °C for 6 h, and tests

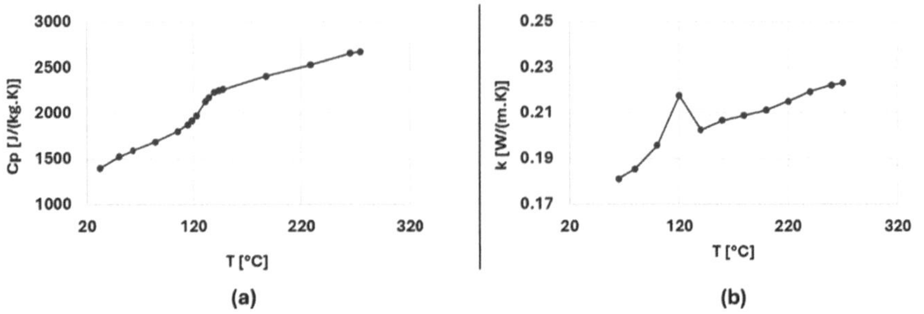

Fig. 1 Thermal properties as a function of temperature of the molded polymer. **a** Heat capacity. **b** Thermal conductivity

Table 1 Important material data (from Moldflow 2024 database)

Property	Unit	Value
Morphology	–	Amorphous
Melt density	g/cm^3	0.92
Solid density	g/cm^3	1.02
MFI (260°/2.16 kg)	g/10 min	43
Shrinkage	%	0.4–0.7
Tensile modulus	MPa	2850
T_g	°C	127

were conducted on an XNR-400C melt flow indexer with a 2.095 mm die diameter and a 21.6 kg load, applying 61.4 MPa pressure, which is higher than in this study's injection molding experiments, validating the use of MFI for T_{no} determination. MFI measurements from 260 °C to 155 °C showed a significant flow slowdown below 165 °C (Fig. 2). T_{no} was estimated between 155 °C and 165 °C, with 160 °C used for this study.

Fig. 2 Curve of MFI values as a function of test temperature

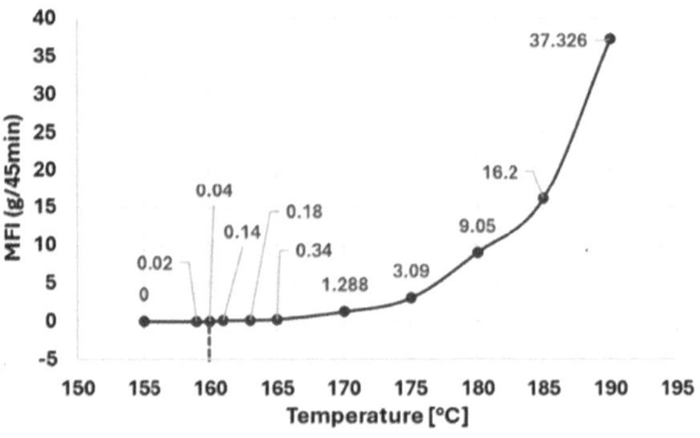

2.2 Mold Insert Preparation

The AISI 316 stainless steel mold insert was polished sequentially with abrasive papers and a 1 μm diamond paste cloth to achieve a mirror-like finish. Three holes, created using a picosecond laser system, were positioned along the injection direction: one at the center and two at ± 2.5 mm (Fig. 3a) to assess replication fidelity relative to the gate.

Two rectangular markings were etched for precise positioning in the injection molding machine. As shown in Fig. 3a, the laser used a spiral pattern (120 μm outer diameter, 5 μm inner diameter, 2 μm spacing) with parameters of 532 nm wavelength, 125 mm/s scan speed, and 75 kHz frequency over 150 passes. Post-machining, the insert was cleaned in heated acetic acid (40 °C) and ultrasonicated for 5 min to remove oxides, followed by a 10-min ultrasonic cleaning in isopropanol to eliminate residual contaminants [15]. To determine the hole depth, initial profilometer measurements were challenging due to difficulty detecting the true bottom. An alternative approach was used: a second insert with identical holes was sectioned laterally, embedded in resin, and polished sequentially. Microscopic examination revealed conical cross-sections (Fig. 3b), likely from laser machining. Polishing continued until the mid-section was exposed, enabling accurate depth measurement and geometry analysis. Figure 3b shows the conical cross-section, while Fig. 3c presents a geometric model for simulation, representing the hole as a cone with a 120 μm large diameter, 28 μm small diameter, and 80 μm height, reflecting the observed shape for numerical modeling.

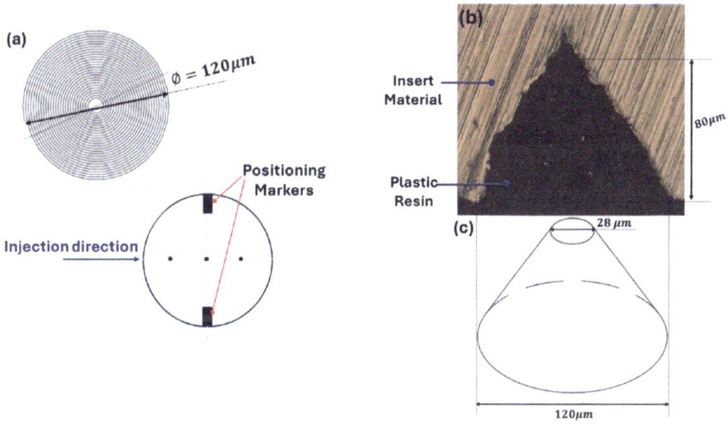

Fig. 3 Schematic of mold insert surface and laser etching strategy. **b** Microscope image showing the conical cross-sectional profile of laser-machined hole. **c** Proposed conical geometry for modeling the hole

Table 2 Injection molding parameters with tested values for mold temperature (T_{mold}) and injection velocity (V_{inj})

T_{mold} [°C]	V_{inj} [mm/s]	T_{melt} [°C]	P_{hold} [MPa]
60–100–120	25–50–100	260	0

2.3 *Injection Molding Process and Replication Measurement*

Injection molding trials were performed on a Wittmann Battenfeld Micro-Power 15 machine using the prepared mold insert with three holes, replicated as pins in discs of 18 mm diameter and 1.5 mm thickness. Injection velocity and mold temperature were varied (see Table 2), with holding pressure set to zero. Five initial shots stabilized the process, followed by three samples collected for analysis.

The pins, labeled Pin 1, Pin 2, and Pin 3 (nearest, central, and farthest from the gate, respectively), were analyzed using a Keyence optical microscope for the 3D acquisition and height measurement. The Replication Ratio RR was calculated as:

$$RR = \frac{H_p}{H_m} \times 100 \tag{1}$$

where H_p is the pin height and H_m is the ideal height of the conical hole (80 μm).

2.4 *Macro-scale Simulation*

The simulation was performed using Autodesk Moldflow Insight 2024. A 3D mesh was generated for the disc and pins, with local refinement (0.01 mm edge length, 10 elements through thickness) around the pins and a coarser mesh (0.1 mm) elsewhere, totaling 1,851,274 elements. The material TOPAS 5013L-10 from the database was assigned, with melt temperature and switch-over matching experiments. Mold temperature and injection velocity were varied to replicate experimental conditions. The "Maximum Volume to Fill per Time Step" was reduced to 0.05% to accurately resolve flow into the pins. The no-flow temperature was set to 160 °C (experimental value), and the heat transfer coefficient was adjusted to 30,000 W/(m^2 K), as recommended by the literature [4, 5, 12, 16]. A "Fill only" simulation was performed to analyze the filling phase.

2.5 *Micro-scale Simulation*

The micro-scale simulation was performed using COMSOL Multiphysics 6.1 with a 2D model to investigate pin filling behavior. A 0.5 mm × 0.5 mm domain, centered on the pin, was extracted from the macro-scale geometry (Fig. 4a). The 2D model, obtained from the disk's longitudinal cross-section (a vertical cut through the pin's central axis), represented the pin as a trapezoidal microcavity (120 μm base, 28 μm top, 80 μm height). A user-controlled mesh, refined near the microcavity and corners, ensured precise flow resolution. The model was divided into two domains: Domain 1, a narrow rectangle on the left filled with molten polymer, and Domain 2, the remaining rectangular region on the right, including the pin, filled with air (Fig. 4b). For Domain 2, air was assigned as a perfect gas, with properties selected from the COMSOL database. For Domain 1, a new material was created in COMSOL to represent the COC polymer. The melt density was assigned a constant value, while thermal conductivity and heat capacity were implemented as interpolating functions using data extracted from Moldflow (see Fig. 1). The Cross-WLF model was implemented to define the polymer viscosity as a function of temperature (T) and shear rate ($\dot{\gamma}$):

$$\eta = \begin{cases} \dfrac{D_1.e^{\frac{-A_1(T-D_2)}{A_2+(T-D_2)}}}{1+\left(\dfrac{D_1.e^{\frac{-A_1(T-D_2)}{A_2+(T-D_2)}}.\dot{\gamma}}{\tau^*}\right)^{1-n}} & if \;\; T > T_{no} \\ \infty(=10^6) & if \;\; T \leq T_{no} \end{cases} \quad (2)$$

where $T_{no} = 160\,°\text{C}$ is the no flow temperature determined previously, n, τ^*, A_1, A_2, D_1 and D_2 are the material-specific parameters from Moldflow database for TOPAS 5013L-10. This conditional function ensures that the viscosity becomes very high when the polymer temperature drops below T_{no} simulating solidification.

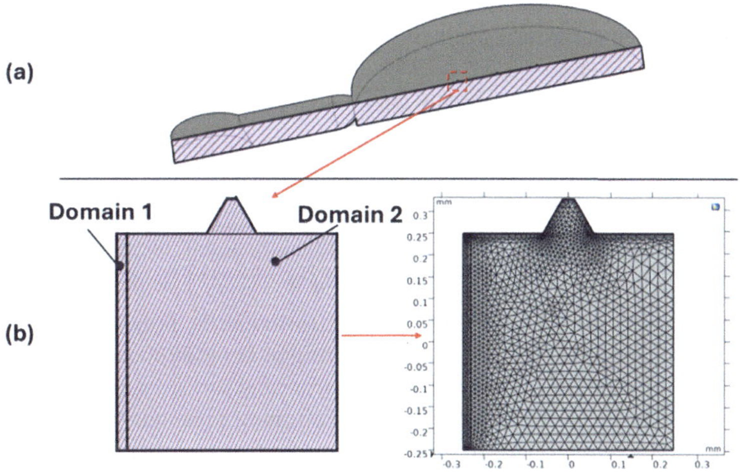

Fig. 4 **a** CAD of the macro-scale model, sectioned to extract the 2D geometry. **b** Zoomed-in view of the 2D model used in COMSOL and the mesh used

During the filling process, multiple physical phenomena, including fluid flow, heat transfer, and material behavior, are intricately involved. To accurately model these interactions, an original framework based on three fundamental governing equations are implemented:

$$\frac{\partial \rho}{\partial t} + \nabla \cdot (\rho \boldsymbol{u}) = 0 \tag{3}$$

$$\rho \frac{\partial \boldsymbol{u}}{\partial t} = -\nabla P + \nabla(2\eta\psi) + \rho g + \nabla\big(\sigma\big(I - NN^T\big)\delta(\Gamma, x)\big) \tag{4}$$

$$\rho c_p\left(\frac{\partial T}{\partial t} + (\boldsymbol{u}.\nabla)T\right) = \nabla \cdot (k\nabla T) + \beta T\left(\frac{\partial P}{\partial t} + (\boldsymbol{u}.\nabla)P\right) + 2\eta\dot{\gamma}^2 \tag{5}$$

where Eq. (3) is the mass continuity equation to ensure conservation of mass of the fluid (polymer or air), relating its density ρ with the velocity field $\boldsymbol{u}$. Equation (4) instead is the momentum equation describing how the velocity field evolves under the influence of forces including the pressure effect $-\nabla P$, the viscosity effect $\nabla(2\eta\psi)$, the gravity effect ρg and the surface tension effect $\Pi = \nabla\big(\sigma\big(I - NN^T\big)\delta(\Gamma, x)\big)$. More in details, P is the fluid pressure, η its viscosity, ψ is the strain tensor, σ is the surface tension at the polymer/air interface, N is the interface normal and $\delta(\Gamma, x)$ is a Dirac delta function localizing the surface tension forces to the interface Γ. Equation (5) instead is the heat equation to account for the conservation of energy, with c_p, k, β, T and $\dot{\gamma}$ are respectively the heat capacity, the thermal conductivity, the thermal expansion coefficient, the temperature and the shear rate of the fluid (polymer or air).

To accurately capture the interface between the molten polymer and air during the filling process, the phase field method was employed. This method introduces a phase field variable ϕ which distinguishes between two phases: $\phi = 1$ represents the polymer phase, and $\phi = -1$ represents the air phase. The evolution of the phase field is governed by the Cahn–Hilliard equation:

$$\frac{\partial \phi}{\partial t} + \boldsymbol{u}.\nabla\phi = \nabla \cdot (\gamma\nabla G) \tag{6}$$

where γ is the mobility parameter controlling the diffusion of the interface. G is the chemical potential, defined as: $G = \frac{3\sigma\delta_i}{2\sqrt{2}}\left[-\nabla^2\phi + \frac{\phi(\phi^2-1)}{\delta_i^2}\right]$, with δ_i and σ are the interface thickness parameter and the surface tension at the polymer/air interface. The phase field method ensures a smooth transition between the polymer and air phases, enabling accurate tracking of the flow front and surface tension effect.

The phase field method is coupled with the governing equations (mass continuity, momentum and heat transfer) through two key mechanisms: materials properties and surface tension. Material properties such as density (ρ), viscosity (η) and thermal properties (k, c_p) are interpolated across the interface using the phase field variable ϕ. For example, the density is calculated as $\rho = \rho_{polymer}\frac{1+\phi}{2} + \rho_{air}\frac{1-\phi}{2}$. Additionally, the surface tension term Π in the momentum equation is derived from ϕ, where the interface normal N and the Dirac delta function $\delta(\Gamma, x)$, which localizes surface tension forces to the interface, are computed as $N = \frac{\nabla\phi}{|\nabla\phi|}$ and $\delta(\Gamma, x) = \frac{|\nabla\phi|}{2}$. This ensures that surface tension effects are accurately modeled at the polymer-air interface. The coupling of these mechanisms ensures that the flow, heat transfer, and interface dynamics are solved simultaneously, providing a comprehensive model of the filling process.

In COMSOL, all these equations were implemented through three physics modules: the Laminar Flow module, the Heat Transfer module, and the Phase Field module. These modules were coupled using the Two-Phase Flow, Phase Field feature to model polymer-air interactions, calculate material properties at the interface, and incorporate surface tension, and the Non-Isothermal Flow feature to couple heat transfer and fluid flow, accounting for temperature-dependent properties and viscous dissipation.

The boundary conditions used in the model are illustrated in Fig. 5, with Fig. 5a showing the thermal boundary conditions and Fig. 5b showing the velocity boundary conditions. At the mold wall (green line in Fig. 5), a no-slip condition was applied for the velocity. This condition was justified based on the macro-scale simulations, which estimated the maximum shear stress at the wall to be 0.09 MPa in the pin's surrounding region for all experimental conditions. According to previous work by Tofteberg and Andreassen [12], the critical shear stress for the COC-steel interface is between 0.1 and 0.18 MPa, similar to that of HDPE. Since the estimated shear stress (0.09 MPa) is below the critical value required for cohesive failure and the onset of wall slip, the no-slip condition was deemed appropriate [17]. For the thermal boundary condition, a convective heat flux with a heat transfer coefficient of 1000 W/(m^2 K) was applied to account for heat exchange between the polymer and mold, critical for simulating cooling and solidification. At the inlet (purple borders in Fig. 5), the velocity field (components along the x- and y-axes) as a function of time was extracted from the macro-scale simulation in Moldflow, ensuring the inlet velocity profile accurately reflected the flow behavior observed at the larger scale. The inlet temperature was set to the melt temperature, also extracted from the macro-scale simulation, ensuring the polymer entered the domain at the correct processing temperature. Two outlets were defined (blue edges in Fig. 5b): one in the main flow and the other at the bottom of the micro-cavity. A zero-pressure condition was applied at both outlets to allow air to escape as the polymer filled the domain. For the main flow outlet, a convective heat flux condition with a heat transfer coefficient of 15 W/(m^2 K) was applied to model heat exchange with the surrounding air, reflecting the limited heat transfer at the outlet. At the polymer–polymer interface (red edge in Fig. 5), a slip condition was applied for the velocity, assuming a fully wetted state to allow smooth flow. For the thermal boundary condition, a convective heat flux with a heat transfer coefficient of 2000 W/(m^2 K) was applied to account for heat exchange between the molten polymer regions, ensuring accurate temperature distribution. The temperature at this interface was set to the melt temperature, extracted from the macro-scale simulation.

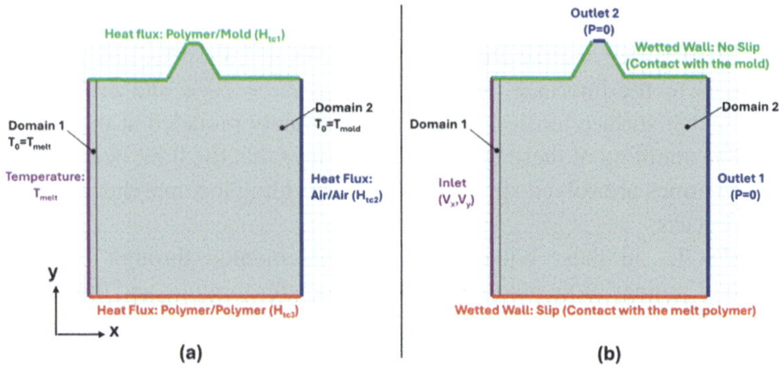

Fig. 5 Boundary conditions setting in COMSOL. **a** Thermal boundary conditions. **b** Velocity Boundary Conditions

The values for the heat transfer coefficients were calibrated through multiple simulations, with the selected values ensuring stability, convergence, and results consistent with experimental observations. Additionally, the surface tension (σ) at the polymer-air interface was set to 0.025 N/m. While the surface tension for solid COC was reported in the literature to range between 0.035 and 0.045 N/m, a lower value of 0.025 N/m was used in this work to account for the decrease in surface tension at the higher melt temperature (260 °C) [18, 19].

3 Results and Discussion

3.1 Experimental Results

The experimental results, presented in Fig. 6, show the replication ratio (RR) for the three pins at different mold temperatures and injection velocities. For all pins and mold temperatures, the RR increased with higher injection velocities. At the lowest mold temperature (60 °C), even with significantly increased injection velocities (up to 250 mm/s, not shown), the RR remained below 65%, demonstrating the limitations of low mold temperatures. A strong correlation was observed between mold temperature and RR, with higher temperatures consistently improving replication. At 100 °C (within the manufacturer-recommended range of 95–115 °C), RR values between 45 and 70% were achieved across all pins, even at the lowest injection velocity (25 mm/s). In contrast, achieving similar RR values at 60 °C required more than doubling the injection velocity, underscoring the importance of selecting an appropriate mold temperature. When the mold temperature exceeded the recommended range (120 °C), the RR plateaued at approximately 85%, even with higher injection velocities, suggesting that excessively high temperatures do not significantly improve replication beyond this point. An example of a pin obtained under these conditions

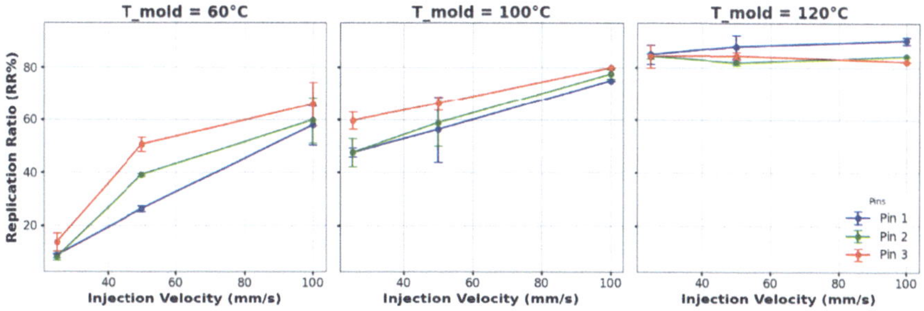

Fig. 6 Experimental Replication Ratio as a function of the distance from the gate (pin 1, pin 2 or pin 3) and injection velocity: **a** $T_{mold} = 60$ °C, **b** $T_{mold} = 100$ °C, **c** $T_{mold} = 120$ °C

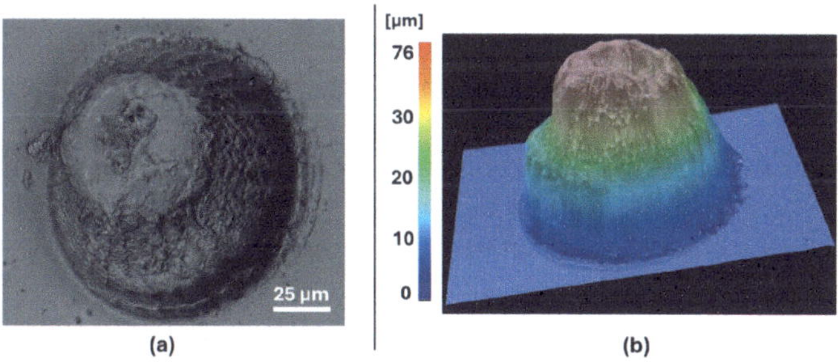

Fig. 7 **a** Real 2D image of a pin obtained at $T_{mold} = 120$ °C and $v_{inj} = 100$ mm/s, captured at a tilt angle of 30°. **b** Corresponding 3D reconstructed image of the same pin

($T_{mold} = 120$ °C and $v_{inj} = 100$ mm/s) is shown in Fig. 7, with Fig. 7a displaying the real 2D image captured using a Keyence microscope, and Fig. 7b presenting the corresponding 3D reconstruction.

At 60 and 100 °C, RR increased with distance from the gate. Pin 1 (closest to the gate, blue curves) showed the lowest RR, while Pin 3 (farthest from the gate, red curves) exhibited the highest. This trend suggests that distance from the gate influences replication, particularly in the absence of a packing phase. Lower replication near the gate is likely due to rapid pressure drop, while improved replication at greater distances may result from flow-induced thermal effects, such as viscous dissipation and longer residence time, which reduce viscosity and enhance filling. These findings align with previous studies emphasizing the importance of holding pressure near the gate and the role of injection speed in replication at the cavity's end. At 120 °C, the trend disappeared, with RR becoming more consistent across all pins regardless of distance from the gate. Higher mold temperatures homogenize the polymer's flow behavior, reducing the pressure differential's impact and diminishing the benefits of increased injection velocity or pressure.

3.2 Numerical and Experimental Results Comparison

In all the macro-scale simulations run in Moldflow, the software predicted complete filling of the pins, even at the lowest injection velocity and mold temperature. This result cannot be attributed to mesh resolution, as a multiscale mesh approach was used, with high-density meshing applied specifically to the microfeature zones. Additionally, simulations employed the heat transfer coefficient recommended in the literature, which typically makes filling more challenging. Despite this, the pins were still fully filled. One possible explanation is the relatively small volume of the pins (0.08% of the total part volume). With a maximum fill volume per time step set to 0.05%, Moldflow may have filled an entire pin in a single time step, treating the microfeature as part of the larger component and neglecting its complexity. Furthermore, Moldflow fails to capture critical microscale effects, such as surface tension and capillary forces, which significantly influence replication fidelity at this scale. As a result, Moldflow is not suitable for accurately simulating microscale filling processes.

All numerical-experimental results comparisons presented in this section focus on Pin 1, the pin closest to the gate, to provide a consistent basis for evaluating replication fidelity under varying process conditions.

At the polymer-mold interface, a static contact angle of ($\theta_s = 135°$) was selected to account for the limited wetting expected for high molecular weight polymers such as COC on metallic mold surfaces [20, 21]. Figure 8 presents a comparison between the real and COMSOL-predicted filling of the pins obtained for a mold temperature of $T_{mold} = 60$ °C at different injection velocities, 25, 50 and 100 mm/s. In each subfigure, the left image shows the 3D image of the real pin obtained using the Keyence optical microscope, while the right image displays the volume fraction of the polymer at the end of the simulation in COMSOL, illustrating the predicted filling. These images provide a visual representation of the filling behavior, allowing for a direct comparison between experimental and simulation results.

From the predicted results, it appears that the COMSOL model has captured the same real profile of the flow interface between the polymer and air, accurately reproducing the shape and curvature of the advancing flow front. This indicates that the model effectively incorporates the surface tension effects and wetting dynamics at the microscale. However, a systematic underestimation of the COMSOL model is evident, as the predicted filling heights remain consistently lower than the experimental values. This discrepancy can be attributed to the use of a static contact angle ($\theta_s = 135°$), which does not fully account for the dynamic contact angle during the filling process. In reality, the dynamic contact angle during the filling process is influenced by flow conditions (e.g., injection velocity, pressure gradients) and temperature-dependent polymer properties [3]. At higher temperatures, the polymer melt becomes less viscous, leading to faster flow front advancement and potentially smaller dynamic contact angles than the static value. This could result in better wetting and higher pin heights in experiments, which the model does not fully capture.

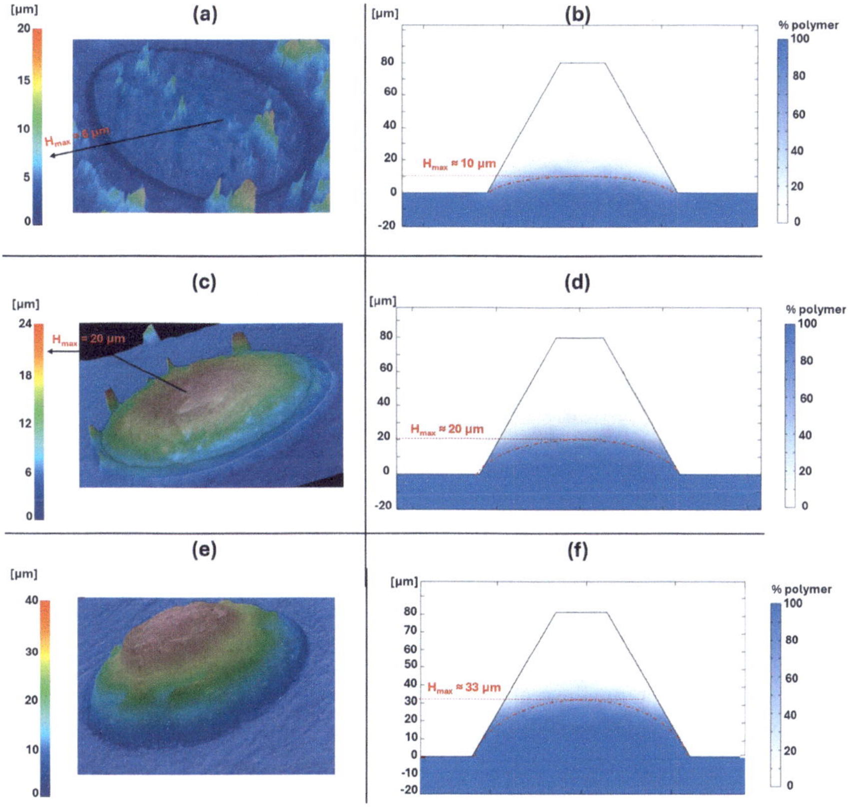

Fig. 8 Comparison of real (left column) and COMSOL-predicted (right column) filling of the pin at $T_{mold} = 60$ °C and **a–b** $v_{inj} = 25$ mm/s **c–d** $v_{inj} = 50$ mm/s **e–f** $v_{inj} = 100$ mm/s

To further investigate the role of wettability in the replication process, additional simulations were performed using a neutral contact angle ($\theta_s = 90$). These simulations aimed to provide a comparative analysis of how different wetting conditions affect the filling behavior and pin height. Figure 9 presents the COMSOL-predicted filling heights for $\theta_s = 90°$ (Fig. 9a), $\theta_s = 135°$ (Fig. 9b), and the real profile of the pin obtained using the Keyence microscope (Fig. 9c) at $T_{mold} = 60$ °C and two injection velocities (left: 50 mm/s, right: 100 mm/s).

- Neutral Wetting Condition ($\theta_s = 90°$):

In this case, the flow front exhibits a pronounced curvature, with the polymer spreading more easily on the mold surface. The dynamic contact angle at the wall is acute ($\theta < 90°$), indicating enhanced wetting due to the no-slip condition and the advancing flow front. This results in a higher predicted pin height compared to both the experimental values and the results obtained with ($\theta_s = 135°$), as illustrated in

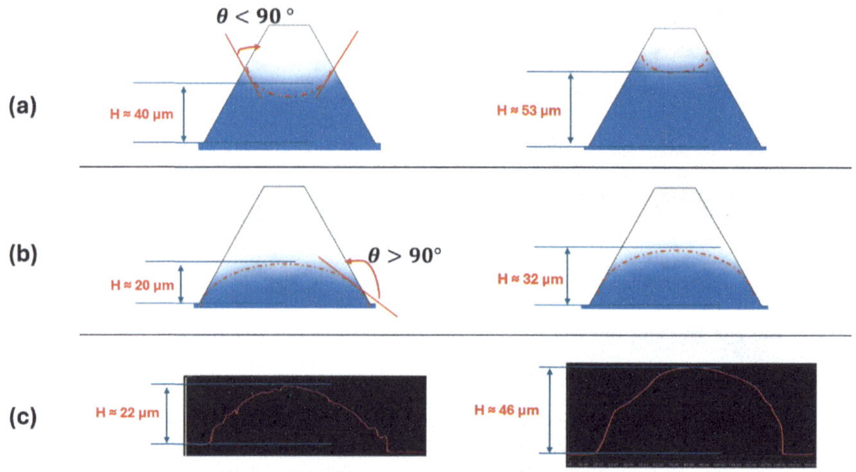

Fig. 9 **a** COMSOL-predicted filling height for a neutral contact angle; **b** COMSOL-predicted filling height for a hydrophobic contact angle; **c** real profile of the pin obtained using the Keyence microscope. Results are shown for $T_{mold} = 60\,°C$ two injection velocities: 50 mm/s (left) and 100 mm/s (right)

Fig. 9a. The overprediction of pin heights suggests that the neutral wetting condition does not fully capture the real-world wetting behavior, particularly the retarding effects of the hydrophobic mold surface.

- Hydrophobic Condition ($\theta_s = 135°$)

In this case, the flow front is flatter, with a sharper transition at the wall. The dynamic contact angle is obtuse, reflecting the retarding effect of the hydrophobic surface. This hinders wetting and reduces the filling efficiency, leading to a lower predicted pin height. Importantly, the shape of the pin obtained with ($\theta_s = 135°$) closely matches the real shape observed in the experiments, as shown in Fig. 9b, c.

Regarding the replicated heights prediction, the bar chart in Fig. 10, comparing experimental and COMSOL-predicted pin heights, demonstrates that the proposed COMSOL model successfully captures the overall trend observed in the experiments. The model with both angles correctly predicts that pin height increases with both injection velocity and mold temperature, confirming that the key physical effects governing the replication process, such as viscous flow, heat transfer, and surface tension, have been effectively incorporated into the simulation.

However, while the numerical results follow the same trend, the COMSOL model with $\theta_s = 135°$ systematically underestimates the replicated pin height in most cases, with the discrepancy becoming more pronounced at higher temperatures. In contrast, the model with $\theta_s = 90°$, while failing to accurately capture the pin shape, predicts heights closer to the experimental values, particularly at higher temperatures. This suggests that the neutral wetting condition better approximates the dynamic wetting

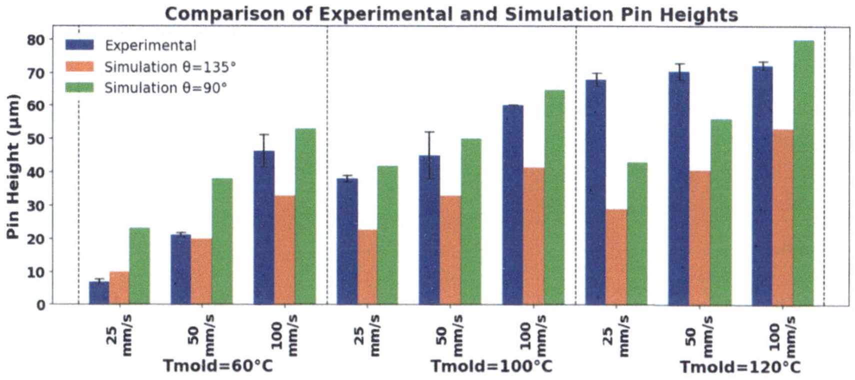

Fig. 10 Comparison of the experimental and numerical height for all the testing conditions

behavior under certain conditions but still does not fully account for the retarding effects of the hydrophobic mold surface.

4 Conclusions

This study explored the replication fidelity of micro-features in injection molding using a multiscale simulation approach, where data such as velocity profiles and temperatures were extracted from macro-scale Moldflow simulations and provided as input to micro-scale COMSOL models, validated by experimental measurements. Key findings include:

- No-Flow Temperature (T_{no}) and Viscosity Condition: In accordance with literature, the no-flow temperature of the polymer was determined to be higher than the transition temperature, and to account for viscosity changes at the microscale, a condition was applied to the polymer viscosity: once the temperature reaches T_{no}, the polymer can no longer flow, and its viscosity becomes infinite.
- Experimental Results: Replication ratio (RR) increased with higher injection velocity and mold temperature, with high RR (~ 85%) achieved at the higher operative temperature. Lower temperatures (60 °C) required significantly higher velocities for comparable RR, with a strong influence of the gate distance.
- Macro-Scale Simulation: Moldflow predicted complete pin filling under all conditions, failing to capture microscale effects like surface tension and capillary forces, highlighting its limitations for micro-feature replication prediction.
- Micro-Scale Simulation: The COMSOL model captured overall trends, but the results varied depending on the contact angle used. Simulations with a hydrophobic contact angle ($\theta_s = 135°$) closely matched the experimental pin shapes but consistently underestimated pin heights. In contrast, simulations with

a neutral contact angle ($\theta_s = 90°$) produced pin heights closer to the experimental values but resulted in different pin shapes due to enhanced wetting.

This work demonstrates the potential of multiscale simulation approaches for optimizing micro-injection molding, providing valuable insights for improving replication fidelity and process design, even if further refinements are needed to ameliorate the model.

Fundings

PNRR–M4C2INV1.5, NextGenerationEU-Avviso 3277/2021-ECS_00000033-ECOSISTER-spk 3. This work was supported by the Emilia-Romagna Region, Italy [PR-FESR 2021–2027—Priorità 1—obiettivo specifico 1.1—Azione 1.1.2—Progetto "CUP E17G22001630003—SAFER"].

References

1. Yang C, Yin X-H, Cheng G-M (2013) Microinjection molding of microsystem components: new aspects in improving performance. J Micromech Microeng 23:093001. https://doi.org/10.1088/0960-1317/23/9/093001
2. Yao D, Kim B (2002) Simulation of the filling process in micro channels for polymeric materials. J Micromech Microeng 12:604–610. https://doi.org/10.1088/0960-1317/12/5/314
3. Choi S-J, Kim SK (2011) Multi-scale filling simulation of micro-injection molding process. J Mech Sci Technol 25:117–124. https://doi.org/10.1007/s12206-010-1025-9
4. Brulez A, Boschard C, Larochette M, Benayoun S (2020) Submillimeter features on injected acrylonitrile butadiene styrene parts: visualization of the flow and numerical simulation. Polym Eng Sci 60:3258–3269. https://doi.org/10.1002/pen.25551
5. Rytka C, Lungershausen J, Kristiansen PM, Neyer A (2016) 3D filling simulation of micro- and nanostructures in comparison to iso- and variothermal injection moulding trials. J Micromech Microeng 26:065018. https://doi.org/10.1088/0960-1317/26/6/065018
6. Lin H-Y, Young W-B (2009) Analysis of the filling capability to the microstructures in micro-injection molding. Appl Math Model 33:3746–3755. https://doi.org/10.1016/j.apm.2008.12.012
7. Shen Y-K, Chang C-Y, Shen Y-S et al (2008) Analysis for microstructure of microlens arrays on micro-injection molding by numerical simulation. Int Commun Heat Mass Transfer 35:723–727. https://doi.org/10.1016/j.icheatmasstransfer.2008.01.013
8. Yu CJ, Sunderland JE, Poli C (1990) Thermal contact resistance in injection molding. Polym Eng Sci 30:1599–1606. https://doi.org/10.1002/pen.760302408
9. Liu Y, Gehde M (2015) Effects of surface roughness and processing parameters on heat transfer coefficient between polymer and cavity wall during injection molding. Int J Adv Manuf Technol. https://doi.org/10.1007/s00170-015-7816-5
10. Kim S, Turng L (2006) Three-dimensional numerical simulation of injection molding filling of optical lens and multiscale geometry using finite element method. Polym Eng Sci 46:1263–1274. https://doi.org/10.1002/pen.20585
11. Kuhn S, Burr A, Kübler M et al (2010) Study on the replication quality of micro-structures in the injection molding process with dynamical tool tempering systems. Microsyst Technol 16:1787–1801. https://doi.org/10.1007/s00542-010-1104-y
12. Tofteberg TR, Andreassen E (2010) Multiscale simulation of injection molding of parts with low aspect ratio microfeatures. Int Polym Proc 25:63–74. https://doi.org/10.3139/217.2318

13. Piccolo L, Puleo K, Sorgato M et al (2021) Modeling the replication of submicron-structured surfaces by micro injection molding. Mater Des 198:109272. https://doi.org/10.1016/j.matdes.2020.109272
14. Mannella GA, La Carrubba V, Brucato V et al (2011) No-flow temperature and solidification in injection molding simulation. Belfast, (United Kingdom), pp 689–693
15. Luo F, Guan Y, Ong W et al (2014) Enhancement of pulsed laser ablation in environmentally friendly liquid. Opt Express 22:23875. https://doi.org/10.1364/OE.22.023875
16. Xu G, Yu L, Lee LJ, Koelling KW (2005) Experimental and numerical studies of injection molding with microfeatures. Polym Eng Sci 45:866–875. https://doi.org/10.1002/pen.20341
17. Anastasiadis SH, Hatzikiriakos SG (1998) The work of adhesion of polymer/wall interfaces and its association with the onset of wall slip. J Rheol 42:795–812. https://doi.org/10.1122/1.550909
18. Yang D, Xu Z, Liu C, Wang L (2010) Experimental study on the surface characteristics of polymer melts. Colloids Surf, A 367:174–180. https://doi.org/10.1016/j.colsurfa.2010.07.008
19. Shin JY, Park JY, Liu C et al (2005) Chemical structure and physical properties of cyclic olefin copolymers (IUPAC technical report). Pure Appl Chem 77:801–814. https://doi.org/10.1351/pac200577050801
20. Zitzenbacher G, Huang Z, Längauer M et al (2016) Wetting behavior of polymer melts on coated and uncoated tool steel surfaces. J Appl Polym Sci 133:app.43469. https://doi.org/10.1002/app.43469
21. Mosaddegh P, Angstadt DC (2008) Micron and sub-micron feature replication of amorphous polymers at elevated mold temperature without externally applied pressure. J Micromech Microeng 18:035036. https://doi.org/10.1088/0960-1317/18/3/035036

White Papers

Adaptive and Sustainable Manufacturing Processes for Profiles and Tubes: Challenges and Innovations

Enrico Simonetto, Antonio Fiorentino, Tommaso Mancia, and Matteo Strano

Abstract Metal profiles and tubes serve diverse industrial needs, from structural support to fluid conveyance, in sectors like automotive, food, energy, furnishing, and appliances. Their production involves a complex supply chain of deformation, welding, and cutting processes, often handled by separate industrial entities, each with specialized skills in machinery, tooling, or manufacturing. It is a relatively traditional sector, where the innovations fostered by the digital and green transitions are crucial to maintain competitiveness and are beginning to sink in. Advanced control systems enhance economic, environmental, and energy efficiency, improve product quality, and reduce waste. Achieving these benefits requires integration across the supply chain and diverse competences. This paper aims at presenting the state of the art, the main innovations, and challenges represented by the use of advanced control systems along the reference supply chain. Particular attention will be paid to the impact of process and machine architecture and the effects on the in-line implementation of new frontier technologies applicable to industrial control such as AI, Digital Twins, sensors, and new concepts of flexible processing and sustainable manufacturing.

Keyword Sustainable processes · Tube forming · Adaptive control · Industry 4.0/5.0

E. Simonetto (✉)
Department of Industrial Engineering, Università degli studi di Padova, Padua, Italy
e-mail: enrico.simonetto.1@unipd.it

A. Fiorentino
Department of Mechanical and Industrial Engineering, Università degli Studi di Brescia, Brescia, Italy

T. Mancia
Department of Theoretical and Applied Sciences, Università degli Studi eCampus, Novedrate, Italy

M. Strano
Department of Mechanics, Politecnico di Milano, Milan, Italy

L. Fratini et al. (eds.), *Selected Topics in Manufacturing*, Lecture Notes in Mechanical Engineering, https://doi.org/10.1007/978-3-031-99501-9_11

1 Introduction

Manufacturing systems continue to evolve in terms of design, architecture, operations, production systems, sensors, and actuators, as discussed by Bowen et al. [1]. These innovations are driven by the introduction of disruptive new technologies in the sector and are pulled by market demands or the emergence of new business models. Recent years have seen pivotal advancements, such as the enhancement of electric drive performance coupled with a reduction in costs, and significant improvements in sensor capabilities—developments largely influenced by the consumer electronics industry, as highlighted by Bleicher et al. [2]. Simultaneously, new artificial intelligence models have emerged. These models enable the collection and analysis of large data sets, reshaping technology in manufacturing sectors, as noted by Kreuzer et al. [3].

As available technologies evolve, so too do customer demands, with increasing expectations for high-performance yet customized products in small, frequently changing batches. According to El Maraghy et al., this shift is often linked to changes in business models developed in response to market needs, without sacrificing the demands for productive efficiency and capital remuneration [4]. Figure 1 highlights how these three aspects are concurrent in the development of new manufacturing paradigms and their temporal evolution. In particular, the figure emphasizes how, within a few years, the most advanced manufacturing systems aim to integrate highly refined control capabilities with the ability to efficiently adapt production to small batches of highly diversified components, as demonstrated by Allwood et al. [5]

In parallel with these technological and market trends, sustainability has become a fundamental driver in the design of both products and processes. According to Hauschild et al. [6] sustainability today encompasses social, environmental, and

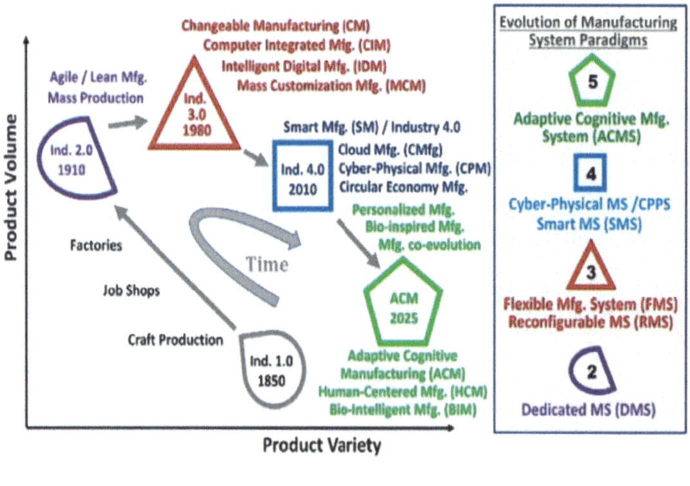

Fig. 1 Manufacturing and manufacturing systems paradigms [4]

economic dimensions, aligning manufacturing strategies with global development goals. Adopting sustainable practices helps companies meet regulatory and societal expectations while improving resource efficiency and reducing waste. This transition offers the opportunity to rethink and redesign manufacturing systems. It encourages innovations that improve process efficiency and flexibility, ultimately increasing value and product quality. However, these processes often require substantial investment in specialized equipment and machinery. Their cost structure is heavily influenced by tooling value and initial ramp-up expenses, as noted by Campi et al. [7]. These often restrict the flexibility of batch sizes and limit the customization of products, posing challenges that run counter to evolving market demands as highlighted by Yang et al. [8].

To address these challenges, efforts can be made not only on the products and organizational systems but also on the manufacturing processes themselves. In this case, rather than changing the basic deformation mechanisms, the paper explores strategic enhancements in control systems. By optimizing these systems and integrating state-of-the-art sensors, actuators, and control systems, metal forming processes can become more efficient and adaptable. This optimization supports the transition from mass production to mass customization, ensuring that the industry remains competitive in a rapidly changing economic and environmental landscape.

As this paper considers a specific manufacturing chain aimed at the production and processing of open and closed profiles, Sect. 2 is dedicated to describing and analyzing this manufacturing chain and the main processes. Section 3 focuses on analyzing the motivations and challenges posed by the integration of control systems in traditional metal forming processes, taking into account not only the benefits but also the costs. Then the main architectures of the control systems and their components, including sensors, actuators, and models are described. Section 4 presents a collection of examples that report the state of the art of the described systems for the main operations involved in the line. Section 5 offers a vision of the future prospects of these systems, ingrained in the forming processes considered.

1.1 Motivations/Scope

This paper aims to identify the main development paths that the reference manufacturing processes, outlined in Sect. 2, can follow to enhance their competitiveness and respond to the technological challenges posed. In particular, it aims to analyze the benefits that the use of process control systems can offer for these processes. It also considers the costs and necessary modifications to processes, equipment, and machinery. Special focus is given to identifying the main technical and practical obstacles.

Regarding control systems, a review of the scientific literature highlights different categories and implementable approaches. As discussed by Allwood et al. [5], these categories may be defined according to several aspects, including: (i) the presence or absence of feedback, (ii) the position of the measurement stage in the process,

(iii) the direction along which the correction propagates, or (iv) the type of algorithm used in the control system.

In the first case, it is possible to identify open-loop systems (see Fig. 2a) or closed-loop control systems. The latter, as described by Polyblank et al. [9], are based on feedback from measurement stages, generally provided by sensors. Even in the second case, it is possible to identify sub-categories based on, for example, the positioning of the feedback steps within the process. It is possible to distinguish between offline approaches, as shown by Fig. 2b, where measurement steps occur upstream or downstream of the process, and in-line approaches (Fig. 2c), where measurements are integrated within the process, as discussed by Wiebenga et al. [10]. In turn, the further downstream the measurement point is moved in the production process, the more necessary it becomes to transition from a feedforward to a feedback control approach.

Considering the third category, a feedforward control, as proposed by Havinga et al. [11] can be used to correct the process parameters for the components in production subjected to external disturbances measured by the sensors, by adjusting the subsequent processing steps, as shown by Fig. 2d. Feedforward approaches, in particular, require effective process models capable of describing the effect of measured disturbances on the properties of the manufactured component. In contrast, feedback-type approaches, as described by Endelt [12], are more effective in responding to unexpected disturbances to correct subsequent components. A further and more detailed classification can be defined in accordance with the fourth point, taking into account the type of algorithm used for control. As discussed by Li et al. [13], control systems can be classified based on response linearity. Aguiar et al. [14], explore frequency-driven designs, while Lim et al. [15], investigate MIMO structures. Finally, Allwood et al. [5] focus on the influence of the underlying process model. Therefore, it is important to note that these approaches, in most cases, are not alternatives to each other, because as specified in Sect. 3, they respond to different problems and must therefore be integrated with each other to increase their effectiveness. An important question to address is: what are the main benefits of using an advanced control system in a manufacturing process? Since a control system allows for the automatic correction of parameters based on measured feedback, the most obvious advantage is increased efficiency. This efficiency gain occurs through a reduction in setup times due to a higher level of automation, and a reduction in scrapped parts both in the setup and running stages by automatically correcting deviations caused by external disturbances, as demonstrated by Tatipala et al. [16]. This can make the production system more stable, even with input materials of lower quality, for example, those with broader tolerances. The implementation of actuators and sensors may also have the additional effect of expanding the productive capacities of the equipment, increasing operational flexibility, as highlighted by Yang et al. [8]. This possibly allowing the production of components with new characteristics, or with higher quality, for example in terms of achievable tolerances. An example can be seen in flexible profiling processes, where the implementation of control systems and additional automation, makes possible to obtain components with a variable cross-section along the axis, usually unachievable with traditional equipment, as shown

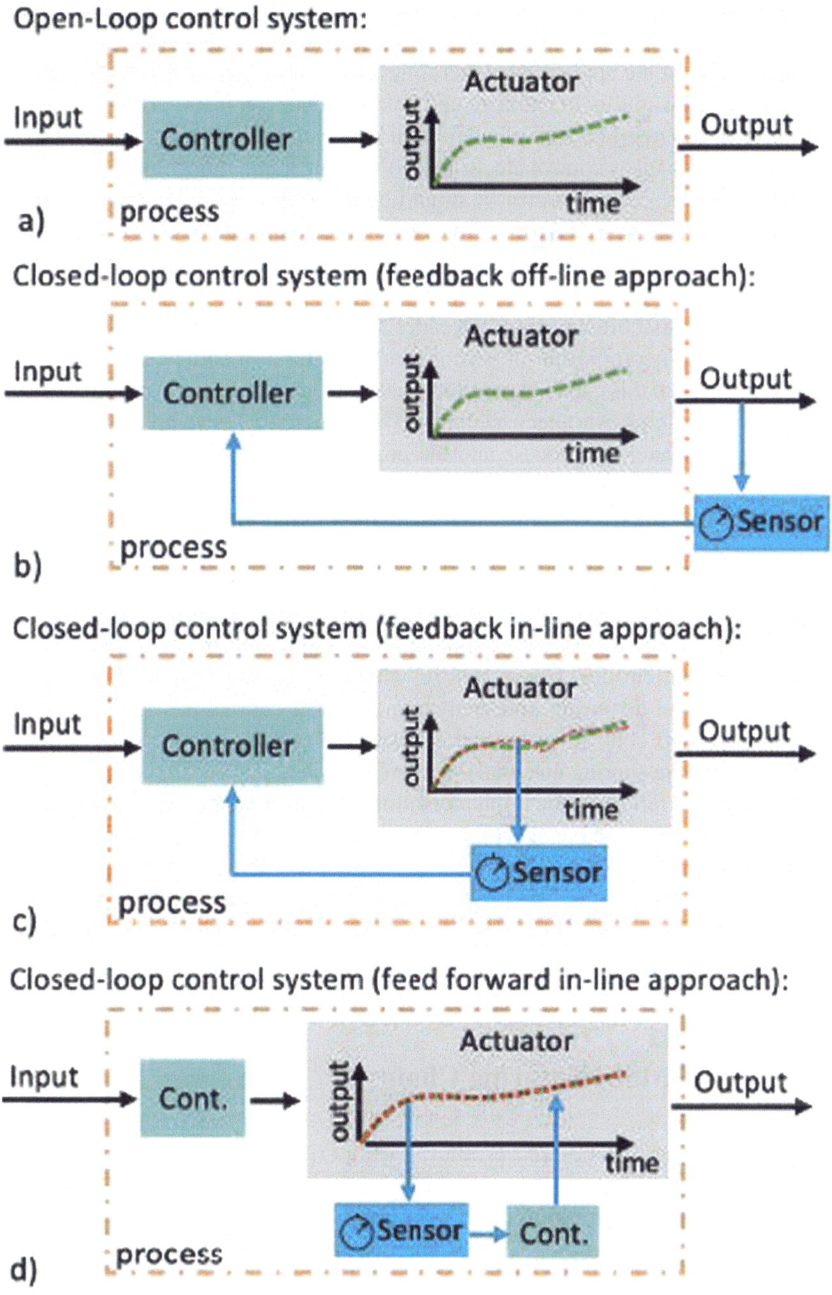

Fig. 2 Control approaches: **a** open loop, **b** closed loop with offline feedback, **c** closed loop with in-line feedback, **d** closed loop with forward in-line

by Woo et al. [17]. However, as noted by Tatipala et al. [16], the implementation of control systems also involves implementation and management costs that can be high depending on the application. Therefore, balancing costs and benefits is a critical aspect, and the ability to estimate and balance them can be crucial in evaluating the success or failure of their implementation. Since control systems require sensors and actuators, as well as control electronics to manage them, this immediately incurs a certain hardware cost. However, it should be noted that for many manufacturing processes, the implementation of sensors and actuators may also require a change in the architecture of the machines and equipment. Since the processes considered are of a conventional type, machines and equipment are the result of a design and optimization process that often occurred decades ago, with architectures not specifically designed for the incorporation of sensors and actuators. In some cases, as seen in examples in Sect. 4, this can lead to a complete redesign of a line or part of it to incorporate control systems. Another aspect involves the potential development of models capable of describing the process and the possible effect of external disturbances, necessary for determining the corrective interventions to be applied.

Many effective models require large-scale data sampling, analysis, and storage, which can result in significant management and storage costs as noted by Santos and Deschamps [17]. This management can involve costs not only for hardware, but also for legal agreements between the involved parties. It should not be underestimated that, as previously described, the machine manufacturer, tools provider, and end user are often different entities. Consequently, while the machine manufacturer would have an interest in acquiring and analyzing as much data as possible to develop more sophisticated control systems, these data are actually produced by different companies, making sharing and analysis more complex.

The success or failure of the implementation of control systems requires a careful evaluation of the benefits, but also of the associated costs. This paper aims to analyze in detail the elements required for the implementation of these systems, to point out the various aspects. Finally, the provided examples will serve to both present the benefits, but also to highlight the issues that hinder their widespread adoption of advanced control system in industrial environments.

2 Reference Manufacturing Chain

The manufacturing of tubes and profiles plays a critical role in industries such as energy [18], automotive [19], aerospace [20], heat exchangers [21], and construction [22], driven by the need for high-performance products and more efficient manufacturing processes. For example, In the automotive industry, Merklein et al. [23] report a growing use of high-strength steel to enhance crash resistance, and Ahmetoglu and Altan [24] describe how tube hydroforming contributes to vehicle weight reduction. The primary materials used in tube manufacturing include steel alloys, aluminum alloys, and other metallic materials, each chosen for their specific properties. Carbon steel is valued for its high mechanical strength and cost-effectiveness, making it ideal for structural applications and gas and liquids pipelines. Stainless steel's superior corrosion resistance makes it ideal for industries with stringent hygiene requirements or chemical exposure. Aluminum alloys, meanwhile, are favored in aerospace and automotive sectors for their lightweight properties. Copper alloys are utilized in applications requiring high thermal conductivity, such as heat exchangers.

Miller [25] provides a comprehensive overview of the manufacturing processes for tubes and profiles, which typically include cutting, shaping, joining, and surface treatments, as illustrated in Fig. 3. Cutting methods vary, including machining, blanking, and unconventional processes such as laser and plasma cutting, which are selected based on production rates, material types, and the required geometry. After cutting, tubes undergo shaping processes, such as bending and forming, where mandrels or the superposition of axial stretching forces are used to maintain the tube's shape and cross-section. For joining operations, Carvalho et al. [26] demonstrate the use of laser welding, while Kumar et al. [27] focus on arc welding techniques. D'Urso et al. [28] introduce more advanced approaches such as friction stir welding, and other studies explore electromagnetic forming to connect both homogeneous and heterogeneous materials. For surface treatments, Pais and Rao [29] review galvanization strategies, whereas Presuel-Moreno et al. [30], analyze anodization techniques. These methods, along with industrial painting, are applied to enhance durability, prevent corrosion, and extend the lifespan of tubes. Moreover, research into green corrosion inhibitors offers an alternative to traditional methods, providing more environmentally friendly solutions. It is worth to notice that this paper focuses on techniques that are akin to sheet metal working, deliberately excluding bulk working processes, such as extrusion, for the sake of clarity and relevance to the specific scope of tube and profile manufacturing, as discussed in Sect. 4.

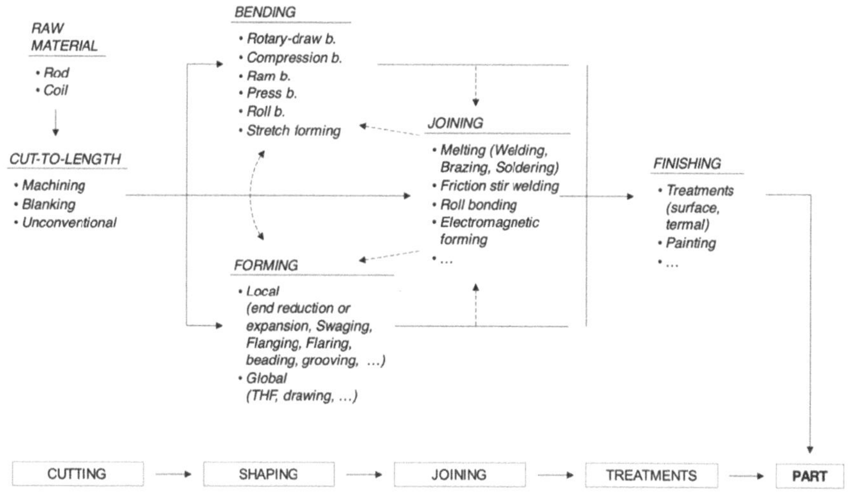

Fig. 3 Tube and profiles manufacturing process chain

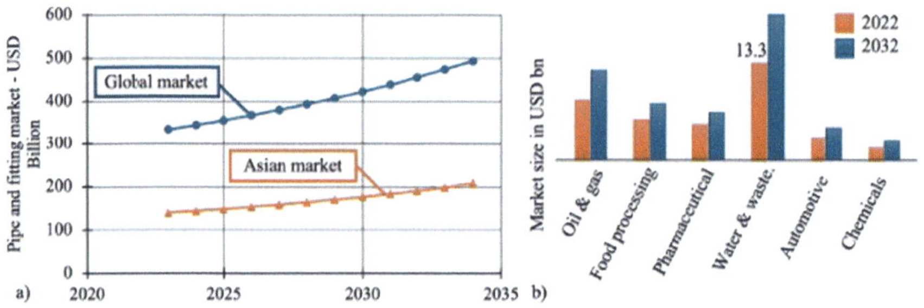

Fig. 4 **a** Global and Asian piping and fittings market size forecast 2024–2034 [31]. **b** Europe Pipe Market, by End User, 2022 and 2032 [34]

The market for pipes and fittings is driven by various end-use sectors, including chemicals, pharmaceuticals, oil and gas, and energy, with piping used for applications such as irrigation, waste management, and water distribution. In 2024, the global market size for pipes and fittings is estimated at USD 345.13 billion, with a projected compound annual growth rate (CAGR) of 3.70% from 2024 to 2034, reaching around USD 494 billion by 2034 [31], as shown in Fig. 4a. The Asia–Pacific region holds the largest market share, accounting for 42% in 2023, and is expected to grow at a similar rate, with an estimated market value of USD 210 billion by 2034. This growth is attributed to increasing population, urbanization, and government initiatives to improve industrialization, construction, and drainage systems. The European market [32] is also experiencing growth, driven by factors such as the oil and gas sector, transportation, construction, and investments in research and development [33], as shown by Fig. 4b. The Italian market, based on data from 2019, shows that a significant portion of steel tube exports and imports occurs at the European level, especially in Western Europe [34]. Italy is a key player in the European market, with 85.8% of its steel tube exports going to other European countries. The Italian steel tube market is also bolstered by a network of 362 companies manufacturing steel tubes, pipes, and fittings [35].

3 Control Systems

This section focuses on the architecture that a control system can assume in profile and tube processing, discussing in particular these system core elements. The first part therefore highlighting in a control systems different approaches can operate at different levels, coexisting to solve different problems. The following paragraphs of the sections are dedicated to discuss and categorize the different types of actuators, sensors, and models used in the reference process chain.

3.1 Architecture

Any manufacturing process is subject to external and internal disturbances. The role of control systems is to adjust the process parameters by correcting any errors due to these phenomena. However, it should be emphasized that, regardless of its performance, a control system cannot compensate for deficiencies in the design phase, such as an incorrect design of the tools in a forming operation. Moreover, the ability of the control system to correct any disturbances depends specifically on the approach with which it was built and the capabilities of its individual components. It is also necessary to highlight that the choice of one approach over another is strongly dependent on the purpose of the control system and the parameters it intends to correct.

Figure 5 presents the hierarchy of control systems in a manufacturing process dedicated to the production of an elongated profile or tubular based part. Figure 5 specifically refers to a single process of a longer process chain necessary to obtain the target part. The initial workpiece for the considered process will thus be an initial semi-finished product, with specific characteristics in terms of both geometry and material properties, within certain tolerances. Similarly, for the target part a series of specifications will be defined, which may be both geometric-dimensional, or concern other properties such as: yield strength, grain size, surface properties, or others. Again, a target value and a tolerance range will be defined. This information will be necessary not only for the control system, but also for the equipment design stage. The first stage, therefore, includes not only the design of the tools but also the definition of the target process parameters. For example, in the case of a bent part, not only the dies will be designed, but also the forming sequence. Another example may concern the design of the terminal cycle for an in-line thermal treatment that a profiled parts should undergo to meet specific yield strength values.

Given these initial inputs, a first form of control presented in Fig. 5 is represented by an offline feed-forward control approach. In this case by measuring a reference property of the initial component, still within the prescribed tolerances, a first adjustment of the process parameters could be made. This approach could be applied, for instance, in a profiling process, where an initial measurement of the input sheet thickness could be used to adjust the gap between the rolls to avoid thickness reductions. It is important to note, that a sensor output can be used directly or indirectly to estimate the control variable. While a direct measurement is relatively simple, it is often not possible, and consequently, it may be necessary to estimate the control variable based on other inputs. In this case, however, it is necessary to use a model that allows estimating this variable based on the measured data. For example, the yield strength of the incoming material cannot be measured directly except with destructive tests. However, it is possible to estimate it based on hardness tests, which require a model to correlate the two quantities.

Once the component is loaded into the machine, the process control system applies the target parameters. Again, the presence of a model is necessary to translate, for example, a forming sequence into tool positions or a thermal path into furnace power.

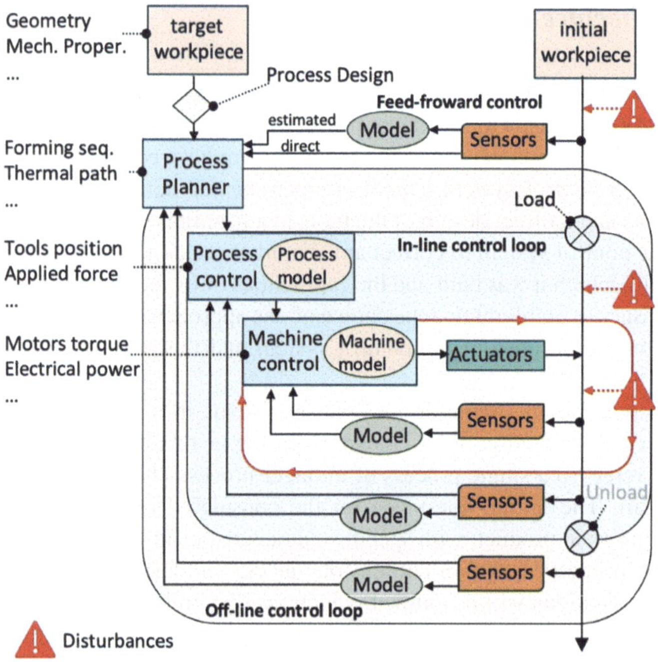

Fig. 5 Control loop architecture for a tube or profile manufacturing process

The process model can vary from a simple kinematic model to a more complex analytical model, or a simulative model as discussed in Sect. 3.3. An additional level of control, called machine control, is responsible for effectively translating the process parameters into machine parameters by controlling the actuators. Keeping the tool path position example, the machine control may use to regulate the electrical motor's speed, velocity, and torque to follow the chosen path.

In an open-loop scheme, the actuator's position is calculated using a kinematic model of the transmission system. However, this makes the system less accurate, unstable over time, and subject to various types of disturbances. One solution is to include a measurement system for closed-loop control of the motor position. In this case, the choice of sensor and control type can significantly impact costs and performance. In the simplest case, a rotary encoder can be used to measure the motor position and consequently the tools positions. However, disturbances such as mechanical backlash, load deformations, and inertial effects could introduce additional errors. These errors could eventually be modelled and compensated for, also based on further measurements such as motor torque or others. Alternatively, a measurement sensor could be used to directly measure the tools position, thus accounting for any backlash and deformations of the transmission system. This second approach, however, generally requires higher costs in terms of sensors, depending on the target accuracy. Clearly defining achievable tolerances is crucial, as this influences the choice of control approach. Additionally, measuring tool position alone does not directly inform about the actual properties of the finished component, which is the true objective of the process. It constitutes only an indirect measurement. For instance, in a bending process, while the achievement of a certain tool position is a given, this does not provide any information about springback. Therefore, if the goal of the control system is to correct springback, it is essential to measure the actual geometry of the component. Based on these in-line measurements of the component, it may be possible, depending on the process, to correct the component in production or eventually adjust the subsequent parts. Whether or not the measured part currently in production can be corrected depends on the process, the position of the measurement point, and the measured equipment. For example, in a rotary draw bending process, once the dies are released and the springback is measured, it is not possible to re-grip the part to adjust the part. However, for an incremental bending operation such as roll bending, this type of approach is possible. Similarly, even where feasible, if the measurement system is positioned offline downstream of the machine, it will not be possible to correct the current part but only the subsequent ones. To automate corrections, process and machine models should, upon defect detection and quantification, calculate the new parameters required to correct deviations. While Fig. 5 shows the case for a single process, Fig. 6 illustrates how a control system could be applied at the process chain level. Specifically, it should be noted that according to the standard approach, the first thing to be designed is the final component, and then the process chain. The final results determine the technical specifications required for initial components. This information propagates backward, creating an opposite flow to that of the physical parts.

A more advanced approach would be to view at the different processes as intermediate steps of a single chain, allowing information to flow in both directions. Measurements taken in-line or offline during a process step could be used with a feedback logic to control the preceding processes. Similarly, the same data could be used in a feedforward logic to adjust the parameters of subsequent processes. For instance, the geometric properties of the component measured during a profiling process could be used by subsequent bending operations to adjust the tool positions, avoiding the appearance of defects. In the same way, formability issues in hydroforming processes following stretch bending could provide important indications for correcting the axial pulling forces applied during bending.

However, it is important to note that the exchange of this data can be significantly more complex than the exchange of simple technical drawings with encoded specifications and tolerances. This approach is further complicated by the fact that different processes are often executed by various actors within the supply chain and are frequently performed at different times and under varying conditions. Some examples of these approaches will be presented in Sect. 4.

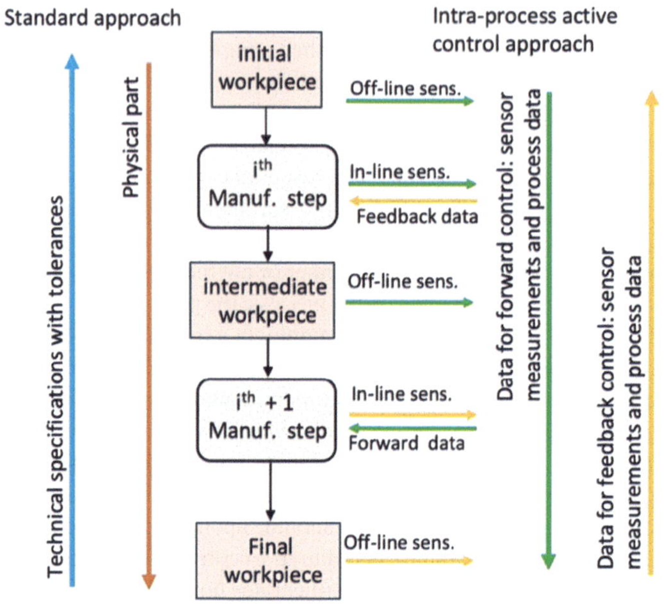

Fig. 6 Control loop for tube or profile manufacturing process chain

3.2 Sensors and Actuators

The implementation of in-line measurement systems in tube forming and processing is crucial for ensuring process control and optimization. This study focuses exclusively on sensors that can be directly integrated within individual production processes, excluding offline measurement solutions.

Perera et al. [36] discuss how measurements in manufacturing processes can be performed either directly or indirectly through estimators that rely on interpretative models to infer relevant parameters from the collected data. To be effectively integrated into production processes, sensors must comply with specific technical and operational constraints. First, the measured quantity must be relevant for process control and accurately reflect the key parameters influencing the forming operation. The measurement bandwidth should be compatible with the process dynamics to ensure reliable data acquisition. Additionally, Galetto et al. [37] emphasize that the measurement volume must align with the component geometry and process characteristics to enable consistent and representative data collection.

For indirect measurements based on interpretative models, real-time processing capabilities are essential to ensure that the acquired data can be processed within the cycle time of the production process. Cao et al. [38] underline the importance of effectively integrating sensors and deploying them in sufficient numbers to ensure adequate coverage of critical process variables. The cost–benefit ratio is another fundamental aspect, as the selection of measurement systems must be justified by a positive balance between the added value obtained and the implementation costs.

Sensors must also operate reliably in harsh industrial environments. Typical forming processes involve vibrations, high temperatures, lubricants, dust, and other disturbances, all potentially compromising sensor performance.

In profiles forming and processing, geometric properties are the primary control targets, as their accuracy and repeatability determine the quality of the final product. Lee et al. [39] highlight that, in specific applications, mechanical and microstructural properties may also be of interest—particularly in cold-forming processes, where material behavior is significantly affected by deformation history.

Measurement approaches can be classified into direct and indirect. Direct approaches measure the target quantity without requiring additional processing, as in the case of optical systems for geometric inspection, displacement sensors, or strain gauges for deformation monitoring. Conversely, indirect sensors acquire correlated quantities that, through an interpretative model, enable the estimation of the desired parameter. Examples include load cells for assessing the stress state of the material or thermal sensors for detecting microstructural variations.

Despite advancements in measurement technology, integrating in-line sensors into tube forming processes presents several challenges. The confined spaces within forming tools and equipment often limit the available installation options, requiring compact and robust sensor solutions. Additionally, the effectiveness of sensors must be evaluated based on the machine architecture. For instance, when measuring force, if the system is hyperstatic, it either requires measurement points equal to the number of constraints or measurements at a few significant locations, allowing the total load to be estimated by considering the system stiffness. This introduces different levels of complexity depending on the specific setup.

Ultimately, the effectiveness of in-line measurement systems depends on their integration with actuators and process models. The collected data must be effectively utilized in process correction strategies, either through direct actuator control or by updating predictive models that optimize process parameters based on real-time information. The ability to leverage in-line measurement data within advanced control frameworks is essential for improving process stability, reducing variability, and enhancing overall production efficiency.

As with sensors, the effect of an actuator on the process variable can be direct or indirect. For example, in a bending process, springback compensation can be achieved directly by increasing the bending angle (overbending) or indirectly by applying an axial tensile force, as occurs in rotary draw bending when increasing the stretching component. In the first case, the actuator directly adjusts the bending angle, while in the second case, it modifies the axial force applied during bending. The latter approach requires a predictive model capable of determining the effect of the superposed stress state to apply an effective correction. This is analogous to how estimators infer a parameter variation based on an indirect measurement.

The effectiveness of actuators in controlling a specific process variable depends on several key factors. First, the type of action that the actuator can apply is crucial, as it determines which process parameter can be modified. For instance, in rotary draw bending, adjusting the bending die rotation directly affects the bending angle, whereas modifying the axial force requires intervention on the pressure die or the carriage system. Another critical factor is the dynamic response of the actuator, in terms of both reaction time and amplitude of variation. Some systems, such as in-line heating for thermal treatments, may have limitations in terms of maximum heating rates, affecting their ability to introduce rapid corrections. Integration within the machine and tooling setup is also a key consideration. Adding an actuator often changes the system degrees of freedom, potentially complicating the overall machine configuration. Additionally, available space can limit the feasibility of integrating new actuators, making compactness and adaptability crucial selection criteria. Another important aspect is the level of interference with other process parameters. For example, in springback control, modifying the axial force may also influence the thickness of the profile, not just the bending angle.

In the last decade, significant advancements in electric motor performance and cost efficiency have led to their increasing adoption, replacing traditional hydraulic actuation in many applications, as highlighted by Kawamoto et al. [40]. Electric actuators offer higher stability and accuracy in movement control while also enabling real-time monitoring of parameters such as position, force, and torque. This provides additional valuable data for process control. However, in systems with multiple degrees of freedom and numerous manual setup adjustments, the transition to fully automated control systems requires a careful cost–benefit analysis before implementation.

3.3 Models

In control systems for tube forming and manufacturing processes, models serve as representations of complex, real-world machinery and processes, allowing engineers to predict, optimize, and control the behavior of these systems. Tube bending, laser cutting, roll forming, and hydroforming processes rely heavily on these models. They help achieve desired outcomes despite complexities and uncertainties related to material deformation, machine dynamics, and variability in production. Control systems require these models to:

- Optimize production quality: predict material deformation, precision in cutting, and other quality parameters for process adjustments.
- Improve process efficiency: minimize resource use by adjusting process parameters in real-time for energy and material savings.
- Enhance adaptability: enable rapid adjustments in manufacturing setups when switching between product specifications or when encountering machine wear or external disruptions.

In this context, a wide range of modeling approaches can be adopted, as summarized in Fig. 7. These include analytical models, which provide fast, explicit solutions but are poorly suited to capturing the complexity of real-world processes, and Finite Element Models (FEM), which offer high fidelity but are computationally

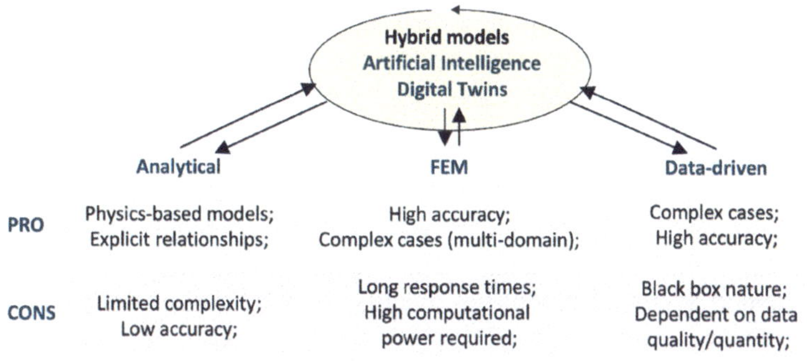

Fig. 7 Overview of modelling approaches for control systems

demanding. Data-driven models represent a third category, leveraging statistical or regression techniques applied to real data. These models can deliver rapid solutions even for complex problems, though their reliability is highly dependent on the quality and quantity of the input data. Finally, hybrid approaches—such as those based on Artificial Intelligence (AI) or the Digital Twin paradigm—have emerged, combining the strengths of the aforementioned techniques to enhance both model accuracy and responsiveness.

Around the beginning of the new millennium, process optimization methods in combination with FEM have become fully mature and accepted in the scientific community and were incorporated in numerous commercial software packages. A great number of diverse methods, approaches, and applications have been proposed. For instance, Dang et al. [41] explore metamodel-based optimization, while Havinga et al. [42] introduce stochastic methods to simulate the uncertainty of real experiments. The most recent and significant challenge is to perform optimal real-time process control and optimization has been addressed by Havinga and van den Boogaard [43]. Any control system requires a model of the observed machine or process, and the extensive availability of metamodels developed for FEM-based offline optimization provides a valuable foundation. Typically, such metamodels or surrogate models are based on the results of Finite Element simulations, as shown by Prates et al. [44], on experimental data, as in the work of Havinga and van den Boogaard [43], or on a combination of both approaches, as demonstrated by Strano et al. [45].

The computational efficiency of metamodels can be even combined with real-world data coming from sensors to build a real-time digital twin which can feed the control system. Given the complexity of tube forming and fabricating processes, no model can realistically be attempted which is based on closed-form analytical solutions. While metamodeling offers significant advantages in efficiency and accuracy, it is essential to consider the potential limitations, such as the need for extensive data for validation. This balance between model complexity and practical applicability remains a critical challenge in the field.

Ingarao et al. [46] classify the most frequently used metamodeling techniques in tube manufacturing into three main categories:

1. Linear Regression and other statistical approximations methods;
2. Methods based on machine learning and other artificial intelligence techniques;
3. Kriging and other numerical interpolation methods.

As an example of the first group, Ben Abdessalem and El-Hami [47], use two different regression metamodels (the response surface method and the least squares support vector regression) to identify the influence of design variables, such as internal pressure and axial displacement, on process outcomes, aiding in the optimization of loading paths in hydroforming.

As an example of method based on AI, a work by Naceur and Arif [48] delves into the modeling and optimization of hydroforming and stamping. It shows that ML-driven metamodels can optimize parameters more efficiently, improving the overall precision and output quality of hydroformed products. By using AI and ML for surrogate modeling, the authors significantly reduced the computational costs associated with traditional finite element methods (FEM), making the manufacturing process more resource-efficient and adaptable to varying requirements. The integration of machine learning (ML) and artificial intelligence (AI) in surrogate modeling for tube manufacturing has advanced significantly, providing enhanced accuracy and efficiency in complex processes like hydroforming, bending, and laser cutting. By implementing ML algorithms, researchers have developed surrogate models that effectively predict process outcomes without the need for extensive physical experimentation or computationally intensive simulations.

Kriging-Based Surrogate Modeling in Tube Bending Processes: An Overview

A much more frequently used technique for metamodeling, rather than regression, is interpolation obtained through gaussian processes (kriging).

A significant application of Kriging surrogate models in tube bending is in predicting springback, a common issue in the rotary-draw tube bending process. In a study by Sözen et al. [49], Kriging was used alongside radial basis functions (RBF) to create a predictive model for springback in rotary-draw bending. The model was trained on finite element (FE) data, enabling it to accurately predict springback across different bending scenarios. This surrogate model approach minimizes the need for extensive physical trials, saving time and resources in tube manufacturing.

Further advancing the field, An et al. [50] utilized a multi-objective genetic algorithm (MOGA) coupled with Kriging to optimize the loading path in multi-stage tube forming, which includes bending and hydroforming processes. By integrating Kriging with MOGA, the study developed an efficient optimization framework for multi-stage forming operations. This method effectively reduced computational demands while providing optimal loading paths, demonstrating the potential of Kriging models to handle the complexity and multiple variables associated with tube bending and forming processes.

Kriging models have also been applied in free bending technology, particularly in cases where tubes with variable curvatures are required. Liu et al. [51] recently proposed a Kriging-based model to determine process parameters for spatial tubes with variable curvature, achieving high precision in parameter selection. By utilizing this surrogate model, the study was able to predict and optimize the parameters for free bending processes, which often involve irregular geometries and require high flexibility in parameter adjustments. This approach showcases the adaptability of Kriging models to accommodate diverse tube bending needs and complex geometries.

The adaptability of Kriging is further evidenced in its application to problems where small sample size is available, e.g., because the computational time is long and accurate modeling can be challenging. Wang et al. [52] introduced a hybrid surrogate model that combines Kriging with Gaussian membership virtual sample generation to overcome the limitations of small sample sizes in tube bending processes. This approach allowed for accurate predictions despite limited data, demonstrating Kriging's robustness in dealing with data scarcity.

Kriging has also been employed to metamodel-based optimization of tube hydroforming, controlling a pulsating water pressure and axial feeding, trying to enhance formability and minimize defects [53]. In conclusion, Kriging-based surrogate modeling has emerged as a powerful tool for enhancing accuracy and efficiency in tube bending processes. From predicting springback and optimizing loading paths to handling small data sets, Kriging models offer versatility and robustness across various aspects of tube manufacturing.

Digital Twins in Tube Forming Processes

Digital Twin (DT) technology has become a transformative tool in modern manufacturing, especially within fields that require precise control, real-time data integration, and predictive modeling.

A recent study by Sun et al. [54] focuses on enhancing the accuracy and efficiency of metal tube bending, a common process in industries like aerospace and automotive. The authors propose a method that integrates real-time physical data with virtual parameters to predict and mitigate common bending defects, such as springback and wrinkling. Traditional methods in tube bending rely heavily on offline data and fixed parameters, which often lead to inaccuracies due to the unpredictable nature of physical deformations. In contrast, the proposed DT-enhanced approach allows the DT system to convert real-time deformation data into a format compatible with virtual modeling. This integration of physical and virtual data provides a dynamic, responsive system that can predict and compensate for springback, significantly enhancing product accuracy. The researchers validate the system with physical experiments on 6061-Al tubes, demonstrating improved prediction and control over previous methods.

Another very recent work by Li et al. [55] delves into real-time full-field deformation monitoring in metal tube bending. The authors address a critical challenge in DT modeling: the delay caused by time lag between data acquisition and model updates, which limits the DT's ability to represent complex deformations accurately. To overcome this, they propose a method using spatio-temporal cascade reconstruction to compensate for these time delays, enabling high-fidelity, real-time deformation monitoring. This approach integrates limited sensor data with theoretical models to create a virtual replica, or DTVM, that maintains accurate, dynamic mapping of the physical tube's deformation. The study emphasizes the necessity of time-delay compensation, especially for industries requiring high precision, where any discrepancy between real-time data and the model could result in costly inaccuracies.

4 Examples and Industrial Cases

The following paragraphs present a series of examples and industrial cases related to the processes outlined in the reference production chain discussed in Sect. 2. It is worth noting that a review of the scientific literature reveals how, for many of the cases and processes considered, only a few contributions thoroughly address the issue of process control. In most instances, despite the objective of developing a control system, contributions tend to focus on a single aspect, such as the development or integration of a sensor, an actuator, or the implementation of a model, without adopting a more comprehensive approach. From one perspective, this highlights how the systematic adoption of these systems often remains limited to laboratory prototypes, with only a few cases achieving commercial success in the industrial sector. However, this can also serve as an incentive, as it suggests significant potential for innovation in this area.

4.1 Flexible Roll Forming

Flexible Roll Forming (FRF) is an innovative metal forming process that represents a significant advancement over traditional roll forming. This technology allows the production of profiles with variable cross-sections along the length of a metal sheet, offering unprecedented flexibility, as shown by Simonetto et al. [56]. Unlike conventional roll forming, which produces uniform cross-sections, FRF enables continuous variations, making it highly suitable for industries requiring complex and customized components, such as automotive, aerospace, shipbuilding, and construction. In traditional roll forming, a long strip of metal passes through a series of fixed contoured rolls, progressively shaping the metal into a consistent cross-sectional profile. This method is efficient for high-volume production as it minimizes downtime, but it lacks flexibility because rolls cannot be modified during the process, as explained by Alvarez [57]. On the contrary, FRF utilizes Computer Numerical Control (CNC) techniques to dynamically adjust the position and tilt angle of the roll axes during the forming process. This capability allows the rolls to move along a path that follows the desired bend line of the profile, thereby producing variable cross-sections. The rolls remain tangential to the deformed flange, preventing unwanted plastic deformation. This enables the FRF process to produce complex profiles with variable cross-sections, meeting specific design requirements across various industries. Recent studies have used finite element modelling (FEM) to predict stress distribution and potential defects in FRF processes. Yan and Li [58] developed FEM-based models to optimize process parameters and improve the overall accuracy of profiles formed using FRF. Building on this, Gao et al. [59] advanced FEM simulations by investigating the influence of local temperature and material properties on formability, further enhancing the reliability and efficiency of the process. However, the FRF process has seen very few cases where it has been implemented with sensor systems that allow in-line process control and enable the assessment of critical issues by providing feedback at the end of the process. One of the most recent research projects is related to the scientific work published by Ren et al. [60]. The study examines a case in which a six-dimensional Digital Twin (DT) model was applied to optimize the roll forming process of "hat" components in a manufacturing environment. The DT model replicates various aspects of the process-physical, virtual, and data-driven-to support real-time monitoring and predictive analysis. By integrating layers for physical layout, digital simulation, and an expert system that combines human and AI insights, the DT enables precise data management and dynamic process control (Fig. 8). In the case study, advanced predictive algorithms such as SAPSO-SVR and the BAT algorithm were used to anticipate and reduce strain variations, resulting in improved product consistency and quality. These improvements reflect the models potential in addressing traditional profiling challenges, such as parameter interdependence and data heterogeneity. Real-time data monitoring of parameters such as torque and material feed allowed early identification of deviations, minimizing downtime and reducing scrap. By also predicting wear and tear, DT facilitated proactive maintenance, thereby extending equipment life. This successful implementation highlights DT's ability to improve quality control and predictive maintenance, with promising implications for extension to broader manufacturing processes where efficient, data-driven production is essential.

4.2 *Sawing and Laser Cutting Process*

The sawing process is a fundamental operation in the metalworking industry, essential for cutting metal sheets, bars, and tubes. Recent advancements in sawing technologies have enhanced precision, efficiency, and safety. These innovations are driven by the integration of advanced simulation techniques and adaptive control systems, which optimize sawing operations for industrial applications.

One significant development in the field is highlighted by Abouridouane et al. [61] who investigated friction behaviour in metal cutting processes. Their study focused on reducing the experimental workload required to analyze friction by directly correlating the apparent friction coefficient with cutting parameters. Using dry orthogonal cutting tests on carbon steel AISI 1045, they observed how variations in cutting speed

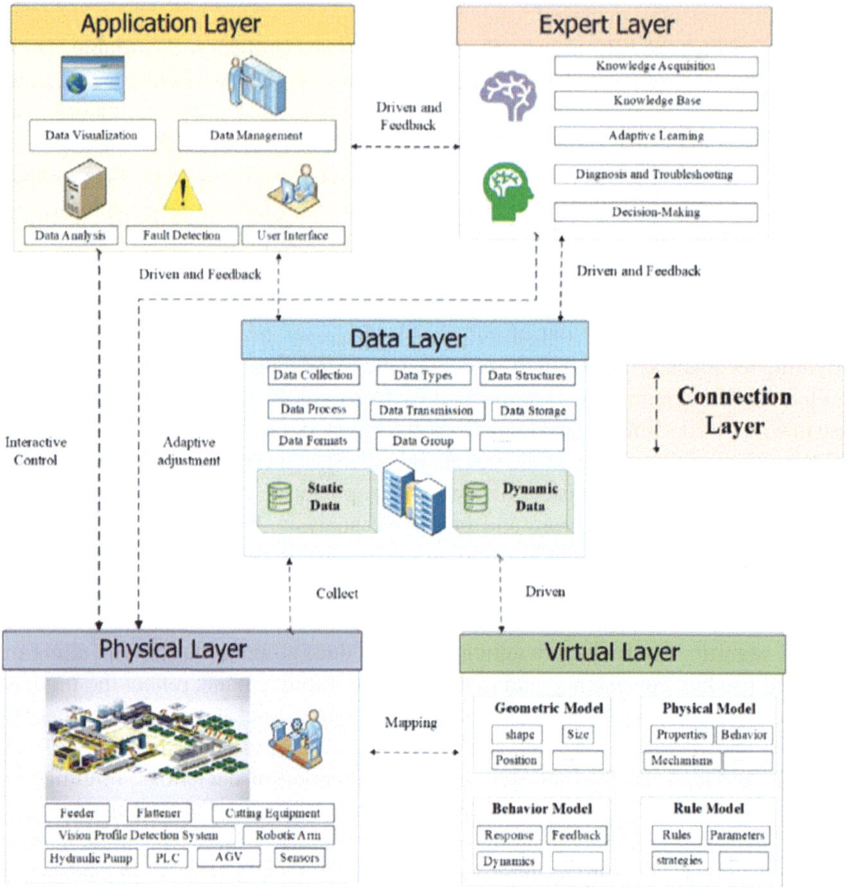

Fig. 8 DT model for roll forming equipment presented by Ren et al. [60]

and uncut chip thickness impacted the transition between sticking and sliding friction. Higher cutting speeds and thicker uncut chips reduced sticking friction, making sliding friction dominant. By developing a validated friction model through finite element simulations, they successfully minimized the number of experiments needed, achieving precise predictions with a drastically reduced workload. This approach not only improves understanding but also enhances the efficiency of sawing processes by leveraging simulation tools. Möhring et al. [62] introduced an innovative method for simulating the sawing process that significantly reduces computational effort. Their approach combines 2D and 3D simulations to balance accuracy and time efficiency, applying a 2D strain state for high cutting width-to-depth ratios. This hybrid simulation strategy achieved more than a 50% reduction in computation time while maintaining strong agreement with experimental results. The cutting forces and chip morphologies predicted using this method were accurate, proving its practical utility in industrial settings where time and resource optimization are crucial. By bridging detailed simulations with real-world applicability, this technique represents a transformative step forward for precision metal cutting. Laser cutting is a widely used technique in the metalworking industry, valued for its precision, efficiency, and ability to handle a variety of materials. Recent advancements in laser cutting technology have further enhanced its capabilities, making it an essential tool in manufacturing. Process monitoring and control in laser cutting are critical to ensuring precision, efficiency, and repeatability. Modern systems often integrate advanced sensors and adaptive control mechanisms to optimize process parameters in real time. For example, Duflou et al. [63] developed a real-time, adaptive monitoring system for laser flame cutting, designed to enhance the quality and efficiency of cutting operations, particularly for thick plates of mild steel. The system uses a combination of optical sensors, such as photodiodes and near-infrared (NIR) cameras, to monitor easily observable physical parameters correlating with cut quality, such as edge striations and dross formation. The adaptive control mechanism automatically adjusts laser power, cutting speed, assist gas pressure, and duty cycle to maintain optimal performance, minimizing defects and maximizing productivity. This integration reduces reliance on manual quality checks while providing robust solutions for high-power CO_2 laser systems. The study also highlighted the capability to generate automated quality reports, facilitating efficient process oversight in industrial settings.

Combining optical sensors with feedback devices allows for real-time analysis of operational conditions and prompt corrective actions. IPG Photonics, for example, integrates high-performance sensors into its laser cutting heads to monitor power and focus, thereby achieving superior cutting quality even at high power levels [64]. Automation also plays a key role, as seen in Prima Power's use of AI algorithms to analyse vast amounts of real-time data [65]. These systems provide adjustments on-the-fly, ensuring product consistency and reducing waste. The integration of advanced technologies such as image processing, optical sensors, and AI systems provides a comprehensive approach to optimizing laser cutting processes. These innovations not only deliver high-quality outputs but also promote sustainability through reduced waste and energy consumption. Consequently, process control in laser cutting represents a technological cornerstone for contemporary industrial applications.

4.3 Bending

Bending operations are a fundamental step in the manufacturing of elongated components with closed or open profiles. Among the most widely used and significant technologies is Rotary Draw Bending (RDB). This process allows for accurate and repeatable bends with excellent geometric accuracy. The high level of constraint applied to the tube contributes to this precision, with a mandrel inside and external support provided by a die, clamping/pressure pads, and a wiper die. While this process is precise and repeatable, it requires expensive, application-specific tools based on profile dimensions, thickness, and bend radius. Due to its industrial importance, this process will be discussed in the following section. To overcome its limitations, several more flexible bending processes are available. However, with increased flexibility comes greater difficulty in maintaining accurate and repeatable process control. Examples of these processes are provided in the Flexible bending processes section.

Although each process has its peculiarities, it is valuable to identify common challenges in bending operations that could benefit from advanced control systems. The most significant challenge is achieving geometric accuracy, specifically in terms of the bend radius and angle, due to the springback phenomenon. Springback depends on several factors: material properties, supply conditions, geometric variations (diameter, thickness), weld seams, profile orientation, tool friction, lubrication, temperature, and other aspects. Other critical control parameters involve the appearance of defects caused by instability phenomena like wrinkling or the application of excessive radial pressure, which can result in ovalization. Such issues can even lead to rupture or localized thickness reduction. In some applications, controlling the final mechanical properties, such as microstructure and yield strength, is also essential. Finally, surface characteristics represent another crucial factor. Non-optimized process parameters can cause imprints or marks, leading to both aesthetic and functional defects, such as poor fittings with seals or gaskets.

Rotary Draw Bending

Focusing on the rotary draw bending process, whose die configuration is shown in Fig. 9a, the control of springback, along with avoiding the onset of wrinkling and fractures, is among the critical issues most frequently analysed in the literature. From a control perspective, the challenge can be divided into the in-process measurement of the tube geometry or key process parameters and the calculation of adjustments to either prevent defects or compensate for them, such as determining the 'over-bend' angle to meet the target bending angle. It is worth noting that corrections must be applied to subsequent components. For example, in the case of wrinkling or fractures, once they form, they cannot be corrected, and adjustments can only be made on the next bend. Similarly, with springback, once the dies are opened, it is not possible to re-bend the current part without introducing additional defects. However, in this case, two process features simplify measurement: (i) springback occurs exactly when the dies open after bending, and (ii) the bend is located between two straight tube sections.

The interest in in-line measurement systems, aimed at reducing traditional trial-and-error correction methods, with automatic systems capable of minimizing setup times and scrap production, has been well documented in the technical literature since the early development of industrial bending machines.

In the case of springback, most applications evaluate it by measuring the angular variation between the straight sections before and after the bending stage, using a variety of methods. Patents such as [67, 68] proposed the use of contact probes to measure the die orientation relative to the clamp die, while [69] introduced the use of a wire sensor.

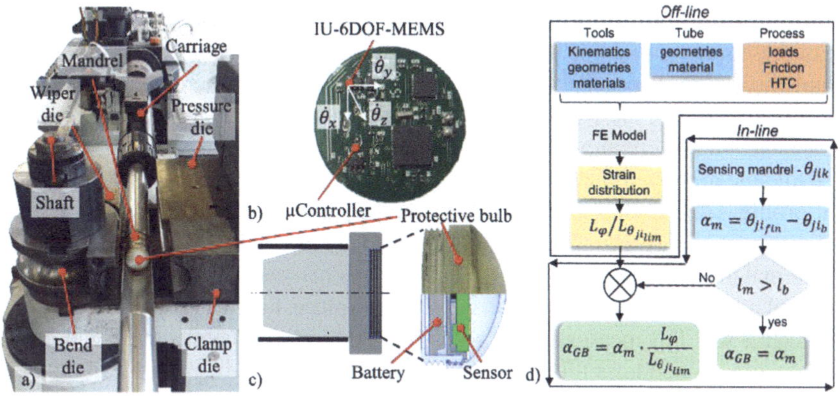

Fig. 9 **a** RDB process, **b** IMU board, **c** IMU probe [66], **d** machine control logic [56]

Since the early 2000s, with the advancement of non-contact sensors and measurement systems, similar approaches have been implemented through direct measurements using laser triangulation [70], digital image processing [71], or a combination of both. For instance, Ha et al. [72] proposed a video system that tracks the positional variation of a laser beam fixed to the tube, as illustrated in Fig. 10a. Although these systems can be mounted directly on the bending machine, they function independently, without requiring active communication with the machine.

Other approaches involve the use of contact sensors embedded in the tools, as described by Saegusa [73], and Watanabe et al. [74]. In these cases, after bending, the machine tools are brought back into contact with the tube to measure the angle based on the angular variation detected by the positioning system of the tools.

However, from an industrial perspective, laser and video measurement systems, as well as contact sensors downstream of the dies, have not seen widespread adoption. This is due to the high cost, complexity, and limitations in covering the entire

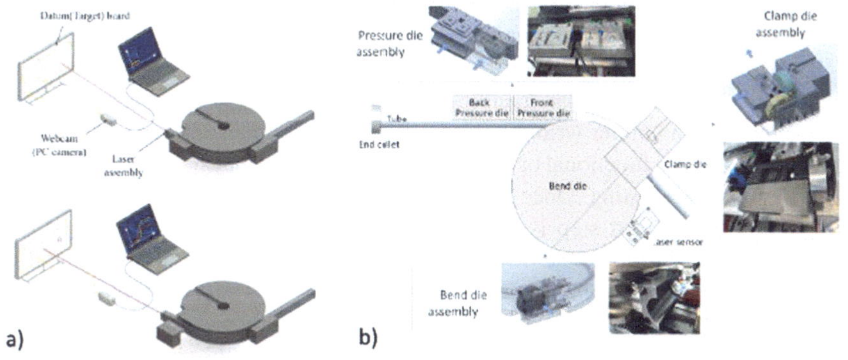

Fig. 10 **a** Laser [72], **b** Force [76] based spring back measurement system

measurement volume for components that develop in 3D space. Additionally, the use of these sensors often necessitates longer straight tube sections, which limits the complexity of the shapes that can be achieved. While sensors integrated into the dies help address some of these challenges, they require the machine to follow a modified work cycle, negatively impacting productivity.

While the aforementioned solutions focus on measuring the outer surface of the bent profile, Ghiotti et al. [66] proposed another approach, which uses an Inertial Measurement Unit (IMU) platform, as described by Simonetto et al. [75] attached to the internal mandrel, as shown in Fig. 9b. This system measures springback based on the angular position of the mandrel inside the workpiece, before and after the process. The accuracy of this measurement can be limited by the system location within the bending zone. To address this, the measurement system can be placed at the tube's extremity, as shown in Fig. 9c, or the total angle can be extrapolated, according to the algorithm presented by Simonetto et al. [56] and illustrated in Fig. 9d.

Another approach is to indirectly determine the springback angle, not through geometric measurements, but by measuring the forming loads. Figure 10b illustrates the integration of load cells into the bending tools to measure the torque and correlate it with the springback angle, as proposed by Ma et al. [76]. This method could enable correction from the very first bent piece by predicting the springback angle based on the measured torque. However, it requires a model capable of accurately correlating the torque with the final angle. This correlation is highly dependent on boundary conditions, making it difficult to implement in an industrial setting. Developing an in-line measurement system that is accurate, cost-effective, flexible, and repeatable continues to be an ongoing research area. Such systems also need to be suitable for standard industrial use.

Wrinkling typically occurs between the wiper and the bend die, in a region that is difficult to access externally, and is often only identified during unloading. However, some authors have proposed in-line measurement systems. Simonetto et al. [77, 78] used an IMU on various machine tools, demonstrating that the mandrel is most sensitive to wrinkling, while vibrations in other components can also indicate these defects or fractures. Borchmann et al. [79] introduced a laser sensor to measure the tube's internal profile via a pocket in the wiper die, later enhancing it with load cells to detect instability boundaries [80]. Additionally, for detecting fractures on the extrados, a thermal camera-based image analysis system was proposed to identify temperature spikes at fracture sites. However, the effectiveness of these systems in detecting minor wrinkling or similar defects, which are often more of an aesthetic issue than a functional one, has yet to be proven, as they have not undergone extensive industrial testing. Once the springback angle or the onset of a defect has been identified, the next step is to calculate the appropriate corrective action. In the first case, standard practice involves bending a sample tube at different angles. Linear regression is then applied to establish the relationship between springback and bending angle through a trial-and-error approach. In the latter case, a sample is bent, and the position of the tools or the applied normal load is manually adjusted. However, it is important to note that small variations in boundary conditions, incoming material, or even the position of the weld line can alter these phenomena. Additionally, as discussed by Ghiotti et al. [66], it is well known that for small bend angles, this springback-bending angle relationship is often non-linear.

To automatically adjust the process parameters, models capable of correlating parameter variations with their output are required, as outlined in Sect. 4. The models that can be used are of various types: analytical, numerical, or based on artificial intelligence (AI) techniques (e.g.: neural networks, deep learning, and fuzzy systems).

Regarding analytical models, the relationship between springback, material parameters, bend angle, and bend radius has been widely investigated and is the subject of extensive literature. Most approaches, however, tend to be generic, without accounting for the specific characteristics of bending technologies. Li et al. [81] propose an analytical model capable of describing the neutral layer shift in the case of a tube subjected to superposed bending and stretching, with reference to the rotary draw bending (RDB) process. Ma et al. [82] extended the model to estimate springback even in the case of materials with different tensile and compressive responses. Although the analytical models, as the one proposed by Welo and Granly [83] enable a swift evaluation of how various variables influence tube bending and can be readily integrated into control systems, they do not consider key factors such as the weld seam, die friction, or the effects of transition zones at the ends of the bent area. A similar limitation affects the instability-focused models developed by Zhao et al. [84]. In this case, models have been developed to calculate the critical axial load based on the ratio between deformed length and bend radius, but instability phenomena are heavily influenced by boundary conditions, often oversimplified in analytical models. Numerical approaches can overcome these accuracy limitations. Yanagimoto et al. [85] demonstrate that finite-element simulations capture boundary-condition effects more realistically, while Zhu et al. [86] use such models to predict springback, Liu et al. [87] analyse wrinkling, and Safdarian [88] investigates fracture behaviour; similar strategies are also applied to cross-sectional ovalization. However, despite their accuracy and precision, these models are highly dependent on input data and often require long computation times, making them less suitable for in-line control systems and more appropriate for the design and optimization phase of the tools. Consequently, the need for reliable and fast models for in-line control systems has led many researchers to explore approaches aligned with what is nowadays considered artificial intelligence. Since the early 2000s, researchers have embraced AI-based strategies for tube-bending applications. Strano [89] introduced fuzzy-logic techniques for automated tooling design, and Lieh and Li [90] applied the same logic to straightforward in-line springback control. Around the same time, Shahin and Elchalakani [91] trained neural networks on experimental data to estimate design loads. Hybrid models that couple neural networks with FEM—first explored by Kami and Dariani [92], to predict wrinkling without costly failure tests—remain influential. This strategy is now widely used to forecast both wrinkling, as shown by Zhang et al. [93] and cross-sectional distortion, as demonstrated by Wang et al. [94] although its main role is still process-optimisation rather than fully fledged control. Recent approaches, like the Graph Neural Network (GNN) proposed by Wang et al. [95], simulate springback in the RDB process in real-time. This method effectively addresses the computational limitations of traditional FEM models without compromising accuracy, achieving maximum errors below 5%. Although FEM models remain crucial for generating training data, this GNN model represents a significant step forward towards the implementation of real-time control systems. An alternative approach to the development of fast simulation models is the one proposed by Borchmann et al. [96], which introduces a system based on Case-Based Reasoning (CBR). This approach involves conducting a sensitivity analysis campaign on the main process parameters to create cases that currently support the operator but can also be implemented in a control system. However, for this approach to be effective on an industrial scale, a significant exchange of data between clients and suppliers along the supply chain is required.

A separate consideration can be made for the Digital Twin of the bending process proposed by Li et al. [97], as illustrated in Fig. 11. In this approach, the Digital Twin integrates data from various sources, both historical and real-time data acquired and processed directly from the machine. Particular emphasis is placed on managing these data from different spatial and temporal domains to make them homogeneous and easily analysable. Li et al. [55] subsequently extended the model—whose declared goal is to predict tube quality in real time, focusing on cross-sectional distortion—to include wall-thickness reduction and stress-distribution predictions. While the use of this tool for springback correction or other defects is not mentioned, the development of DT certainly seems like a promising direction for addressing these issues.

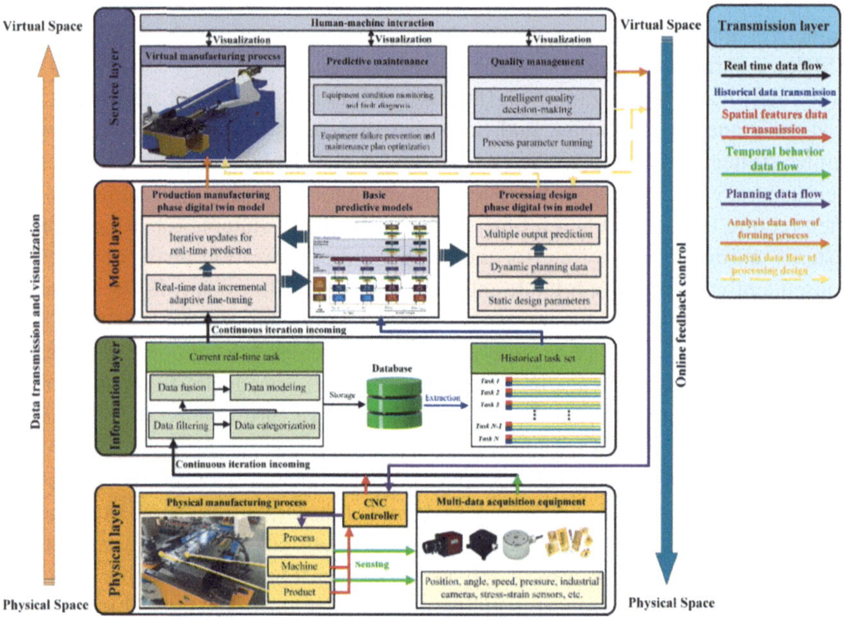

Fig. 11 Digital Twin of the tube bending machine [97]

In the final section, special attention is given to actuators. Albertelli and Strano [98] trace the early-2000s shift from hydraulic to electric drives, which cut energy use, boosted accuracy, and increased process speed. Yet the literature still lacks examples of equipment that adds new degrees of freedom for true in-line process control.

Interest is instead shifting toward flexible tooling. Kaleem et al. [99], accelerate small-batch production by 3D-printing pressure dies, while Li et al. [100] design diameter-adjustable mandrels that can adapt to either the tube's exterior or interior. A further frontier is thermal actuation. Ghiotti et al. [101] show that superimposing a localized thermal field can markedly reduce springback and tailor microstructure in high-strength-steel bends; Simonetto et al. [102] refine the concept using induction and Joule heating. Extending the idea, Li et al. [103] demonstrate that warm-bending strategies widen the bending window for several tube materials.

Flexible Bending Processes

Beyond RDB, there are numerous other bending processes used in technical practice that, in many cases, share the same control issues. This paragraph focuses on flexible bending processes aimed at achieving complex shapes while using standard tools. As highlighted in Sect. 2, the cost of tooling is often decisive in determining the minimum batch size for economically viable production. For this reason, the development of flexible technologies that minimize tooling costs is becoming increasingly attractive. A clear example of industrial success is the three-roll push bending process, the diagram of which is shown in Fig. 12a. These processes, which start from standard tools for a given tube diameter, achieve different bend radii and angles by adjusting the roller position. A similar principle underpins the Torque-Superposed Spatial (TSS) method introduced by Chatti et al. [105], presented in Fig. 12b, and the free-form bending approach detailed by Maier et al. [106] in Fig. 13. In the first case, the tool setup consists of three pairs of rolls that guide and transport the profile through the bending process, while a roll-based guiding system (bending head) defines the bending curve in the horizontal plane. The 3D bending contour is controlled by a superimposed torque, which has two functions: (i) to change the bending plane by twisting the profile, or (ii) to compensate for an undesirable twisting moment. Meanwhile, the second uses a bending die, orientable along three axes, through which the tube is pushed by a feeder. In a parallel with the RDB process, it is important to note that, an all these flexible processes, the tube is not constrained except in the bending area, and the springback does not occur at a precise and well-defined moment, but continuously as the tube exits the dies. Beyond the process limitations, while on one hand, operational flexibility increases, on the other hand, these technologies require more advanced control systems, as small variations in process parameters can lead to significant deviations of the workpiece parameters. In general, the greater the degrees of freedom of the system and thus its flexibility, the more complex and demanding the process control requirements become [8]. These processes find their main application in the production of freeform three-dimensional components, where the primary geometric parameters can be extremely challenging to define and measure accurately.

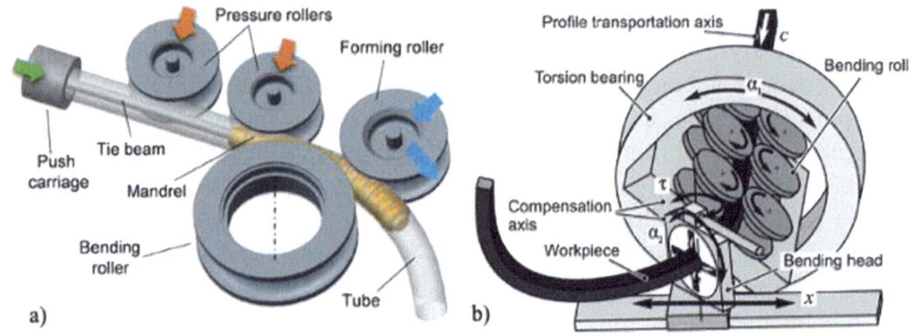

Fig. 12 Principles of flexible tube bending process: **a** three-roll-push (TRPB) bending [104] and **b** torque superposed spatial (TSS) bending [105]

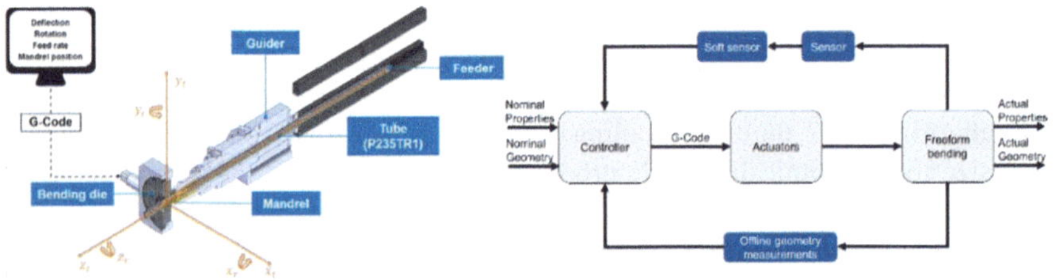

Fig. 13 **a** Principles of freeform tube bending process and **b** closed-loop control algorithm [106]

In three-roll push bending (TRPB), most studies determine the bending radius for simple planar bends with contact probes, as shown by Kawaguchi et al. [107] or with laser systems—either one-dimensional [108] or two-dimensional [109] dimensional. While Simonetto et al. [104], expanded the approach by embedding an inertial sensor in the mandrel; this not only measures the radius but also detects wrinkles, a capability they refined in a subsequent study in [77]. However, while the former systems are generally too bulky and expensive, in the latter case, a non-standard mandrel is required. An even more complex case is represented by TSS and free bending, specifically developed to obtain 3-dimensional profiles, which, however, entails a significant increase in the volume and measurement parameters. An attractive solution is the adoption of vision systems. Wang et al. [110] demonstrate that offline stereo-vision setups achieve high precision in measuring bend radii, yet Staupendahl et al. [111] show that the analysis time of such systems remains too long for integration into automatic control loops. On the other hand, in-line vision systems, as proposed by Scandola et al. [112] are limited to simple geometries and have only been applied at a laboratory level.

An alternative approach is to control the workpiece geometry through indirect measurements. This approach, proposed by Staupendahl et al. [113] for the case of TSS and illustrated in Fig. 14, involves measuring the applied bending and twisting torque. These process torque, through a semi-analytical model, are used to calculate

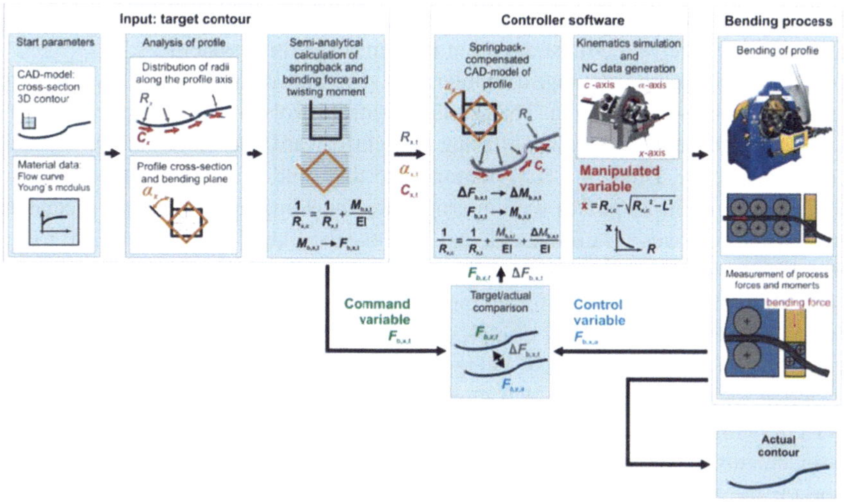

Fig. 14 Concept of indirect closed loop control in TSS, based on force measures [113]

the radii of curvature and thus to control the resulting geometry in a closed-loop system. Obviously, in this case, the reliability of the data available for the force to geometry transfer function is of fundamental importance to ensure the accuracy of the results. While geometric parameters are generally the most significant, they are not the only parameters of interest. For example, Maier et al. [106] developed a soft sensor for the free-bending process that controls residual stresses by linking them to the bending-die position and the hardness of the incoming material. Ismail et al. [114] hypothesizes the possibility of closing the control loop through in-line residual stress readings using a Barkhausen noise sensor. However, since residual stresses are correlated with tool position as well as the part geometry, it seems difficult to adjust one variable independently from the other. Similarly to the rotary draw bending process, a potential control system for actively modifying mechanical properties could involve the introduction of an actuator capable of applying a thermal field. Within the ProTuBend project [111], this approach is explicitly presented to produce high-strength steel profiles.

4.4 *Forming Process*

Once the tubes are formed through bending processes, which were previously discussed and are responsible for changing the shape in space, they then undergo a series of processes that alter their cross-sectional geometry. These include flanging, hydroforming, drawing, tube end forming, reduction and expansion, swaging, flaring, beading, and grooving. Among these, tube drawing and hydroforming are the processes that receive the most attention in terms of control systems.

In traditional tube drawing, a tubular rod is pulled through a die to adjust or change its cross-sectional area. The primary control involves adjusting axial tension to alleviate pressure on the dies, offering straightforward but limited flexibility and control. Similarly, Li et al. [115] demonstrate that cold drawing of ribbed tubes, carried out in multiple passes with intermediate annealing, achieves high-precision forming and removes rib-groove defects. To close the loop, Li et al. [116] add force sensors that track drawing loads and feed this information back to optimise process parameters, although local deformation control and overall system sophistication remain challenging. For more complex forming requirements, advanced systems like dieless drawing are utilized. Dieless drawing leverages fuzzy control systems integrated with vision sensors that monitor the tube outer radius and dynamically adjust drawing speed to improve dimensional accuracy. Despite these technological advancements, challenges such as oscillations in the profile persist as discussed by Supriadi and Manabe [117].

Hydroforming research shows a similar progression. Strano et al. [118] formulate time-dependent loading curves that can be determined adaptively during FEM-based planning. Manabe et al. [119] extend this idea by embedding fuzzy controllers and sensors in the die, dynamically adjusting the filling trajectory and sharply reducing wrinkles and bursts, while Endelt [12] confirms the benefits of real-time feedback from finite-element simulations.

Nakamori et al. [120] push the concept further for Y-shaped joints, combining high-accuracy contact sensors with adaptive fuzzy logic to refine material flow and improve dimensional precision.

Although most of these contributions concentrate on T- or Y-shaped parts—geometries often chosen for academic clarity—they collectively highlight a clear movement toward smarter sensing and control strategies that could be transferred to a wider range of complex tubular components.

4.5 Integrated Processes

The tube manufacturing industry has seen substantial advancements through the integration of subsequent steps in the process chain, resulting in enhanced accuracy, efficiency, and cost-effectiveness. Traditionally, tube production involves discrete processes such as cutting, bending, etc., each of which is completed in sequential stages and with very little automation between subsequent steps. However, manufacturers are increasingly adopting integrated methods that combine multiple processes in a single workflow, effectively reducing setup times, labor costs, and potential sources of error, as presented in the concept shown by Fig. 15.

One notable example of process integration in tube manufacturing is an approach that combines laser cutting and bending operations into a single, seamless process. This integrated method addresses a common challenge in tube manufacturing: the

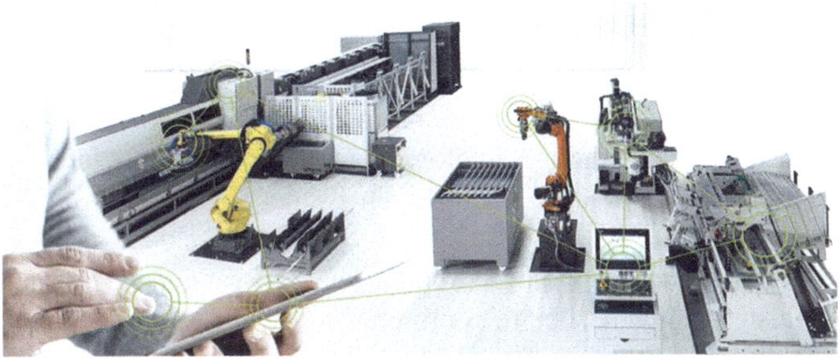

Fig. 15 Connected production line (e.g., hole position compensation in cutting—bending stages) (courtesy of BLM)

misalignment of holes and slots after bending. When a tube is bent, the material elongates, causing any pre-cut holes to shift out of alignment. Implementing precise software algorithms that anticipate the distortion caused by bending, enable to automatically adjust the positioning of holes and slots to ensure that they align correctly after bending.

The tube manufacturing industry is advancing with the development of integrated cells that automate and connect laser cutting, storage, and bending processes. Traditionally completed as separate steps, these processes are now unified, reducing manual intervention, improving precision, and enhancing efficiency. Central control software coordinates each stage, allowing real-time monitoring and seamless workflow adjustments, making production faster and more consistent.

The process begins with a laser cutting unit programmed for high-precision cuts based on CAD files. Sensors ensure alignment and accuracy, and the unit's integration with subsequent stages prevents the need for manual transfers, reducing setup time and potential errors.

After cutting, tubes are transferred to a robotic storage system that organizes them efficiently, ensuring readiness for the bending process. The storage unit automation reduces downtime, allowing for a smooth and continuous workflow. Tubes are then moved to the bending unit, which works in sync with the cutting unit. The software anticipates elongation from bending, so holes or slots are pre-adjusted for accurate alignment in the finished product. The central control software is key, orchestrating each step and enabling data-sharing across units. This software monitors operations and makes adjustments as needed, enhancing product quality and minimizing waste.

Looking forward, the future of process integration in tube manufacturing will likely involve greater use of artificial intelligence (AI) and machine learning. These technologies could enable machines to learn from each production cycle, refining their processes over time to improve accuracy and efficiency further. Additionally, as more manufacturers adopt integrated systems, the cost of these technologies may decrease, making them accessible to a broader range of companies.

5 Vision and Perspectives

The adoption of advanced control systems in profile and tube manufacturing is supported by several established technological trends, particularly in sensors, actuators, and computational models, each driven by developments in key industrial sectors.

The evolution of sensor technologies has been largely influenced by the consumer electronics industry, which has led to continuous improvements in performance and cost reduction. These advancements enable wider integration of high-precision inline measurement systems, supporting real-time monitoring and closed-loop control strategies that enhance process efficiency and stability.

In parallel, the electrification of industrial systems, also driven by the transformation of automotive powertrains, has spurred the development of more efficient and compact electric actuators. The increasing replacement of hydraulic and pneumatic systems with electric alternatives offers higher precision, faster response times, and improved energy efficiency, making real-time adaptive control more effective.

At the computational level, the rise of AI-driven models is shaping new control strategies, improving process predictions and enabling real-time optimization of operating parameters. Advances in computational power allow the deployment of digital twins that dynamically integrate real-time data with predictive simulations, enhancing process stability and adaptability.

Despite these advancements, several challenges still hinder widespread implementation. Fragmentation within the manufacturing sector creates obstacles in data ownership and interoperability, as different companies manage machines, sensors, and production separately. The absence of standardized data-sharing protocols further complicates cross-system integration.

Another major challenge is the need for specialized expertise. While automation reduces manual intervention, it requires skilled personnel to design, implement, and manage AI-driven control systems. Advanced control systems need experts skilled in both data analysis and manufacturing processes. Understanding constraints, machine capabilities, and production dynamics is essential to fully exploit these control technologies. This increasing reliance on data analysis and digital infrastructure poses a barrier, particularly in traditional manufacturing environments. In this context, training and education provided by higher education institutions play a critical role. These institutions must equip engineers and technicians with a comprehensive skill set that integrates data science with process engineering knowledge, enabling effective real-world applications. Additionally, architectural constraints remain a limiting factor. Many existing machines were not designed for extensive sensor integration or real-time control, making retrofitting complex and costly. The required investments for system upgrades or new equipment can slow adoption, especially in capital-intensive industries.

Nevertheless, the alignment of technological progress with industry demands is driving the transition toward more digitalized, adaptive, and sustainable manufacturing systems. Overcoming barriers related to data integration, workforce specialization, and system adaptability will be crucial for enabling widespread adoption of advanced control solutions, ultimately enhancing production efficiency and competitiveness.

6 Conclusions

The manufacturing of profiles and tubes involves a complex sequence of forming, cutting, and joining operations, often managed by different entities along the supply chain. In this context, advanced control systems represent a fundamental enabler for improving process efficiency, product quality, and resource optimization, offering solutions to long-standing challenges related to variability, precision, and sustainability.

This paper has analyzed the state of the art of control systems in the reference production chain, highlighting their key components—sensors, actuators, and computational models—and their integration within industrial processes. The discussion covered different control architectures, from open-loop systems to fully adaptive solutions, showing how modern approaches leverage in-line measurement, predictive modeling, and AI-based optimization to enhance process stability and flexibility.

Through case studies, the paper has also examined the main industrial applications, demonstrating how control systems can mitigate common issues such as springback in bending, dimensional accuracy in profiling, and defect formation in hydroforming and cutting operations. Practical implementations reveal both benefits and challenges. Key issues include integrating systems into existing machine architectures and evaluating cost–benefit trade-offs when introducing new control elements. The discussion on trends and challenges has highlighted key factors influencing the adoption of advanced control systems. While technological progress in sensors, actuators, and AI-driven models favors their implementation, structural barriers remain. These include sector fragmentation, which complicates data sharing and process integration, the demand for specialized expertise, and the constraints imposed by legacy equipment, often designed without real-time adaptive control in mind.

In conclusion, while advanced control systems are becoming increasingly central to modern manufacturing strategies, their widespread adoption still requires overcoming technical, organizational, and economic challenges. Future developments should focus on standardizing data exchange protocols, improving the interoperability of control architectures, and fostering multidisciplinary competencies to support the transition toward more autonomous and adaptive production systems. By addressing these aspects, the industry can fully exploit the potential of advanced control, achieving higher precision, efficiency, and sustainability across the entire production chain.

Acknowledgements The authors express their gratitude to Blm Group S.p.A., Condevo S.p.A., Gasparini S.p.A., and Ricos Srl for their valuable discussions, insights, contributions, and guidance during the preparation of this manuscript.

References

1. Bowen DT, Russo IM, Cleaver CJ et al (2022) From art to part: learning from the traditional smith in developing flexible sheet metal forming processes. J Mater Process Technol 299
2. Bleicher F, Biermann D, Drossel WG et al (2023) Sensor and actuator integrated tooling systems. CIRP Ann 72:673–696
3. Kreuzer T, Papapetrou P, Zdravkovic J (2024) Artificial intelligence in digital twins—a systematic literature review. Data Knowl Eng 151
4. ElMaraghy H, Monostori L, Schuh G, ElMaraghy W (2021) Evolution and future of manufacturing systems. CIRP Ann 70:635–658
5. Allwood JM, Duncan SR, Cao J et al (2016) Closed-loop control of product properties in metal forming. CIRP Ann Manuf Technol 65:573–596
6. Hauschild MZ, Kara S, Røpke I (2020) Absolute sustainability: challenges to life cycle engineering. CIRP Ann 69:533–553
7. Campi F, Favi C, Mandolini M, Germani M (2019) Using design geometrical features to develop an analytical cost estimation method for axisymmetric components in open-die forging. In: Procedia CIRP. Elsevier B.V., pp 656–661
8. Yang DY, Bambach M, Cao J et al (2018) Flexibility in metal forming. CIRP Ann 67:743–765. https://doi.org/10.1016/j.cirp.2018.05.004
9. Polyblank JA, Allwood JM, Duncan SR (2014) Closed-loop control of product properties in metal forming: a review and prospectus. J Mater Process Technol 214:2333–2348
10. Wiebenga JH, Weiss M, Rolfe B, Van Den Boogaard T (2012) A numerical approach to robust in-line control of roll forming processes. In: 14th International conference on metal forming, METAL FORMING 2012
11. Havinga J, van den Boogaard T, Dallinger F, Hora P (2018) Feedforward control of sheet bending based on force measurements. J Manuf Process 31:260–272. https://doi.org/10.1016/j.jmapro.2017.10.011
12. Endelt B (2022) In-process feedback control of tube hydro-forming process. Int J Adv Manuf Technol 119:7723–7733. https://doi.org/10.1007/s00170-022-08683-6
13. Li Y, Liu D, Li D et al (2024) Multisource disturbances rejection control for linear systems with unmeasured states under process and measurement disturbances. J Franklin Inst 361:106737
14. Aguiar APVDA, Ramos ETG, Acioli G, Barros PR (2024) Frequency data driven optimal design of different PI control structures for a multivariable process. In: IFAC-PapersOnLine. Elsevier B.V., pp 376–381
15. Lim Y, Venugopal R, Ulsoy AG (2010) Multi-input multi-output (MIMO) modeling and control for stamping. J Dyn Syst Measur Control Trans ASME 132:1–12. https://doi.org/10.1115/1.4001332
16. Tatipala S, Wall J, Johansson C, Larsson T (2020) A hybrid data-based and model-based approach to process monitoring and control in sheet metal forming. Processes 8:89. https://doi.org/10.3390/pr8010089
17. Santos S, Deschamps F (2024) Industry 4.0 enablers provide a typical architecture for industrial manufacturing data. In: Procedia CIRP. Elsevier B.V., pp 37–42
18. Vieira RD, Arroyo F, Idarilho GNN (2024) Line pipe manufacturing processes. Handbook of pipeline engineering, pp 311–365
19. Asnafi N, Lassl G, Nilsson T (2003) Tubular hydroforming of automotive side members with extruded aluminium profiles. J Mater Process Technol 142:93–101. https://doi.org/10.1016/S0924-0136(03)00467-9

20. Liu QWC (2010) Pipe-assembly approach for aero-engines by modified particle swarm optimization. Assem Autom 30:365–377
21. Aresti L, Florides GCP (2018) A review of the design aspects of ground heat exchangers. Renew Sustain Energy Rev 92:757–773
22. Seda M (2009) Three strategies for combating the corrosion of steel pipes carrying desalinated potable water. Desalination 238:166–173
23. Merklein M, Wieland M, Lechner M et al (2016) Hot stamping of boron steel sheets with tailored properties: a review. J Mater Process Technol 228:11–24. https://doi.org/10.1016/j.jmatprotec.2015.09.023
24. Ahmetoglu M, Altan T (2000) Tube hydroforming: state-of-the-art and future trends. J Mater Process Technol 98:25–33
25. Miller GG (2003) Tube forming processes: a comprehensive guide. SME Society of Manufacturing Engineers, Dearborn, Michigan (US)
26. Carvalho SM, Baptista CARP, Lima MSF (2016) Fatigue in laser welded titanium tubes intended for use in aircraft pneumatic systems. Int J Fatigue 90:47–56. https://doi.org/10.1016/j.ijfatigue.2016.04.018
27. Kumar A, Vincelli J, Gupta MC, Sapp M (2010) A study on laser cleaning and pulsed gas tungsten arc welding of Ti-3Al-2.5V alloy tubes. J Mater Process Technol 210:64–71
28. D'Urso G, Longo M, Giardini C (2013) Microstructure and mechanical properties of friction stir welded AA6060-T6 tubes. Key Eng Mater 554–557:977–984. https://doi.org/10.4028/www.scientific.net/kem.554-557.977
29. Pais M, Rao P (2021) An up-to-date review on industrially significant inhibitors for corrosion control of zinc. J Bio Tribocorros 7:1–13
30. Presuel-Moreno F, Jakab MA, Tailleart N et al (2008) Corrosion-resistant metallic coatings. Mater Today 11:14–23
31. (2025) Piping and fittings market size to hit USD 493.97 Billion by 2034. https://www.precedenceresearch.com/piping-and-fittings-market
32. Europe Pipe Market Size, Share, Growth|Report, 2034. https://www.marketresearchfuture.com/reports/europe-pipe-market-6222
33. Europe Pipe Market Size, Share & Industry Trends Analysis By 2030. https://www.databridgemarketresearch.com/reports/europe-pipe-market?srsltid=AfmBOoocb5QHnthzvL0spElC_6MsTd8aIuu269VywziBqbm05QNOT1fd
34. Italy: share of export value of steel tubes by macro-region 2014–2019|Statista. https://www.statista.com/statistics/1027249/distribution-of-value-of-steel-tubes-exported-from-italy/
35. Italy: steel tubes, pipes and fittings: producers by legal form|Statista. https://www.statista.com/statistics/537388/steel-tubes-pipes-and-fittings-producers-by-legal-form-italy/
36. Perera YS, Ratnaweera DAAC, Dasanayaka CH, Abeykoon C (2023) The role of artificial intelligence-driven soft sensors in advanced sustainable process industries: a critical review. Eng Appl Artif Intell 121
37. Galetto M, Mastrogiacomo L, Maisano D, Franceschini F (2015) Cooperative fusion of distributed multi-sensor LVM (large volume metrology) systems. CIRP Ann Manuf Technol 64:483–486
38. Cao J, Brinksmeier E, Fu M et al (2019) Manufacturing of advanced smart tooling for metal forming. CIRP Ann 68:605–628
39. Lee J, Kim D, Quagliato L et al (2017) Change of the yield stress in roll formed ERW pipes considering the Bauschinger effect. J Mater Process Technol 244:304–313. https://doi.org/10.1016/j.jmatprotec.2017.01.022
40. Kawamoto K, Ando H, Yamamichi K (2018) Application of servo presses to metal forming processes. In: Procedia manufacturing. Elsevier B.V., pp 31–38
41. Dang VT, Labergère C, Lafon P (2019) Adaptive metamodel-assisted shape optimization for springback in metal forming processes. Int J Mater Form 12:535–552
42. Havinga J, Mandal PK, van den Boogaard T (2020) Exploiting data in smart factories: real-time state estimation and model improvement in metal forming mass production. Int J Mater Form 13:663–673. https://doi.org/10.1007/s12289-019-01495-2

43. Havinga J, van den Boogaard T (2015) Inverse identification of process variations for thin steel sheet bending. Key Eng Mater 651–653:1363–1368. https://doi.org/10.4028/www.scientific.net/KEM.651-653.1363
44. Prates PA, Marques AE, Oliveira MC, Fernandes JV (2019) Comparing metamodeling techniques for variability analysis in sheet metal forming processes. In: Esaform, p 120003
45. Strano M, Semeraro Q, Iorio L, Sofia R (2018) Hierarchical metamodeling of the air bending process. J Manuf Sci Eng 140:1–34
46. Ingarao G, Marretta L, Di Lorenzo R (2015) A comparison between three meta-modeling optimization approaches to design a tube hydroforming process. In: Key engineering materials
47. Ben Abdessalem A, El-Hami A (2014) Global sensitivity analysis and multi-objective optimisation of loading path in tube hydroforming process based on metamodelling techniques. Int J Adv Manuf Technol 71:753–773
48. Naceur H, Arif W (2021) Modeling and Optimization in manufacturing by hydroforming and stamping. In: Modeling and optimization in manufacturing. Wiley, pp 13–47
49. Sözen L, Guler MA, Bekar D, Acar E (2012) Investigation and prediction of springback in rotary-draw tube bending process using finite element method. Proc Inst Mech Eng C J Mech Eng Sci 226:2967–2981
50. An H, Green D, Johrendt J, Smith L (2013) Multi-objective optimization of loading path design in multi-stage tube forming using MOGA. Int J Mater Form 6:125–135
51. Liu L, Wu J, Wang M et al (2024) A novel acquisition strategy of process parameters for spatial tubes with variable curvature formed by free bending technology. Int J Adv Manuf Technol 132:4069–4086. https://doi.org/10.1007/s00170-024-13568-x
52. Wang Z, Gu Y, Zhang S et al (2024) A transferred hybrid surrogate model integrating Gaussian membership virtual sample generation for small sample prediction: applications in metal tube bending. Eng Appl Artif Intell 129:107560. https://doi.org/10.1016/j.engappai.2023.107560
53. Bucconi M, Strano M (2018) Metamodeling and optimization of the THF process with pulsating pressure. In: AIP conference proceedings. American Institute of Physics Inc.
54. Sun C, Wang Z, Zhang S et al (2022) Digital-twin-enhanced metal tube bending forming real-time prediction method based on multi-source-input MTL. Struct Multidiscipl Optim 65
55. Li J, Wang Z, Zhang S et al (2025) Digital-twin virtual model real-time construction via spatio-temporal cascade reconstruction for full-field plastic deformation monitoring in metal tube bending manufacturing. Robot Comput Integr Manuf. https://doi.org/10.1016/j.rcim.2024.102860
56. Simonetto E, Ghiotti A, Bruschi S (2021) In-process measurement of springback in tube rotary draw bending. Int J Adv Manuf Technol 112:2485–2496
57. Alvarez W (2006) Roll form tool design: fundamentals. Industrial Press
58. Yan Y, Li Q (2010) FEM modeling and mechanics analysis of flexible roll forming. Appl Mech Mater 44–47:132–137
59. Gao JF, Li Q, Zhao W (2013) Thermal stress analysis for local heating variable cross-section roll forming. Adv Mat Res 683:599–603
60. Ren Y, Dong J, He J et al (2024) A novel six-dimensional digital twin model for data management and its application in roll forming. Adv Eng Inform 61:102555
61. Abouridouane M, Bergs T, Schraknepper D, Wirtz G (2021) Friction behavior in metal cutting: modeling and simulation. Procedia CIRP 102:405–410. https://doi.org/10.1016/j.procir.2021.09.069
62. Möhring HC, Menze C, Drewle K et al (2023) A novel approach for simulating a sawing process with reduced simulation time. CIRP J Manuf Sci Technol 42:72–80. https://doi.org/10.1016/j.cirpj.2022.12.019
63. Duflou JR, Fallahi Sichani E, De Keuster J, Kruth J-P (2009) Development of a real time monitoring and adaptive control system for laser flame cutting 1203
64. (2024) IPGphotonics. https://www.ipgphotonics.com/it/products/beam-delivery/laser-processing-heads/laser-cutting-heads. Accessed 13 Feb 2025
65. (2021) Power line—Prima industrie. https://www.primapower.com/files/primapower/attachments/Power%20Line_lug21_IT.pdf. Accessed 6 Feb 2025

66. Ghiotti A, Simonetto E, Bruschi S, Savio E (2021) Enhancing the accuracy of in-process springback measurements of complex tube bending processes using cost-effective embedded sensors. Meas Sci Technol 32
67. Clarke KW (1967) Metal forming machine
68. Blurton-Jones TJ (1999) Tube bending apparatus and method
69. Morikawa SKY (1974) Device for detect a bend angle upon precisely bending a metallic pipe
70. Wamemiinde R, Berndt D (2009) Method and device for determining
71. Ghiran et al. (2009) Bending apparatus and method of bending a metal object
72. Ha T, Welo T, Ringen G, Wang J (2022) A strategy for on-machine springback measurement in rotary draw bending using digital image-based laser tracking. Int J Adv Manuf Technol 119:705–718. https://doi.org/10.1007/s00170-021-08178-w
73. Saegusa S (1994) Tube bending machine
74. Watanabe et al (1994) Bend correction apparatus and method
75. Simonetto E, Ghiotti A, Bruschi S (2015) Feasibility of motion-capture techniques applied to tube bending. In: Key engineering materials. Trans Tech Publications Ltd, pp 1128–1133
76. He X, Welo T, Ma J (2024) In-process, real-time monitoring of forming forces in rotary draw bending process. Int J Adv Manuf Technol. https://doi.org/10.1007/s00170-024-14370-5
77. Simonetto E, Ghiotti A, Bruschi S (2017) Dynamic detection of tubes wrinkling in three roll push bending. In: Procedia engineering. Elsevier Ltd., pp 2316–2321
78. Simonetto E, Ghiotti A, Bruschi S, Gemignani R (2017) Dynamic detection of instability defects in tube rotary draw bending. In: Procedia manufacturing. Elsevier B.V., pp 319–328
79. Borchmann L, Frohn-Sörensen P, Engel B (2020) In situ detection and control of wrinkle formation during rotary draw bending. In: Procedia manufacturing. Elsevier B.V., pp 589–596
80. Borchmann L, Heftrich C, Knoche J et al (2023) Control of material flow using measuring methods for wrinkle and crack detection during rotary draw bending. In: Procedia CIRP. Elsevier B.V., pp 857–862
81. Li H, Ma J, Liu BY et al (2018) An insight into neutral layer shifting in tube bending. Int J Mach Tools Manuf 126:51–70
82. Ma J, Li H, Fu MW (2021) Modelling of Springback in tube bending: a generalized analytical approach. Int J Mech Sci 204
83. Welo T, Granly B (2010) A new adaptive bending method using closed loop feedback control. Transactions of nonferrous metals society of China (English Edition) 20:2111–2117
84. Zhao C, Han Z, Du B et al (2020) Wrinkling prediction of aluminum alloy tubes during reduced diameter compression forming. Int J Adv Manuf Technol 106:65–75
85. Yanagimoto J, Banabic D, Banu M, Madej L (2022) Simulation of metal forming—visualization of invisible phenomena in the digital era. CIRP Ann 71:599–622. https://doi.org/10.1016/j.cirp.2022.05.007
86. Zhu YX, Liu YL, Yang H, Li HP (2013) Improvement of the accuracy and the computational efficiency of the springback prediction model for the rotary-draw bending of rectangular H96 tube. Int J Mech Sci 66:224–232
87. Liu N, Yang H, Li H et al (2015) An imperfection-based perturbation method for plastic wrinkling prediction in tube bending under multi-die constraints. Int J Mech Sci 98:178–194
88. Safdarian R (2020) Investigation of tube fracture in the rotary draw bending process using experimental and numerical methods. IntJ Mater Form 13:493–516
89. Strano M (2005) Automatic tooling design for rotary draw bending of tubes. Int J Adv Manuf Technol 26:733–740
90. Lieh J, Li WJ (2005) On-line fuzzy logic control of tube bending. In: Intelligent systems in design and manufacturing VI. SPIE, p 59990D
91. Shahin M, Elchalakani M (2008) Neural networks for modelling ultimate pure bending of steel circular tubes. J Constr Steel Res 64:624–633. https://doi.org/10.1016/j.jcsr.2007.12.001
92. Kami A, Dariani BM (2011) Prediction of wrinkling in thin-walled tube push-bending process using artificial neural network and finite element method. In: Proceedings of the Institution of Mechanical Engineers, Part B: Journal of Engineering Manufacture, pp 1801–1812

93. Zhang S, Yuan Y, Wang Z et al (2022) A hierarchical prediction method based on hybrid-kernel GWO-SVM for metal tube bending wrinkling detection. Int J Adv Manuf Technol 121:5329–5342. https://doi.org/10.1007/s00170-022-09691-2
94. Wang C, Wang Z, Zhang S et al (2023) Reinforced quantum-behaved particle swarm-optimized neural network for cross-sectional distortion prediction of novel variable-diameter-die-formed metal bent tubes. J Comput Des Eng 10:1060–1079. https://doi.org/10.1093/jcde/qwad037
95. Wang Z, Wang C, Zhang S, et al (2024) Towards high-accuracy axial springback: mesh-based simulation of metal tube bending via geometry/process-integrated graph neural networks. Expert Syst Appl 255
96. Borchmann L, Kuhnhen C, Frohn P, Engel B (2019) Sensitivity analysis of the rotary draw bending process as a database of digital equipping support. In: Procedia manufacturing. Elsevier B.V., pp 592–599
97. Li J, Wang Z, Zhang S, et al (2024) Task incremental learning-driven digital-twin predictive modeling for customized metal forming product manufacturing process. Robot Comput Integr Manuf 85
98. Albertelli P, Strano M (2017) Tube bending machine modelling for assessing the energy savings of electric drives technology. J Clean Prod 154:83–93. https://doi.org/10.1016/j.jclepro.2017.03.212
99. Kaleem MA, Steinheimer R, Frohn-Sörensen P et al (2024) Additive manufacturing of polymeric pressure die for rotary draw bending process. Int J Adv Manuf Technol 134:1789–1804. https://doi.org/10.1007/s00170-024-14221-3
100. Li J, Wang Z, Zhang S et al (2023) A novelty mandrel supported thin-wall tube bending cross-section quality analysis: a diameter-adjustable multi-point contact mandrel. Int J Adv Manuf Technol 124:4615–4637. https://doi.org/10.1007/s00170-023-10838-y
101. Ghiotti A, Simonetto E, Bruschi S (2021) Insights on tube rotary draw bending with superimposed localized thermal field. CIRP J Manuf Sci Technol 33:30–41. https://doi.org/10.1016/j.cirpj.2021.02.012
102. Simonetto E, Venturato G, Ghiotti A, Bruschi S (2018) Modelling of hot rotary draw bending for thin-walled titanium alloy tubes. Int J Mech Sci 148:698–706. https://doi.org/10.1016/j.ijmecsci.2018.09.037
103. Li G, He Z, Ma J et al (2021) Springback analysis for warm bending of titanium tube based on coupled thermal-mechanical simulation. Materials
104. Ghiotti A, Simonetto E, Bruschi S, Bariani PF (2017) Springback measurement in three roll push bending process of hollow structural sections. CIRP Ann Manuf Technol 66:289–292
105. Chatti S, Hermes M, Tekkaya AE, Kleiner M (2010) The new TSS bending process: 3D bending of profiles with arbitrary cross-sections. CIRP Ann Manuf Technol 59:315–318. https://doi.org/10.1016/j.cirp.2010.03.017
106. Maier D, Stebner S, Ismail A et al (2021) The influence of freeform bending process parameters on residual stresses for steel tubes. Adv Ind Manuf Eng 2
107. Kawaguchi et al (1988) Roller bending apparatus equipped with a curvature measuring unit
108. Ingvarsson (2008) Machine for bending off long products and a method to control such a machine
109. Caporusso et al (2010) Machine for continuously bending an elongated workpiece at predetermined radii
110. Wang X, Liu J, Liu S et al (2018) Accurate radius measurement of multi-bend tubes based on stereo vision. Measurement (Lond) 117:326–338
111. Staupendahl D, Becker C, Hermes M et al (2014) Flexible and cost-effective innovative manufacturing of complex 3D-bent tubes and profiles made of high-strength steels for automotive lightweight structures EUR 26688 EN
112. Scandola L, Böhm V, Maier D et al (2024) An inline point-tracking approach for the real-time monitoring of the free-form bending process. Adv Ind Manuf Eng 9
113. Staupendahl D, Chatti S, Tekkaya AE (2016) Closed-loop control concept for kinematic 3D-profile bending. In: AIP conference proceedings. American Institute of Physics Inc.

114. Ismail A, Maier D, Stebner S et al (2021) A structure for the control of geometry and properties of a freeform bending process. In: IFAC-PapersOnLine. Elsevier B.V., pp 115–120
115. Li W, Liu S, Feng P et al (2023) Application of shaped tube as preform to eliminate the forming defects of ribbed tube in multi-pass drawing. J Manuf Process 92:247–261. https://doi.org/10.1016/j.jmapro.2023.02.052
116. Li W, Wang B, Liu S et al (2024) Cold precision forming and microstructure evolution of ribbed tube for nuclear fuel cladding by multi-pass drawing. J Mater Process Technol 326
117. Supriadi S, Manabe K (2013) Enhancement of dimensional accuracy of dieless tube-drawing process with vision-based fuzzy control. J Mater Process Technol 213:905–912
118. Strano M, Jirathearanat S, Altan T (2001) Adaptive FEM simulation for tube hydroforming: a geometry-based approach for wrinkle detection. CIRP Ann 50:185–190. https://doi.org/10.1016/S0007-8506(07)62101-5
119. Manabe KI, Chen X, Kobayashi D, Tada K (2014) Development of in-process fuzzy control system for T-shape tube hydroforming. In: Procedia engineering. Elsevier Ltd, pp 2518–2523
120. Nakamori T, Shukuno K, Manabe K (2017) In-process controlled Y-shape tube hydroforming with high accurate built-in sensors. In: Procedia engineering. Elsevier Ltd., pp 43–49

Machining of Critical Aerospace Components: Challenges and Case Studies

Niccolò Grossi, Paolo Albertelli, Rachele Bertolini, Alessandra Caggiano, Gianni Campatelli, Matteo Cragnolini, Andrea Lelio Maurizio, Luigi Nele, Paolo C. Priarone, Antonio Scippa, Marco Sortino, Giovanni Totis, and Sandro Turchetta

Abstract The aerospace industry in Italy is an increasingly important contributor to economy, with a highly skilled workforce and the presence of companies operating globally. Machining processes constitute a fundamental part of the manufacturing supply chain, with a global market size exceeding 22 billion euros in 2020. However, evolving industry demands for components with increasingly stringent requirements necessitate optimized and reconfigurable machining processes. This White Paper

N. Grossi (✉) · G. Campatelli · A. Scippa
Department of Industrial Engineering, University of Florence, Firenze, Italy
e-mail: niccolo.grossi@unifi.it

P. Albertelli
Department of Mechanical Engineering, Politecnico Di Milano, Milano, Italy

R. Bertolini
Department of Industrial Engineering, University of Padova, Padova, Italy

A. Caggiano
Department of Industrial, Electronic and Mechanical Engineering, University of Roma Tre, Roma, Italy

M. Cragnolini
Laboratorio BS, Raveo, Italy

A. L. Maurizio
Pietro Rosa TBM, Maniago, Italy

L. Nele
Department of Chemical, Materials and Production Engineering, University of Naples Federico II, Napoli, Italy

P. C. Priarone
Department of Management and Production Engineering, Politecnico Di Torino, Torino, Italy

M. Sortino · G. Totis
Polytechnic Department of Engineering and Architecture, University of Udine, Udine, Italy

S. Turchetta
Department of Civil and Mechanical Engineering, University of Cassino and Southern Lazio, Cassino, Italy

L. Fratini et al. (eds.), *Selected Topics in Manufacturing*, Lecture Notes in Mechanical Engineering, https://doi.org/10.1007/978-3-031-99501-9_12

investigates advanced methodologies for optimizing chip removal processes in the manufacturing of aerospace components and outlines key challenges and potential solutions through case studies, aiming to provide practical insights for industrial players. The paper focuses on critical components such as mirrors, aeroengine airfoils, and aerospace structures addressing issues like deflection/deformation, vibrations and surface quality/ integrity.

Keywords Machining processes · Aeronautical & aerospace · Stiffness · Vibration · Surface Integrity

1 Introduction

The aerospace sector stands as a beacon of technological progress and innovation, a tradition spanning over a century. This expansive industry encompasses various domains, including commercial aviation, defense and military applications, as well as space exploration and satellite endeavors. Its global reach fosters collaboration among nations, stimulating exports and driving economic growth. In 2022, the Aerospace and Defense (A&D) sector reported a revenue of $741 billion, underscoring its substantial contribution to the global economy [1]. Moreover, the year 2023 saw a resurgence in demand for products within the A&D sector. The aerospace industry experienced a recovery in domestic commercial aviation, with revenue passenger kilometers exceeding pre-COVID-19 levels in most nations, resulting in heightened demand for new aircraft as well as aftermarket goods and services [2]. In the United States a major role has been played by the defense sector, due to emerging geopolitical tensions, coupled with a focus on military modernization, fueled strong demand throughout 2023 [3]. And the demand for A&D goods and services is anticipated to remain strong throughout 2024, encompassing both civilian and military sectors. As highlighted by Deloitte in its "Aerospace and defense industry Outlook", the A&D companies are facing major challenges that will require the development of new solutions and approaches [4]. These are:

- Talent development: A&D companies are addressing workforce challenges amid rising demand and changing workforce expectations;
- Supply chain: global supply chain complexity is driving a multitiered solution;
- Digital transformation: The industry can unlock growth and efficiency through digital technologies;
- New product introduction: evolving customer preferences and growing emphasis on sustainability are fueling product innovations;
- Defense and commercial spending: industry spending is helping drive growth and innovation.

Several of these challenges will necessitate a novel organizational structure for companies, whilst others will be closely associated with the manufacturing aspect, specifically, the demand for innovative products/materials and the requirement to implement digital models throughout the design and manufacturing process.

In Europe, despite the Covid-19 pandemic, the aerospace industry persevered, maintaining its position as a global leader with €238 billion in revenue in 2021, providing employment for 879,000 individuals [5]. While large enterprises predominantly reside in a few Member States of European Union (E.U.) such as France, Germany, Italy, and Spain, the industry boasts a network of dynamic small and medium enterprises (SMEs) throughout the E.U., many of which are world leaders in their respective fields [6]. To maintain competitiveness, companies invest significantly in research and development, striving to create cutting-edge products that meet the rigorous demands of the aerospace sector that strives to overcome the modern challenges, one among them fuel efficiency. These activities are supported by numerous European Union initiatives, one of the most significant being the "Clean Sky Initiative", which supports the development of new, cleaner technologies for the A&D sector.

Aerospace parts manufacturing emerges as a pivotal sub-sector, with a global market size estimated at $913.13 billion in 2023. North America currently dominates this sector (51.7%), yet Europe's market size, reaching €273.3 billion in 2023, is poised for growth surpassing that of the U.S. by 2030, driven by increased defense spending, by some European countries, among them Italy [7]. Of the first 10 companies of the sector the majority is from U.S., two are French (Airbus, Safran), and one Italian (Leonardo) [8]. Furthermore, European industry has played a significant role in developing novel services that rely extensively on space infrastructures, encompassing telecommunications, navigation, and Earth observation. Various sectors of the economy, including transport, telecommunications, media, and public entities, benefit from these capabilities, consequently stimulating innovative downstream activities.

Italy has a strong heritage in the aerospace sector, holding the distinction of being the third nation, following the U.S.S.R. and the U.S., to successfully launch a satellite into orbit [9]. As a founding member of the European Space Agency (ESA), Italy ranks as the third-largest contributor, investing €589.9 million in 2021 [10]. The nation hosts numerous specialized aerospace clusters and districts, strategically positioned across 11 regions, collectively forming the Italian National Aerospace Technology Cluster (CTNA). According to the CTNA, the Italian aerospace industry ranks seventh in the world and fourth in Europe in terms of revenue (€16.4 billion in 2018) and is also one of 9 countries with a space agency with a budget of more than $1 billion per year [10]. The Italian aerospace sector contains 514 active business, that employ nearly 47,300 people combined. According to the European Space Agency (ESA), Italy is the 6th leading global space power, with total space industry revenue amounting to €2.23 billion-14% of total revenue- in 2018 (about 600 businesses that employ around 7000 people [9]).

In terms of activities, the aerospace sector is highly heterogeneous and expansive, encompassing a diverse range of fields from aircraft construction and assembly to the design, testing, and implementation of space services, as well as the transportation of goods and passengers. The Italian aerospace industry comprises 543 companies (Istat, 2020), with the majority (89%) engaged in the manufacturing, repair, and maintenance of aircraft, spacecraft and related devices, along with satellite telecommunications.

The national context is characterized by a significant concentration of large entrepreneurial entities, which are complemented by numerous Small and Medium-sized Enterprises (SMEs) primarily situated in five regional districts (Campania, Lazio, Lombardy, Piedmont and Puglia). According to extant industry studies (SRM, 2019; ICE 2019), these SMEs are highly specialized, representing micro-centers of excellence, valorized on an international scale through continuous collaboration with large national companies, research centers, and universities. These synergies have enabled the sector to attain a prominent position in the fields of aeronautics and space on a global scale. This industrial landscape, predominantly comprised of SMEs, is characterized by a strong propensity for innovation. However, it lacks the scale to conduct structured R&D activities, necessitating the development of research centers capable of supporting the activities of this industrial ecosystem.

This sector is also a significant contributor to the Italian economy, and the revenue associated with R&D companies is substantially higher than other sectors, as reported by CESI [11]. The multiplication factor is 2.6, approximately 71% higher than the mean value of other Italian sectors. As for Deloitte [12], the sector is going to grow dramatically also due to the new product required by the market, that in the near future could open new large market opportunities due to their widespread application, such as electric-powered aircraft, urban air mobility, and the new space economy.

Regarding Italy, these emerging market demands are projected to result in a turnover increase of 7–8 billion in 2025–2030 [13]. The regions most involved in A&D sectors have already established policies to support such growth. It is particularly noteworthy to examine the Campania Region, where priority development lines indicate that the primary sector to support growth is Smart Manufacturing. This sector places special emphasis on advanced design, advanced manufacturing processes, digitalization of manufacturing processes, advanced maintenance. Among these innovation lines, manufacturing processes play a predominant role. The aerospace industry is one of the most technologically advanced industries in the world and has very high precision requirements, dimensional and geometrical tolerances, to ensure efficiency, but also safety. Although several manufacturing processes are involved in the aerospace components supply chain, and the sector is prone to the adoption of innovative manufacturing processes (e.g., Additive Manufacturing [14]), machining is by far the most implemented solution for achieving such stringent requirements. Aerospace machining market has reached globally the value of $21.5 billion in 2020 for metal machining and $1.9 billion for aerospace machine tools [15]. Aerospace machining plays a crucial role in enhancing the performance, reliability, and safety of modern aircrafts. Its importance stems from its capacity to manufacture components with high-level mechanical properties, tolerance and surface finish. The rapid advancement of the aerospace sector is intricately linked to the high-quality and efficient manufacturing of aerospace parts and components.

1.1 Aim and Scope

The scenario highlights how the aerospace sector stands as one of the cornerstones of advanced engineering, characterized by a strong emphasis on innovation and research. Due to precision, accuracy, performance, and safety in aerospace components, machining processes still play a strategic role. Moreover, the increasingly stringent requirements of the machined parts require continuous process optimization (e.g., optimal parameters selection, special machining strategies and fixturing). This investigation is still often conducted using standard trial and error approaches, thus providing neither optimized nor reconfigurable solutions, leading to high number of scraps, and hence high costs. As example, an analysis by Boeing [16] assessed that rework and scrap costs due to part distortions exceeded $290 million across four aircraft programs, also revealing that thin-walled aerospace parts have a 47% likelihood of geometrical or dimensional non-conformities after machining. The scientific community, often in close collaboration with industry, has investigated and provided structured solutions and predictive tools to address machining issues. This white paper is therefore targeted to some of the latest and innovative strategies for the machining of aerospace components. It outlines key challenges and potential solutions also through significant case studies, aiming to provide practical insights for industry players. Since the "machining for aerospace components" is a broad topic, this paper focuses on conventional machining processes and their variants, such as turning (including Single Point Diamond Turning (SPDT) [17]), milling (including multi-axis milling, mirror milling [18] and Ultra-Precision Fly Cutting (UPFC) [19]) and drilling (including Vibration-Assisted Twist Drilling (VATD) [20]). These processes are essential for manufacturing a diverse array of aerospace components, ranging from conventional ones, such as aircraft engine parts and aerostructures, to specialized components designed for space applications. The literature sources show that the most critical aerospace components that must undergo machining operations may have the following requirements.

1. complex geometries and thin-walled parts characterized by a low structural stiffness, posing challenges due to their high static deflections and vibrations;
2. tight tolerances that can be difficult to achieve using non-optimized conventional machining processes;
3. high-quality surface finish, reaching the nanometric scale for some components, presenting challenges both in manufacturing and metrology;
4. high-performance materials such as titanium, nickel-based alloys, and composites, known for their low machinability.

This work is dedicated to some representative categories of such critical aerospace parts: telescope mirrors, aero-engine airfoils, and aerospace structures. These categories are presented in Table 1 together with the specific components, materials and machining processes commonly used. Section 2 introduces such components and discusses their criticalities. From a product quality perspective, the machining processes must overcome issues related to distortions, vibrations and surface integrity. Such issues are discussed in Sect. 3 along with potential methodologies/solutions as well as open challenges and future trends. Finally, Sect. 4 presents manufacturing case studies that highlight the practical implementation of solutions applied to the machining of aerospace components and materials.

A schematic of the structure of this work is provided in Fig. 1, highlighting the topics analyzed and the ones addressed in the case studies.

Table 1 Investigated critical aerospace components

Category	Component	Materials	Investigated processes
Telescope mirrors	Metal mirror	Pure nickel, aluminum, Mg alloys	SPDT, milling, UPFC
	Non-metal mirror	Glass ceramic, Si, AMC	
Aero-engine airfoils	Blade	Cast Al alloys, Ti aluminide, nickel alloys, cobalt alloys, AMC	Milling, Turn-milling
	Impeller	TiAl alloys, nickel alloys	5-axis milling, SAM
	Blisk	TiAl alloys, titanium matrix composite, nickel alloys	
Aerospace structures	Fuselage panel	Cast Al alloys, HTD Al alloys, AMC CFRP	Mirror milling, Drilling
	Aircraft skin	Cast Al alloys, HTD Al alloys, non-oxide matrix CMC, AFRP	Mirror milling, Drilling
	Wing frame	PMC, CFRP	Milling, Drilling

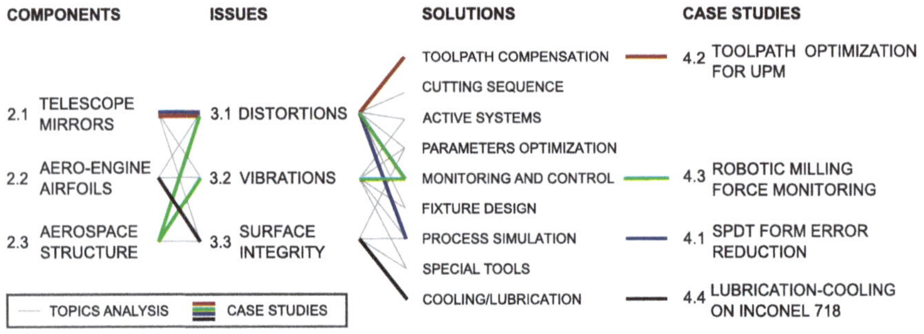

Fig. 1 Structure of the work, highlighting the connection between the selected aerospace components, their investigated issues and the identified solutions categories. The topics of the case studies are highlighted with thicker lines

2 Critical Components and Materials

2.1 *Telescope Mirrors*

Space optical systems, particularly astronomical telescopes, are crucial for advancing our understanding of the universe. Mirrors are the main component of many space telescopes. They combine the stringent mass and stiffness requirements typical of space components, with the extreme requirements in terms of surface finish needed for optical performance. Fabricating these mirrors requires unique technologies due to their different curvatures and specifications, which directly impact telescope performance. The mirrors consist of a substrate and often require optical coatings. The substrate's material selection is critical, considering factors like stiffness, thermal stability, and environmental resistance. Common materials include various glasses, silicon carbide (SiC), aluminum and beryllium. Glass, fused silica, and Zerodur® are cost-effective choices for commercial applications. However, weight limitations with glass have led to the development of segmented and active mirrors. SiC is increasingly used in manufacturing large-aperture mirrors due to its excellent physical and mechanical characteristics, such as low density, high strength, chemical stability, and low coefficient of thermal expansion. However, the material's extreme hardness makes it difficult to machine, causing rapid tool wear and inducing surface shape errors. Steadily increasing interest is being put in the development of fully metallic mirrors due to their easier machinability, improved thermal behavior and lower cost with respect to vitreous ones, as well as the possibility to build the entire telescope with the same material [21]. A typical material of choice for these kinds of mirrors is aluminum, which poses several advantages in the operative phase. On the other hand, such material is soft, hence very susceptible to deformations.

Optical coatings are commonly adopted for achieving desired optical performance (e.g., reflectivity). Coatings typically include multiple layers of materials like silver, aluminum, nickel, gold, or beryllium. Mirrors can have different shapes depending on their spectral band requirements, with aspheric and freeform mirrors gaining popularity for their design flexibility. The production of such components usually consisted of three main machining phases: (i) rough traditional machining, (ii) ultra-precision machining, (iii) post-polishing. The latter is a very slow process (i.e., removal rate tenths of micron per minute) [22], thus requiring starting surface errors in the order of microns. For this reason, it is critical to reduce as much as possible the surface errors coming from the previous phase (ultra-precision machining). Single Point Diamond Turning (SPDT) is one of the most used ultra-precision machining processes, which is especially indicated for the axial-symmetric surfaces that characterize most mirrors' reflective surfaces.

Such a process is an analogue to traditional turning, using a very small diamond tool and extremely precise machine tools. Using this process, it is possible to obtain excellent surface finishes, with roughness in the order of $1 \div 10$ nm and shape error down to $0.1 \div 1$ μm RMS [23, 24], particularly when working on vitreous mirrors. Ultra-Precision Fly Cutting (UPFC) is an alternative, especially in case of non-axial-symmetric surfaces. UPFC employs a single point cutting rotating tool and it is generally used to generate optical quality surface finishes on flat workpieces producing surfaces with sub micrometric form error and nanometer roughness [19].

Several issues arise in the manufacturing of mirrors mainly related to the very strict requirements in terms of tolerance and surface roughness. Mechanical vibrations occurring during the machining process is one factor compromising the surface roughness [25]. However, the most important problem is related to the deformation of the workpiece, affecting the tolerances and causing optical aberration. Such deformation is the result of several concurrent causes:

- Cutting forces generated during the process [26]. For standard cutting parameters, such forces in ultra-precision machining are very low, and their influence is significant only in case of very aggressive parameters or very low-stiffness components;
- Clamping/Fixturing that induces a deformation on the reflective surface of a mirror [21] due to screw tightening [27] mounting operation [28] and coupling tolerances. Their influence is very case-dependent [29, 30] and the careful design of the support system and clamping torque could drastically reduce this issue [31];
- Centrifugal forces affect the turning operations (i.e., SPDT) deforming the machined workpiece. To reduce such an effect, the reduction in rotational speed is a generally applicable approach. Additionally, the increase of supporting points can be applied, however this could even be detrimental [32];
- Gravity could deform the mirror during machining affecting its surface, and this is especially valid in SPDT where the direction of gravity acceleration varies with respect to a reference frame integral with the workpiece [33].

Deformation is critical especially for mirrors made of aluminum because of its low elastic modulus. Additionally, the difficult polishing characteristics of aluminum alloy materials pose a major challenge for existing polishing techniques [34]. Another issue is tool wear [35], especially in SiC mirror polishing for which advanced technologies such as ultrasonic vibration-assisted fixed-point grinding [36], and compensation strategies [37, 38] have been proposed to tackle this issue. Finally, the brittleness of SiC also poses a challenge in machining, affecting the mirror surface integrity [39]. A summary of the main issues is provided in Table 2.

Table 2 Telescope mirrors main issues

Main Issue	Description	Relevant application
Distortions	Cutting forces, clamping forces, centrifugal forces and gravity affect the stringent requirements of tolerance and roughness	SPDT and polishing of metallic mirrors
Vibrations	Vibrations coming from spindle and guideways, stability of tool and workpiece	SPDT of metallic mirrors
Surface integrity	Mirror surface integrity is critical for silicon carbide polishing	Polishing of SiC mirrors

2.2 Aero-Engine Airfoils

Gas turbines serve as the workhorses of various industries, including power generation, aviation, and oil and gas. Central to the efficiency and performance of gas turbines are turbine and compressor blades. These components must withstand high temperatures, extreme pressures, and corrosive environments while maintaining optimal aerodynamic profiles. Due to their complex geometry, high dimensional accuracy and superior mechanical properties, engineering and manufacturing of such blades are non-trivial tasks and require advanced manufacturing techniques. These components are classified based on several factors, including their application, construction and aerodynamic characteristics [40]. Axial blades, commonly used in aircraft engines and power generation turbines, have a straight flow path parallel to the axis of rotation. The geometry of blades is critical to their performance.

Blade profiles are meticulously designed to optimize aerodynamic efficiency, minimize flow losses, and withstand mechanical stresses. The airfoil shape, camber, twist, and chord length are tailored to specific operating conditions and performance requirements (Fig. 2a). Typically, blades in a turbine are divided into two categories: vanes (i.e., stator blades) and rotor blades. In addition to aerodynamic and heat-induced stresses, rotor blades must also withstand significant centrifugal forces.

For this purpose, a meticulous control of their shape and of their surface condition is required. Modern CFD techniques are extensively used to determine blading and annulus (i.e., inner and outer ring) shapes. These techniques have implied a large increase in the complexity of all blading, a heavy reduction of dimensional tolerances and strict requirements in terms of surface finish. Materials used for turbine and compressor blades must have high-temperature resistance, mechanical strength, and corrosion resistance. The most used materials are listed below.

- High alloy steels and particularly stainless steels such as AISI 420, 410, 316, 321, 17-4PH, 13-8Mo, 15-5PH, Greek Ascoloy and more. There are also cast-specific grades, which are commonly cast in molds to produce the blade blank,

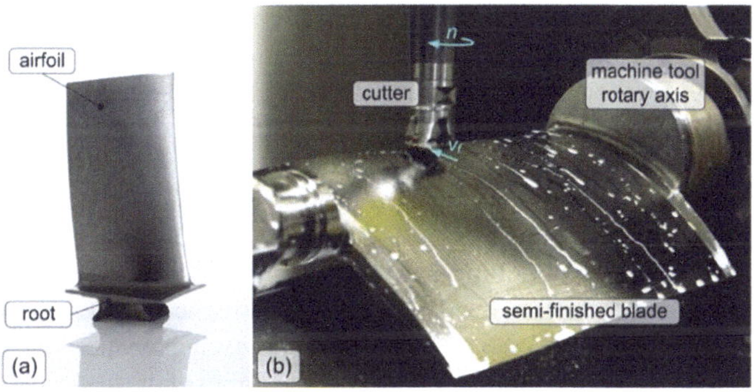

Fig. 2 **a** Typical turbine blade geometry and nomenclature and 5-axes milling of a turbine blade **b** Courtesy of Pietro Rosa TBM

used especially for the hot section (turbine) of the engine, sometimes including internal cooling and aerodynamic cavities inside the thickness of the blade;

- Nickel-based superalloys are renowned for their high-temperature strength and oxidation resistance. Common wrought Nickel superalloys comprise IN718, A-286, IN625, X-750, Incoloy 901, Waspaloy, Udimet 700, Nimonic 80-A.;
- Titanium alloys offer excellent strength-to-weight ratios and corrosion resistance. Most important examples are Ti6Al4V, Ti6-2-4-2, Ti8-1-1, Ti10-2-3, Ti6-2-4-6;
- Other advanced materials such as gamma-Titanium Aluminides and Ceramic Matrix Composites (CMCs), which are particularly suitable for these high-temperature applications due to their exceptional thermal stability and lightweight properties.

The manufacturing process of turbine blades typically involves casting or forging as the primary methods for shaping the initial blade blanks. These methods impart the desired grain structure and mechanical properties to the blades. Theoretically, additive manufacturing techniques could be used for producing near-net-shape blades. Nevertheless, additive manufacturing tends to cause internal defects and residual stresses, which are hardly tolerated by the aerospace sector. However, thanks to process optimization and to other significant advances in this field, Laser Powder Bed Fusion (L-PBF) and Electron Beam Melting (EBM) technology are being increasingly utilized for producing turbine and compressor blade. 3D printed turbine blades for aeronautical applications can be found in the two low-pressure turbine stages of the GEnX motor. However, after the primary production processes, secondary machining operations are required to achieve the final dimensions and surface finish, especially for blades of medium to large size. For this purpose, 5-axes milling on modern CNC milling machines is mostly adopted (Fig. 2b) [41, 42].

Several challenges persist in the manufacturing of turbine and compressor blades. One significant challenge is controlling and reducing tool wear, especially when machining hard-to-machine materials like nickel-based superalloys and ceramics. Stiff machine tools, innovative tool materials, coatings and cooling strategies can mitigate tool wear and extend tool life. Also, the ceramic composites are difficult-to-cut materials because of their brittleness, abrasiveness, and thermal conductivity during machining. Developing optimized cutting parameters, tool geometries, and machining strategies tailored to specific ceramic compositions is crucial for enhancing process efficiency and product quality.

Moreover, achieving strict requirements in term of tight dimensional tolerances, surface finish and surface integrity is very challenging in blade machining. For example, machining accuracy will depend on the effective material allowance left by the previous semi-finishing operation, and it will be severely affected by the relaxation of internal stresses during machining. Another source of inaccuracy will be the tool quasi-static deflection, while excessive relative tool-workpiece vibrations may greatly affect surface quality. In particular, self-excited chatter vibrations may frequently arise in this field, because of the high flexibility of the blade, of the cutting tool or of both elements. Therefore, special cutting tools, devices and methodologies

Table 3 Aero-engine airfoils main issues

Main Issue	Description	Relevant application
Distortions	Internal stress relief and quasi-static deflection under cutting loads of tool and thin-wall workpiece affect the accuracy of part	Milling of thin-walled airfoils/blades or of shrouded impellers using slender tools
Vibrations	Chatter vibrations arising because of flexibility of both workpiece and tool	
Surface integrity	Difficult machinability of challenging materials. Stringent requirements in terms of surface quality	Milling of blades made of ceramic composite or nickel-based alloys

have been developed in the last decades to avoid or suppress chatter occurrence, especially when milling blades.

Surface integrity is another problem affecting blade machining. The blade surface after the final passes must exhibit a very good morphology and microstructure at the microscopic level, free from detached, reattached, or distorted grains, and without intergranular cracks. This requirement extends to all tools generating the final surfaces and all surfaces within 0.5 mm with reference to the finished geometry, thus including most semi-finishing operations. Additionally, for most new blade geometries any final polishing after milling is prohibited, unless it is fully automated (such as mass finishing or vibro-tumbling), preventing the masking of any superficial defects. This necessitates extremely stable and carefully controlled cutting conditions to avoid even the smallest surface defect. This issue is closely related to the evaluation of tool life. It is essential to predetermine tool life to avoid using a worn cutting edge for these demanding milling operations. A summary of the main issues is provided in Table 3.

2.3 *Aerospace Structures*

The evolution of modern aerostructures has been largely driven by the need for fuel efficiency, weight reduction, and improved aerodynamic performance. Aerostructure materials must support aircrafts for long-term use (60,000 flight hours), handling both static weight and additional service loads. These materials need to balance low density for weight reduction with mechanical properties able to stand extreme temperature conditions (from −30 to 370 °C), high humidity or dry environments, and UV radiation [43]. The design and selection criteria for aerostructure materials are defined by the specific application. For example, aircraft wings must endure bending from static weight, dynamic loads from maneuvers and additional stresses from the landing gear and control surfaces during taxiing, take-off and landing. The wing upper surface undergoes compression in flight and tension during taxiing, with the opposite for the lower surface, thus demanding materials with high tensile and compressive strength [44]. The fuselage faces high cabin pressure and shear loads, requiring materials with high tensile and shear strength [45]. The need to face the challenge of aerodynamic performance requirements while reducing weight has prompted extensive research and development into lightweight metal alloys and composite materials, with a primary focus on high-strength aluminum alloys, titanium alloys, and carbon fiber reinforced polymers (CFRP), which offer a combination of low density, high strength, and excellent fatigue resistance [45].

Aluminum-based alloys, such as 2024 Al alloy, are commonly utilized in fuselages due to their moderate yield strength (324 MPa), good fracture toughness (37 MPam1/2), and high elongation rate (21%). These alloys are widely employed in aerostructures due to their availability, lightness, strength and manufacturability, and retain strength in thicker sections due to their lower quench sensitivity. The 7010 and 7050 Al alloys, known for their high strength, stress corrosion cracking resistance and toughness, are ideal for heavy plate applications. The 2024 Al alloy, offering high strength and excellent fatigue resistance, is also commonly used. New generation Aluminum–Lithium (Al–Li) alloys, such as Alcoa 2099 and Constellium AIRWARE™ 2050, have been recently used in modern aircrafts like the Airbus A320 NEO and A330 NEO [46].

Titanium alloys, with their medium density, good strength-to-weight ratio, oxidation resistance at intermediate temperatures, and low coefficient of thermal expansion, are increasingly used in modern aircrafts such as the Airbus A350XWB and Boeing B787. These alloys are compatible with carbon fiber composites, reducing risks of galvanic corrosion and thermal expansion issues. The most widely utilized alloys include alpha (α), beta (β), and alpha–beta (α/β) Ti–6Al–4 V. Beta tempered Ti–6Al–4 V is chosen for damage tolerance performance, while Ti–3Al–2.5 V is used in anti-crash frames. For landing gear applications, Ti–10 V–2Fe–3Al has replaced high-strength low-alloy steel in the Boeing 777, and Ti–5Al–5 V–5Mo–3Cr (Ti-5553) is used in the Boeing 787 [47].

Lightweight composite materials, such as carbon fiber reinforced polymers (CFRP) are increasingly used in aircraft structures their high strength (3450–4830 MPa), high elastic modulus (224–241 GPa), and high-temperature capability (290–345 °C) [48]. CFRP components like panels, stringers, frames, spars, skins, and panels provide reduced operational costs, environmental impact, and fatigue and corrosion-free properties. Modern aircrafts like the Airbus A350 XWB and Boeing 787 Dreamliner feature a structural weight composition of roughly 50% composites, 20% aluminum, 15% titanium, and 6–10% steel [49].

Machining of aerostructure components involves a range of processes tailored to the advanced materials employed which, while offering superior performance, often pose significant challenges in terms of machinability. The high precision and stringent quality requirements of the aerospace industry necessitate advanced machining techniques able to ensure the integrity and performance of each component [46]. Machining processes for aluminum alloys require an accurate tool selection, where carbide tools with coatings such as titanium nitride (TiN) or diamond-like carbon (DLC) are often used to enhance tool life and performance. Coolants are employed to manage heat generation and prevent thermal damage to the workpiece. Additionally, advancements in cutting tool geometries and coatings contribute to further improve the machinability of aluminum alloys, enabling higher cutting speeds and prolonged tool life. High-speed machining (HSM), allowing for higher cutting speeds and feed rates, is employed to enhance productivity and achieve superior surface finishes [46]. Titanium alloys present significant challenges in machining due to their low thermal conductivity, high strength and tendency to work-harden which generate high cutting forces, rapid tool wear, and poor surface finish [46]. Due to the low thermal conductivity of titanium alloys, the heat generated during cutting is not efficiently dissipated, leading to high temperatures at the cutting zone which can cause thermal damage to the workpiece and accelerated tool wear.

Machining of composite materials, particularly CFRP, involves unique challenges due to their anisotropic and inhomogeneous nature. Conventional machining processes can cause delamination, fiber pull-out, thermal damage and poor surface finish [50, 51]. With reference to machining of metal-alloy thin-walled components used in aircraft structures, such as those located in aircraft fuselages and wings, significant challenges are related to surface integrity issues, machining deformation, and machining chatter. The combined effects of cutting forces and heat can lead to surface integrity issues ranging from surface deterioration defects and metallurgical alterations affecting material properties to residual stress causing part deformation due to uneven stress distribution and low part stiffness [41]. Poor surface quality reduced geometric precision and potential residual stresses occur. Moreover, machining chatter, i.e., self-excited vibration occurring during the cutting process, poses significant challenges to ensuring reliable clamping and stable machining and can lead to reduced dimensional accuracy and degraded surface quality, substantially impacting the workpiece assembly and performance. Chatter can also induce additional vibration in the cutting tool, causing increased tool wear and reduced tool lifespan. In this context, fixtures play a vital role in machining by ensuring positioning, clamping, vibration suppression, protection, and cooling, directly impacting the part quality. To enhance rigidity and minimize deformation, both external holding devices, which add support points for stiffness, and internal structures with reinforcements or fillers are utilized [52]. Among the machining processes used in aircraft manufacturing, drilling stands out as the most employed technology. As a matter of fact, in the assembly of a modern aircraft, several discrete aerostructures need to be connected, and mechanical joining techniques by means of rivets and bolts are widely employed. For instance, Boeing's 737 and Airbus's A320 families rely on approximately 600,000 and one million mechanical joints, respectively. The challenges associated with machining aerospace aluminum alloys, titanium alloys, and composite materials become even more complex when drilling multiple materials together, as in the case of hybrid stacks for aircraft assembly. Accordingly, efforts must be made to enhance drilling process quality, productivity and efficiency [46, 51]. Aircraft assembly lines utilize drilling operations including three key processes: drilling, reaming, and countersinking. In some cases, drilling and reaming are combined into a single operation known as One-Shot Drilling (OSD) [49]. A general goal when working with aerospace components composed of multiple layers stacked together is to achieve "One-up drilling", which involves drilling through all the layers in a single operation instead of drilling each layer separately and sequentially. However, this is challenging because each layer may necessitate different cutting speeds, feeds, and tool geometries to ensure the desired hole quality, accuracy, and finish are achieved consistently across all layers. Materials, assembly complexity and stringent quality standards in the aviation industry impose significant demands on cutting tools, susceptible to rapid tool wear and potential failure that can compromise the integrity and reliability of aircraft components.

Regarding surface integrity issues when drilling titanium alloys, burr formation at the tool entry and exit, as well as poor surface roughness, are major concerns. Burr height at the exit increases with cutting speed, while surface roughness is primarily affected by increased feed rate. Poor surface roughness can lead to microcracking, reducing fatigue resistance and increasing susceptibility to early fracture [53]. While 'rolled back' or 'leaned-back' burrs are typically observed in dry processing, uniform burrs without attachments were predominant when wet drilling Ti–6Al–4 V [54]. Moreover, the use of helical point drills with larger helix and point angles has been found to reduce burr size. This aligns with the findings in [55], where spiral drills with a 135° point angle produced detached caps with small residual burrs, whereas web-thinned conventional twist drills with a 118° point angle generated larger, non-uniform crown-shaped exit burrs. When drilling aluminum alloys, the high ductility of these alloys causes long chip formation and burrs at drilled hole entrance and exit due to plastic deformation. Two types of burrs, uniform and crown, can form at the exit of drilled aluminum plates, depending on the workpiece material properties, drilling parameters and cutting forces [56]. Entrance hole burrs result from material plastic flow, while exit hole burrs form due to material deformation caused by high compression rates at the hole center. Reducing burr height and width during drilling is essential to improve hole quality, minimize difficulties in fastener positioning and riveting, and lower manufacturing costs [56]. Burrs, typically harder than the original material due to strain hardening, impact part accuracy, surface quality, and assembly, sometimes resulting in rejection. High cutting temperatures and large drill point angles (120°–140°) exacerbate burr formation by increasing localized deformation and cracking. As burrs hinder assembly and handling, an additional deburring process is required, which is costly and time-consuming [57]. When stacked aluminum sheets are drilled together for aircraft assembly, interfacial burr formation is a common issue. Burrs develop at the interface between the sheets, requiring de-stacking and deburring before final assembly and fastener installation. Since drilling stacked sheets is frequent in aircraft manufacturing, minimizing or eliminating burrs can reduce costly, non-value-added operations. Research studies identified drill point angle, clamp type, and clamping distance as the most significant factors influencing interfacial burr formation [58]. Different clamping conditions affect the deflection of the workpiece during drilling, which is responsible for increased burr formation [59]. A summary of the main issues is provided in Table 4.

Table 4 Aerospace structures main issues

Main issue	Description	Relevant application
Distortions	Thin-walled structures could deform under the effect of cutting forces, clamping forces and residual stresses	Metal-alloy thin-walled components used in aircraft structures, e.g., fuselages and wings
Vibrations	Chatter can lead to reduced dimensional accuracy and degraded surface quality, tool wear and reduced tool lifespan	
Surface integrity	The combined effects of cutting forces and heat can induce surface integrity issues (e.g., surface deterioration defects, metallurgical alterations, residual stress) Drilling of titanium and composite materials can lead to burr formation and delamination	Drilling components made of composites Machining difficult-to-cut materials (e.g., titanium alloys)

3 Main Issues and Methodologies

3.1 Distortions

Machining distortion or deformation is the result of the combination of several factors that cause the differences in shape and size compared to the desired ones on the part after machining [60]. When aerospace components are machined, part distortion is a common issue that results in unexpected non-conformance parts, rework costs, and delays. First, several aerospace components present high flexible features, likely to be subjected to distortions. This is the case of thin components or elements used in aircraft structures, such as aircraft fuselages, wings, blades, engines casings, and thin fuel tanks, classified as Aerospace Thin-walled Components (ATWCs) [41]. However, even for aerospace parts that are not included in such category, distortions could be an issue. This happens in case of parts with very stringent tolerance, such as optical surfaces (e.g., telescope mirrors) in which unacceptable optical aberration could arise with very small distortions (i.e., order of nanometers). Finally, material is a crucial element for several reasons: *(i)* elastic modulus affects distortion, *(ii)* machinability of the material influences cutting forces, one of the main factors causing distortions, *(iii)* residual stresses of the bulk depend on manufacturing processes used before machining. In the aerospace sector, both soft material such as aluminum alloys and difficult-to-cut materials, such as nickel and titanium alloys are vastly employed. Four are the main sources of machining distortion: residual stresses, cutting loads, clamping loads, other loads.

Residual stresses include both Bulk or blank-initial Residual Stresses (BRS) and Machining-Induced Residual Stresses (MIRS) [61]. The first are the inherent tensions in the material blanks resulting from the part's processing history (e.g., rolling, forging, quenching, casting, and extrusion) thermo-mechanical loading and metallurgical changes, which leave the material with residual stresses. Stress-relieving treatments are frequently used however, total stress-free states are hardly reached.

Since in thin-walled aerospace components massive removal of material is generally achieved (90–95%) significant residual stresses are released causing the distortion of the part. BRS are considered one of the main causes of part distortion in the aerospace industry, especially for thin-walled parts [62]. On the other hand, MIRS are caused by the machining process itself due to the generation of high temperature, forces and strain fields in the cutting zone. The profile of MIRS (i.e., penetration and shape) strongly depends on machining condition (i.e., forces and heat) and material proprieties [63]. MIRS and BRS together determine the material's stable energy state, resulting in part deformations once the part is removed from the fixture following the final finishing process. Approaches aiming at tackling these issues for aerospace applications are mainly three: prediction, measurements and control. Predictive approaches are both numerical and analytical, with the latter much more employed and generally exploiting Finite Element Method (FEM). However computational efficiency and materials characterization are still open issues [62]. Analytical formulations are far more efficient, but they neglect some critical aspects of residual stresses such as tool wear or cutting fluid and are still limited to simplified scenarios (e.g., straight turning) [64]. Residual stresses are measured by traditional approaches such as layer removal or hole drilling [65] or more complex approaches estimating residual stressed by coupling models with part distortions measurements [66]. Such results are then used to evaluate predictive models or propose control strategies. Literature shows mainly three mitigation/control techniques aiming at optimizing: *(i)* location of the part in the stock [67], *(ii)* toolpath and cutting sequence [68], *(iii)* clamping loads using active fixture [69].

Cutting loads include both cutting forces and thermal load (i.e., heat). Cutting forces interact with the workpiece and the tooling system generating an elastic deformation that produce a dimensional error, generally called surface-form error [70]. This error generally depends on static deflection but could also be caused by forced vibrations [71]. Surface-form error is critical in case of using slender tools or machining thin-walled workpieces, in this last scenario resulting error is more difficult to be predicted since stiffness changes with position and material removal [72]. In addition, heat produced during the machining process generates temperature gradients and thermal expansion stresses, leading to thermal distortions [73]. This aspect is becoming even more critical with the recent trend of adopting dry cutting or at least reducing the usage of cutting fluids. Several works are dedicated to cutting loads modeling. Analytical cutting forces models have been proposed with high-level of generalization and accuracy. FE modeling of the cutting process is also possible but generally implemented for investigating different effects than distortions [74]. Thermal loads models are more complex to validated, since measurements is not straightforward. Thermal modeling approaches are mainly two: temperature estimation based on power consumption [75] and positioning heat sources on the shear plane and chip-tool contact generating the estimated heat [76]. Cutting loads are then applied to numerical or analytical models of workpiece and tool to predict distortions and deflections [72]. Off-line mitigation strategies in aerospace application include, in order of complexity: *(i)* parameters optimization, *(ii)* toolpath compensation and *(iii)* cutting sequence selection. The first approach modulates feed and spindle speed (i.e., when forced vibrations are significant) to meet tolerance of the part, toolpath is not changed. In toolpath compensation instead, distortions simulated by predictive models [77] are used to modify the actual toolpath to minimize part errors, thin-walled aerospace blade and casing are the common tested components. Finally cutting sequence optimization entails the generation of the toolpath and the definition of machining allowance based on the desired tolerance [72]. In addition, on-line strategy can also be followed using sensors to measure the distortion and adaptive strategy to achieve deflection reduction [78]. As for cutting forces, also clamping forces cause elastic distortions of the workpieces that once machined are released leading to geometric and dimensional errors [79]. Since clamping forces are essential to hold the part and act against cutting forces, optimal value and optimal position of clamps should be designed [80]. Off-line methodologies to reduce such distortions entail the definition of the optimal layout and pre-load based on simulation-derived data [81]. On the other hand, smart clamping is a promising on-line approach [82].

Other loads include special loads not comprised in the previous categories, such as centrifugal forces, gravity forces and aerodynamical forces that could be significant only in very specific scenarios, such as SPDT of optical surfaces. The centrifugal force acting on a machined workpiece during SPDT has been evaluated by Zhang et al. [32]. In their study they minimized such effect by reducing the spindle speed and inserting in the system multiple support points. While the reduction in rotational speed is a generally applicable result, the increase of supporting points is very case dependent and requires additional investigation. During the turning operation, the direction of gravity acceleration varies with respect to the workpiece reference system. Gottini et al. [33] included such effect in their predictive model, showing a significant influence on the SPDT of an aerospace mirror case study. The rotation of the workpiece and its supports may generate some aerodynamic forces on the system. However, they can be considered significant only for high spindle speeds, rarely adopted in SPDT.

One of the main open challenges of distortion control and mitigation methods in the aerospace sector is their implementation: industry demands for general and automatic approaches. Most of the approaches are tested for aluminum alloy components only, and the ones focused on other materials are mainly dedicated to ATWCs, very few works are dedicated to non-thin wall parts such as optical surfaces. In addition, research efforts are needed for improving simplified analytical models (especially for residual stresses) and comprehensive approaches that could tackle multiple sources of distortion. Recent works are dedicated to automation and integration of different methodologies, however, while some improvements have been achieved in terms of predictive approaches, control methodologies require additional investigation. Planning of the processes seems to be the more practical solution to the issue compared to using advanced equipment or on-line approaches. Finishing non-uniform no and toolpath optimization based on distortions seem to be promising concepts. Artificial Intelligent (AI) based techniques are growing and could take advantage of progresses of other fields that share the distortion issues, such as AM. These technologies paired with increased availability of data could be key enablers for advanced techniques that could act on the basis of data trend. However, data appear not enough for tackling such

Table 5 Summary of distortion issue in aerospace components

Causes	Main solutions	Relevant components
Residual stresses	Toolpath compensation Cutting sequence (location of the part in the stock) Active system (smart clamping)	Thin-walled components made of difficult-to-cut materials (e.g., airfoils)
Cutting loads (i.e., cutting forces and thermal loads)	Parameters optimization Toolpath compensation Cutting sequence Monitoring and control (adaptive strategy)	Thin-walled components (e.g., aerostructures and airfoil) or slender tools applications
Clamping loads	Fixture design (optimal layout and clamping forces) Active system (smart clamping)	Thin floors components (e.g., aerostructures) Double supported thin-walled parts (e.g., airfoils) Telescope mirrors
Other loads (e.g., centrifugal forces, gravity)	Monitoring and control Toolpath compensation Process simulation	Tight tolerance components (e.g., space applications)

issue that comes from multiple factors, therefore the data-based approaches should be paired with physic-based models. Table 5 shows a summary of the distortion issue as discussed in this section, along with the main solutions adopted.

3.2 Vibrations

Every machining operation is affected by relative dynamic vibrations between the tool and the workpiece, which can be classified as free oscillation, forced vibrations and self-excited chatter vibrations [83]. Free oscillations are unimportant here, while forced vibrations are negligible in turning—where cutting conditions are almost stationary—but they play a significant role in milling because of the periodic cutting forces exchanged between tool and workpiece. Forced vibrations can be attenuated (but not extinguished) by varying the spindle speed, such that the main harmonics of the cutting forces do not match the mechanical resonances of the machining system. Lastly, self-excited chatter vibrations consist in large, undesired vibrations having a different origin. They ruin the machined surface quality, accelerate tool wear and damage machine tool elements such as the main spindle bearings. The onset of chatter vibrations does mainly depend on the selected spindle speed n (rpm) and depth of cut a_p (mm). Accordingly, the (n, a_p) plane can be split into two regions, one stable and one unstable, as illustrated in Fig. 3. This is the so-called stability lobes diagram (SLD), which can guide the selection of optimal cutting parameters to enhance the material removal rate while assuring chatter-free conditions [84].

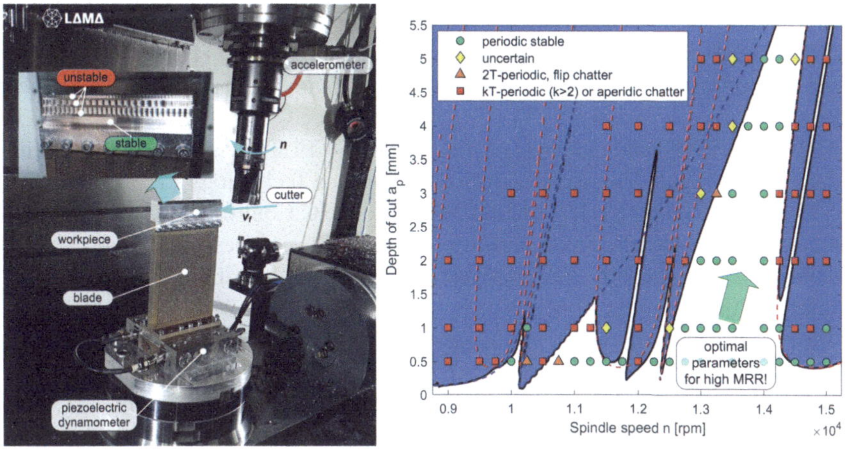

Fig. 3 Experimental setup for assessing the stability of a thin-walled, blade-like workpiece *(left)*. Obtained predicted stability lobes versus experimental points *(right)* [88]

Main chatter cause is the regenerative mechanism. In milling, it can be basically defined as the influence of the previous tooth vibration on the effective chip thickness perceived by the current tooth passing through the same angular position. In other words, the undulation left by the previous tooth passage on the machined surface affects the current tooth-workpiece engagement. Therefore, cutting forces can be altered, thus further influencing vibrations. This loop can reinforce itself causing large vibrations and forces, which tend to grow up exponentially until the loss of contact between tooth and workpiece occurs, and the process can start again. This chaotic regime is responsible for the severe problems outlined above. Thus, it must be absolutely avoided. Passive methodologies are mostly based on the integration of special mechanical elements into the tooling system or into the workpiece-fixture system for enhancing their stiffness or their damping [85]. Another passive solution in milling consists in the adoption of cutters with complex geometries, such as an irregular pitch between subsequent teeth [86], different or variable helix, or even the more exotic "serrated" cutting edge geometries [87].

Tuned mass dampers can be also installed inside special cutting tools or on the machine tool structure to counteract against vibrations [89]. Passive solutions are preferred in industry because they are relatively simple to implement and cheap. Alternatively, semi-active approaches can be applied for chatter suppression such as the spindle speed variation technique, consisting in the continuous modulation of the spindle speed around the nominal value, for disturbing chatter onset [90]. Active approaches do also exist, which are based on the real-time, closed-loop monitoring and compensation for vibrations by using specific sensors and actuators [91]. Both semi-active and active approaches are rarely used in industry, including those of the aerospace sector, because they are expensive, complex and require radical modifications of the machine tool architecture.

Hybrid techniques consist in the combination of sensor-based chatter detection techniques plus real-time adaptation/regulation of cutting parameters [92]. In this perspective, in the last 15 years some commercial solutions have been developed, such as: Okuma NAVI (Chatter Avoidance Control); Mazak SMOOTH TECHNOLOGY (Chatter Control); DMG MORI—CELOS with Vibration Control; FANUC—iHMI with Vibration Control (Fine Surface Technology); HAAS—Vibration Detection System (VDS); HEIDENHAIN—TNC Control with Dynamic Efficiency and Dynamic Precision and a few others. Nevertheless, automatic chatter detection and mitigation in real industrial scenarios is still difficult and not widespread as research literature may claim.

Last but not less important, predictive/preventive methods based on preliminary calculation of the SLD have a great potential for process optimization, especially when milling turbine blades or other thin-walled aeronautical structures [41, 42]. Nevertheless, a lot of input data are required to perform such calculations: nominal tool geometry and milling kinematics, dynamic compliances of tool and workpiece, cutting pressure of the workpiece material, etc. Some of these data are generally unknown and cannot be predicted with sufficient accuracy through CAE simulations. To the authors' knowledge, CUTPRO by MAL Inc. was one of the first commercial software tools capable of calculating stability lobes diagrams (SLDs). It is currently used by USA aerospace industry. Additionally, Third Wave AdvantEdge by Third Wave Systems offers similar simulation capabilities. In the last years a few CAM software products have incorporated features aimed at improving stability and reducing vibrations, but with no explicit capability of computing the SLD. However, according to authors' direct experience with such commercial software, their predictions are generally accurate and useful only in relatively simple case studies, but they can be rather inaccurate when considering more complex case studies.

Possible reasons of this unsatisfactory performance are unmodelled milling dynamics and the presence of relatively large uncertainties affecting the model coefficients used for SLD calculation. This may happen when no direct measurements can be carried out on the considered experimental setup, or in the presence of noisy data. Under such circumstances, deterministic SLD are inadequate. Robust or probabilistic SLD should be computed instead, by using special algorithms which are capable to take into account such uncertainties [93].

In addition, it is worth noting that most of the chatter prediction algorithms developed in the last 30 years are case-dependent, i.e., specialized to solve a given type of machining operation. Specifically, when addressing 5-axes finish milling on turbine/compressor blades or other thin-walled aeronautical structures, the calculation of the SLD is even more challenging because material removal does continuously alter the modal parameters of the workpiece. Moreover, the dynamic compliance does considerably vary from point to point on the blade, thus influencing the SLD. Stability boundaries do also depend on a greater number of factors (e.g., the tilt angle) that need to be adequately chosen [42], in comparison to the more trivial face milling case.

In conclusion there is still a lack of a good, general-purpose commercial software for SLD calculation in industry. However, thanks to the promises offered by

Table 6 Summary of vibration issue in aerospace components

Causes	Main solutions	Relevant components
Forced vibrations	Parameters optimization (e.g., spindle speed) Special tools and Fixture design (increasing the damping/stiffness of the machining system by using passive approaches)	Milling of thin-walled parts or when adopting slender tools
Chatter vibrations	Parameters optimization (based on SLD calculation) Special tools (e.g., variable helix, crest tools) Active systems (e.g., active damping) Monitoring and control (real-time parameters regulation, spindle speed modulation)	Milling of thin-walled parts Milling with slender tools Boring/internal turning Turning a slender workpiece

the implementation of a digital twin in next generation machine tools [94], it is expected that effective solutions for SLD calculation and preventive cutting process optimization will be soon available, especially for the aerospace sector.

Table 6 shows a summary of the vibration issue, as discussed in this section, including the main solutions adopted.

3.3 Surface Quality and Integrity

Field and Kahles first introduced the concept of surface integrity in the 1960s [95]. Surface integrity encompasses the surface layer's geometric characteristics and physical and mechanical properties. The geometric features of the surface are typically described by surface roughness, waviness, and knife pattern direction. Meanwhile, the physical and mechanical properties include residual stress, hardness, metallographic structures, and micro-structural changes. Surface integrity is crucial for aerospace components' operational quality, service life, and reliability [96]. This is especially relevant for difficult-to-machine metals such as nickel-based superalloys and titanium alloys, which constitute a significant portion of aircraft structural and engine components.

Machining these alloys is challenging due to their high toughness and work-hardening behavior [97, 98]. Numerous studies in the literature address issues related to machining nickel-based alloys, such as rapid tool wear and undesirable alterations to machined surfaces [99]. These alloys easily form a work-hardened layer in response to machining-induced strain loading on the subsurface. Additionally, their low thermal conductivity often increases temperatures at the tool cutting edge, causing adhesion of the workpiece material to the cutting edge and accelerated tool wear due to hard abrasive particles in the alloy's structure. The high localized heat, temperature gradients, and stress-induced pressures can also cause microstructural changes, creating a white layer or surface damage that can detrimentally affect the performance of the machined part.

The machining process significantly influences machined parts' surface integrity, including the process parameters, the tool, and cooling conditions [100]. Ensuring high surface integrity is crucial, as industrial standards prohibit severe plastic deformations and thermally abused layers on part surfaces. Surface integrity of aerospace alloys is significantly influenced by machining parameters, as evidenced by several studies on various alloys such as titanium, and nickel-based alloys [96]. The choice of tool material significantly affects cutting temperature, cutting force, and surface roughness, all critical components of surface integrity. While carbide tools are commonly used for machining nickel-based superalloys, their poor heat resistance limits cutting speeds and industrial efficiency. Ceramic tools, in contrast, offer high hardness, strength, and stability, making them a preferable alternative. Ceramic tools do not require cutting fluid, thus avoiding environmental pollution and operator risks from high-temperature fluid volatilization [101]. High-speed dry milling of Inconel 718 with SiAlON ceramic tools greatly improves productivity, reaching speeds up to 1000 m/min. Initially, fresh ceramic tools produce a smoother surface finish (Ra = 0.52 μm), but worn tools result in a rougher surface and high tensile residual stresses. Additionally, the process forms a 1–2 μm thick white layer due to rapid heating and cooling. Despite these issues, SiAlON tools are recommended for roughing, followed by carbide tools for finishing to enhance surface integrity [102]. Tool geometry is also critical; positive rake tools are recommended to reduce cutting forces and improve chip evacuation. Sharp cutting edges and a high rake angle can help mitigate the effects of work hardening.

Various sustainable cooling/lubrication methods, including cryogenic cooling with liquid nitrogen (LN_2), minimum quantity lubrication (MQL), and high-pressure coolant (HPC) lubrication have been studied for their effects on machining performance and surface integrity. Cryogenic cooling with LN_2 reduces tool wear, surface roughness, cutting temperature, and energy consumption, improving the machining characteristics of nickel-based superalloys [103]. Compared to dry cutting, MQL enhances surface integrity and tool life, with additional benefits including reduced cutting force, improved surface finish, lower cutting temperature, extended tool life, and reduced white layer thickness [104].

Recent advances in MQL for titanium machining indicate that its cooling effect is limited due to low flow rates (10–150 ml/h), making friction reduction the primary benefit [105]. To boost MQL cooling efficiency, several studies have developed a hybrid cryogenic fluid and MQL (cryo-MQL) method. Specifically designed nozzles and CFD simulations have optimized cryo-MQL systems, though performance was sometimes lower compared to conventional cutting [106]. Studies comparing cryo-MQL with other lubrication methods underscore its industrial potential, demonstrating the versatility of supercritical CO2-MWF jets in turning, drilling, and tapping [107]. Tests in turning steel and milling nickel-based alloys revealed promising results for tool life and surface quality, although conventional cooling still excelled in milling [108]. HPC lubrication, on the other hand, enhances tool life, reduces cutting forces, and improves chip fragmentation, all of which contribute to better surface integrity during machining [109].

The impact of harsh machining conditions, such as excessive tool wear without coolant, on surface integrity was evaluated during drilling operations on four nickel-based superalloys: Alloy 718, Waspaloy, Alloy 720Li, and RR1000 [110]. Regardless of the material, white layers with ultra-fine grains were formed. These layers were thicker in RR1000 (10 μm) and Alloy 720Li (4 μm) compared to Waspaloy (3 μm) and Alloy 718 (2 μm). Additionally, RR1000 and Alloy 720Li exhibited higher levels of plastic deformation and tensile residual stresses in their SPD layers. This explains why the aerospace industry discards tools before reaching their wear threshold to prevent compromising the surface integrity of safety–critical components due to potential cutting-edge damage [46].

In addition to costly experimental approaches, significant efforts have been made in conducting numerical simulations to improve product quality [111]. A physically based constitutive model to predict surface integrity during machining, considering different cutting parameters and cooling conditions, was developed in Ref. [112]. The model integrates dynamic recrystallization, hardening, and recovery effects to simulate the microstructural changes and mechanical behavior of Waspaloy. Similar results were found in Ref. [113], where a physics-based constitutive model integrated into FE simulations was used to predict surface modifications during machining. A less effort-intensive approach to evaluating the surface integrity of aerospace alloys was proposed in Ref. [114], specifically through an innovative Pendulum-based Cutting Test (PbCT) method. As shown in Fig. 4, this method involves the deceleration of the pendulum from the entrance to the exit of the workpiece, which is used to investigate the influence of varying cutting speeds on chip morphology and surface integrity. This technique allows for continuous mapping of surface integrity and cutting mechanism transitions across different cutting speeds within a single test, contrasting with traditional approaches that require multiple tests. The method links cutting energy consumption with surface integrity, enhancing the understanding and optimization of machining processes for advanced materials.

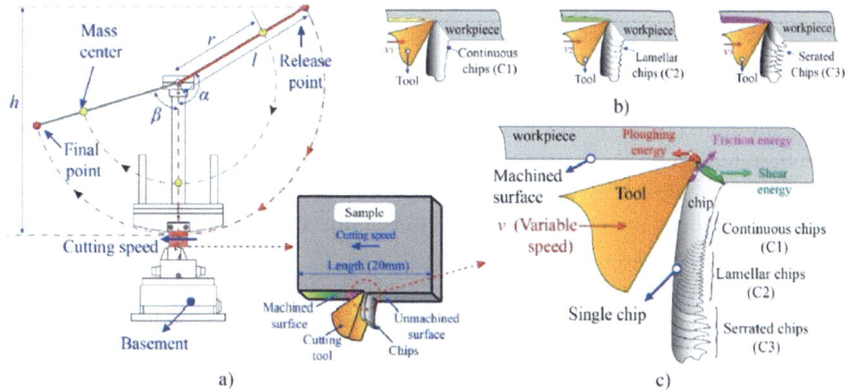

Fig. 4 Schematic of PbCT method (**a**); cutting conditions by varying cutting speeds (**b**); an insight of continuous variation of morphology chip in a single PbCT (**c**) [114]

Metal additive manufacturing (AM) presents substantial benefits for the aerospace industry, enabling cost and lead-time reductions, novel materials and design solutions, mass reduction, and component consolidation for enhanced performance [115]. Although currently limited to non-structural components, AM is increasingly applied to functional aerospace applications, including propulsion systems and structural components [116]. NASA's advancements at the Marshall Space Flight Center (MSFC) highlight the successful design and manufacturing of various components using powder-bed fusion, laser-directed energy deposition, arc-based deposition, and laser-wire cladding, achieving significant cost and schedule reductions through these AM processes [117].

AM components often have poor surface quality, necessitating post-processing for better surface finish and dimensional accuracy [118]. Machining titanium alloys produced by additive manufacturing is particularly challenging due to increased anisotropy, strength, and hardness, which reduce machinability. The microstructural features resulting from the additive process greatly influence surface integrity and tool wear during machining, making it essential to understand their impact. In Ref. [119], the microstructural differences in Alloy 718 produced by Laser Based Powder Bed Fusion (LB-PBF) and conventional wrought methods, and their effects on tool wear were examined. Electron Beam Powder Bed Fusion (EB-PBF) material, with columnar grains and a strong [001] texture, has more δ-phase precipitates and oxide inclusions, leading to higher tool wear due to heat generation and abrasive inclusions. LB-PBF material, with mixed textures and smaller grains, shows higher kernel average misorientation (KAM) values and slower wear progression, especially after heat treatment. Wrought material, with equiaxed grains and weaker texture, exhibits moderate wear and significant adhesion of workpiece material to the tools. The study in Ref. [120] examined the effect of anisotropy on tool wear during the end milling of Ti6Al4V produced via Laser (LB-PBF) with different build-up orientations. Key results included a gradual decrease in tool life with increasing build-up orientation, a correlation between tool wear and surface roughness, and anisotropic effects on cutting mechanisms and chip formation.

In conclusion, the surface integrity of aerospace alloys, especially nickel-based superalloys and titanium alloys, is crucial for the operational quality, service life, and reliability of aerospace components. High-performance cutting tools and sustainable cooling/lubrication methods, such as cryogenic cooling and minimum quantity lubrication, have been developed to improve surface integrity and extend the life of high-value parts. Besides numerical simulation, innovative methods like the PbCT, enhance the understanding and optimization of machining processes. AM offers significant benefits but requires careful post-processing to achieve the desired surface quality. Ongoing research and development, particularly in numerical simulations and advanced characterization techniques, are essential for maintaining high reliability and performance in aerospace components, contributing to industry advancements. Table 7 shows a summary of the issue discussed in this section, along with the main solutions adopted.

Table 7 Summary of surface integrity issue in aerospace components

Causes	Main solutions	Relevant components
Nickel-based superalloys and titanium alloys machining	Parameters optimization Process simulation (physics-based model integrated into FE simulations) Special tools (e.g., coatings) Cooling/lubrication (e.g., cryogenic, MQL)	Nickel-based and titanium components (e.g., airfoils, aerostructures)
Metal additive manufacturing components machining		Special components (e.g., titanium brackets, repaired airfoils)

4 Case Studies

In this section four case studies are presented to show the potential application of technologies and solutions discussed in Sect. 3. Table 8 reports the main characteristics of the case studies, as already shown in Fig. 1.

4.1 *Form Error Reduction in SPDT of a Space Component*

Ariel is an ESA M-class mission aiming to the survey of a sample of over 1000 planets using a fully aluminium Cassegrain off-axis telescope [21]. The main mirror of such telescope is M1, an elliptical mirror characterized by a parabolic reflective surface. To achieve the mission performance objectives, it is crucial that M1 remains free from deformation throughout both the assembly process and operational phases. For this purpose, three flexible components (i.e., flexure hinges) are placed between the back of the mirror and the Telescope Optical Bench.

Table 8 Case studies

Section	4.1	4.2	4.3	4.4
Description	Form error reduction in SPDT of a space component	Toolpath optimization for ultra-precision machining of mirrors	SMARTROBODYN: smart dynamometer for robotic milling	Effect of lubrication-cooling conditions on surface integrity of Inconel 718
Component	Telescope mirror	Telescope mirror	Aerospace structure	Engine Airfoil
Material	Aluminum alloy	Aluminum alloy	Lightweight alloys	Inconel 718 alloy
Main issue	Distortions	Distortions	Distortions – Vibrations	Surface integrity
Solution	Process simulation	Toolpath compensation	Monitoring and control	Cooling/lubrication

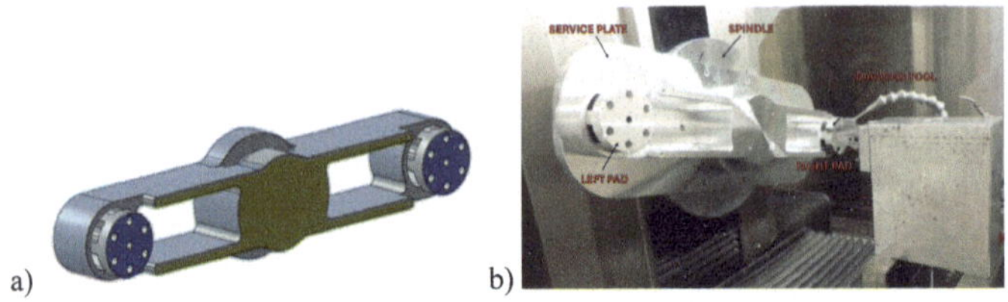

Fig. 5 **a** Flexure hinge, **b** machine set-up

The design of these components is shown in Fig. 5a with the interface surfaces to M1, called here pads, in blue.

During the operational phase, the double blades of these flexure hinges allow for radial flexibility in the back plane of M1, mitigating the deformations induced by thermal gradients. Moreover, to reduce the impact of the mounting of such components to the mirror, a planar tolerance of 2 μm was required on the interface pads. To achieve such a tight tolerance these surfaces are manufactured using Single Point Diamond Turning (SPDT).This process is generally performed using a trial-and-error approach, adjusting the parameters to achieve the desired surface quality. However, in case of strict tolerance, relying solely on expertise may not yield the required surface quality within a short timeframe. This case study shows how applying a simulation-based approach can minimize trial-and-error procedure, streamline production, and provide valuable insights to efficiently identify and address error sources in achieving the target requirements. The initial manufacturing setup of the flexure hinge was based only on the best practices and expertise of the machine operators.

The component was machined on a Nanotech 650 Multi-Axis Freeform Generator machine at the Laboratorio BS facilities, using low tightening torques, slow rotation speeds, small feed rates, and cutting depths to minimize forces during the process, as generally recommended. A wide support plate, machined with SPDT, was used to cover the full surface area of the component thus increasing the stiffness of the system. The flexure is fixed to the plate on three zones: the central area and the two lateral areas corresponding to the pads (Fig. 5b). After machining, the pads surfaces were measured with a Coordinate Measurement Machine (CMM). The results indicate an out-of-plane deviation and slightly tilted surface with a common zone planar flatness tolerance of 18.4 μm, significantly exceeding the specified requirement of 2 μm. The main sources of errors identified for this case study were: *(i)* tightening forces, *(ii)* centrifugal forces, *(iii)* gravitational forces, iv) coupling tolerance. Additional effects (e.g., thermal effects, slide tilt errors, vibrations) were neglected since they were not considered significant in this case. To support the actions aiming at improving the quality of the targeted surfaces, a two-step approach is followed: (1) the contribution of tightening, centrifugal and gravitational forces will be predicted using simulations, (2) influence of coupling tolerance will be assessed by measuring the interface surfaces. For the first step, Finite Element Analyses were performed in MSC Nastran according to the modelling techniques proposed by Gottini et al. [33]. Flexure hinges and service plate were modelled using first order hexahedral mesh, tightening was included applying the preload force (corresponding to 2 Nm) to the screws, gravity was simulated by combining the results of analysis on the two orthogonal directions and centrifugal force was computed and applied radially using a specific Nastran card (RFORCE). Results are shown in Fig. 6. The simulation shows that centrifugal, gravity and clamping forces are responsible of a deviation of about 0.3 μm, suggesting that addressing these effects does not significantly contribute to the reduction of the errors. Coupling tolerance are hence investigated by measuring the contacting surfaces. Since the service plate surfaces are machined with SPDT while flexure hinges' ones with milling, the effect of the first is neglected.

Multiple points on the component coupling surfaces were measured using a CMM machine and results show an out-of-plane deviation of about 0.13 mm, critical value that probably induced a significant deformation during the tightening process. Based on the observation a new setup for the second trial was defined. Since gravity, centrifugal force and tightening contribution is low, no specific actions (e.g., reduction of spindle speed, preload adjustment) are pursued. On the other, to reduce the contact area and minimize the impact of high form deviation of contact surface, a new connection between plate and component was defined, without the need of reworking the main component. Only the central area of the service plate was used to connect plate and flexure hinges, removing the connection to the lateral zones. Service plate was modified accordingly, as shown in Fig. 7. The new configuration was simulated to check the influence of gravity, centrifugal force and tightening effects using the

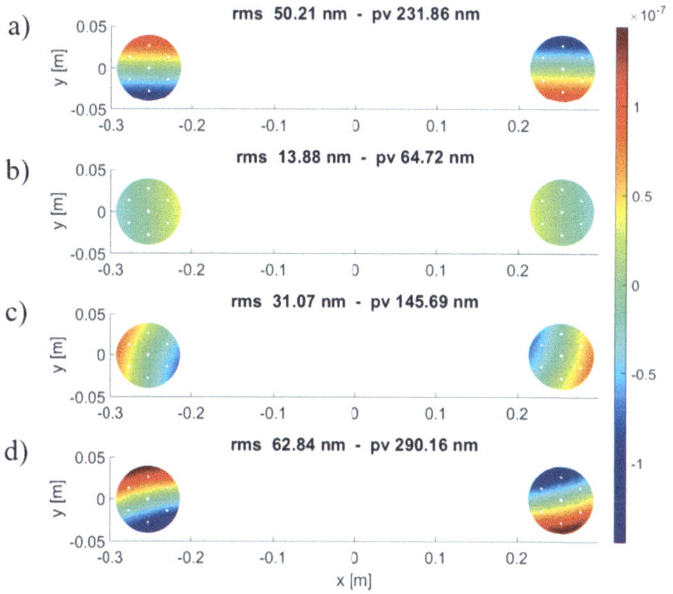

Fig. 6 Analyses results on the pad surfaces machined using the first machining attempt, **a** gravity, **b** centrifugal force, **c** tightening, **d** combination of the three

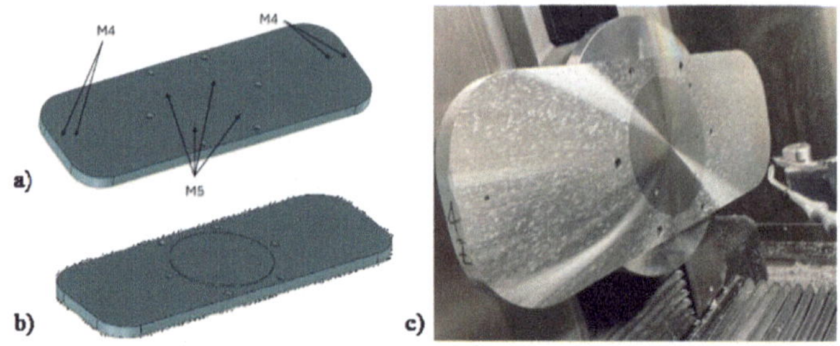

Fig. 7 **a** service plate for the first machining trial **b** modified service plate for the second trial **c** SPDT machining of the modified service plate

same approach previously presented. Results show that these effects are still negligible with a total impact of about 0.13 μm on the surface form error, even lower than in the previous configuration. The proposed new machining setup was used to machine the component using the same cutting parameters.

The machined surfaces of the pads were then measured using a CMM. Measurements revealed that all the points of the surfaces are within a planar tolerance of 1.5 μm, well below the value obtained previously (18 μm), meeting the specification requirements (2 μm) on the second trial. The proposed simulation-based approach allows to reduce trial-and-error iterations, improving manufacturing efficiency and precision in the production of the space component using SPDT. A combination of simulation and measurements supported the identification of the causes of form deviation, suggesting targeted actions to improve the resulting surface quality.

4.2 *Toolpath Optimization for Ultra-Precision Machining of Mirrors*

This case study is dedicated to defining an effective methodology for the compensation of quasi-static dimensional errors affecting a mirror-like surface obtained through SPDT. The methodology was developed and validated on several benchmarks made of aluminum alloys. Among them, an axisymmetric reflector made of aluminum AL6068, with an external diameter of 130 mm, $R_a < 10$ nm and very strict dimensional tolerances was considered, see Fig. 8a, b.

Another component was a free form reflector made of aluminum AL5083, size about 240 mm × 160 mm, $R_a < 30$ nm and very strict dimensional tolerances, see Fig. 8c, d. In the last decades similar methodologies have already been developed, as shown in the recent works [121]. However, their practical implementation in real industrial environments is not straightforward for many reasons. In the current case, a relatively simpler, semi-empirical method of compensation a posteriori (i.e., after semi-finish) is under development. The main principles and ad-hoc solutions behind the developed algorithm are described in the following.

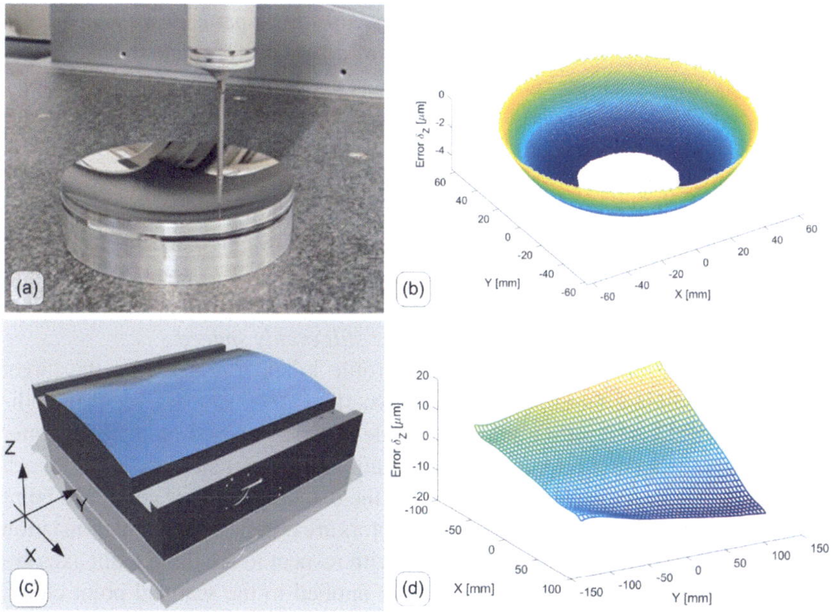

Fig. 8 Case studies investigated for developing the toolpath compensation strategy. Axisymmetric reflector (**a**) and the dimensional error between the semi-finished part and the nominal target (**b**). Free-form reflector (**c**) and the associated dimensional error between the semi-finished part and the nominal target (**d**)

The main assumption here was that the machined surface can be expressed as a function $z_e(x, y)$, which can be decomposed as in Eq. (1):

$$z_e(x, y) = z_n(x, y) + \mathrm{d}z_m(x, y) + \mathrm{d}z_r(x, y) \tag{1}$$

where $z_n(x, y)$ is the ideal, nominal target surface, $\mathrm{d}z_m(x, y)$ are deviations due to macroscopic machining errors such as part mechanical deformations and thermal distortions, tool inaccuracies, linear axes trajectory errors, spindle eccentricity, $\mathrm{d}z_r(x, y)$ are further deviations due to local and generally random effects.

It was assumed that the function $\mathrm{d}z_m(x, y)$ represents a repeatable and systematic form error, greater than the random tiny deviations $\mathrm{d}z_r(x, y)$. One further hypothesis is that the effect produced by $\mathrm{d}z_m(x, y)$ can be significantly reduced by introducing a tool path distortion of opposite sign and comparable magnitude.

Based on these hypotheses, in order to minimize the systematic errors, the main steps for part geometry correction were decided. Firstly, the unknown function $\mathrm{d}z_m(x, y)$ is estimated from the CMM scan of the machined part. Secondly, the toolpath is recalculated in order to partially compensate for the deformation measured in the initial part. Eventually, the former steps are iterated until the desired tolerance is achieved.

To this aim, main steps of the algorithm are:

- *Measurement of semi-finished surface*: after semi-finish, part surface is inspected by using a CMM, by touching the vertexes of a very uniform and evenly spaces grid (along X and Y directions, grid spacing si about 0.05 mm)
- *Generation of the target point cloud*: the target point cloud having a higher resolution than the scanned point cloud (about 10 time more points) is derived from a mathematical model.
- *Importing initial point clouds*: this procedure can be complex due to the size of the files (hundreds of MB), the number of points (millions), and the variety of available formats.
- *Undersampling*: the scanned point cloud is undersampled to obtain a reduced *but representative set of points, which should be still evenly spaced.*
- *Collimation*: the undersampled scanned profile is superimposed on the target profile. The collimation is performed in two steps: *(i)* coarse barycentric collimation: the barycenter of the scanned profile is superimposed on the barycenter of the reference point cloud; *(ii)* fine collimation using distance vectors: for each point of the undersampled scanned profile, the distance from the reference point cloud is calculated. Once all the distance vectors are obtained, the rigid translation vector and the barycentric rotation angles with respect to the three main axes are calculated. The translation and rotation are applied to the scanned point cloud. This process is repeated until the magnitudes of the translation and rotation fall below predefined thresholds.
- *Calculation of the mathematical distortion model*: for all the points of the scanned profile, the distance vector with respect to the reference point cloud is calculated again using the collimation algorithm. A set of vectors is obtained, one for each scanned point. Finally, a 10th order polynomial model is interpolated on the z component of the distortion vectors, by considering x and y as main independent coordinates, as shown in Eq. (2).

$$\mathrm{d}z_m(x, y) \approx f(x, y) \tag{2}$$

- *Reference profile compensation*: the z coordinate of all points in the reference point cloud is adjusted using the expression in Eq. (3):

$$z_n'(x, y) = z_n(x, y) - k \cdot f(x, y) \tag{3}$$

where the value of the coefficient $k \approx 0.5$ is a good first guess value.

The above-described calculation procedure is very demanding, especially for point clouds composed of tens of thousands or millions of points, as in the case of free-form surfaces. With the aim of reducing to elaboration time, most critical calculations were performed with modern high-performance NVIDIA GPUs—models RTX 4070 or RTX 4080. This choice was crucial to reduce the elaboration time from several hours to tens of minutes. According to the obtained preliminary results, the presented method is effective for enhancing dimensional accuracy. In most of the considered case studies, the required tight tolerances were achieved, although some further improvements are necessary.

4.3 SMARTROBODYN: Smart Dynamometer for Robotic Milling

To machine large or complex aerospace components or thin-walled structures, such as aeronautical fuselages and wings, robotic milling can be the best industrial solution, thanks to its high operational versatility [122–124]. In addition, robotic milling of light alloys (such as aluminum, magnesium and titanium alloys) is an advanced machining technique of increasing industrial interest for many strategic sectors including aerospace, automotive, energy and others. Nevertheless, robotic milling applications are hindered by the excessive flexibility of the robotic arm and by other problems, which do considerably limit the overall performance of this technique. In order to enhance such performance, real-time cutting force measurement would be very effective. A raw estimate of the quasi-static cutting force components can be obtained from the torque signals extracted from robotic arm joints, whenever they are available. Nevertheless, this method fails in the presence of small cutting forces—such as in finish milling—and it is also inadequate for characterizing their high-frequency harmonics, which would be necessary for advanced applications like detection and mitigation of undesired chatter vibrations.

To address this need, further and more advanced sensors are required. For example, a plate dynamometer can be installed between the worktable and the workpiece. However, this solution is costly and not practical for most industrial applications. On the other side, integrating a dynamometer between the end effector of the robotic arm and the spindle housing would be a promising solution. This was recently done by a few researchers, see Refs. [125] and [126]. Unfortunately, their solutions are constrained by a limited frequency range and are based on multiple calibrations.

The identification of dynamometer transmissibility requires a dedicated modal analysis that has to be carried out in the final experimental setup. Subsequently, a numerical filter can be generated and used for an accurate estimation of the effective input forces [88]. In robotic milling there is a further complication, because the final dynamometer transmissibility depends on the type and pose of the arm. As a consequence, multiple calibrations at different poses would be necessary to compute an effective filter. This will likely discourage real industrial applications of this type of sensors and filtering techniques.

In this field piezoelectric dynamometers should be preferred because of their high stiffness, satisfactory dynamic bandwidth and outstanding signal-to-noise ratio [127]. However, their response is influenced by the robotic arm configuration, as the other types of sensing devices. Furthermore, they are possibly too expensive in comparison to robot price.

There is currently no device that can be integrated into a robotic milling system, capable of estimating the 3D dynamic cutting forces and tool vibrations over a wide frequency range, without weakening the machining system, having a moderate cost. To address these needs, an innovative sensorised device has been recently developed at the LAMA FVG laboratory of the University of Udine (Fig. 9). It was founded on the concepts of Universal Dynamometer and Universal Inverse Filter presented in

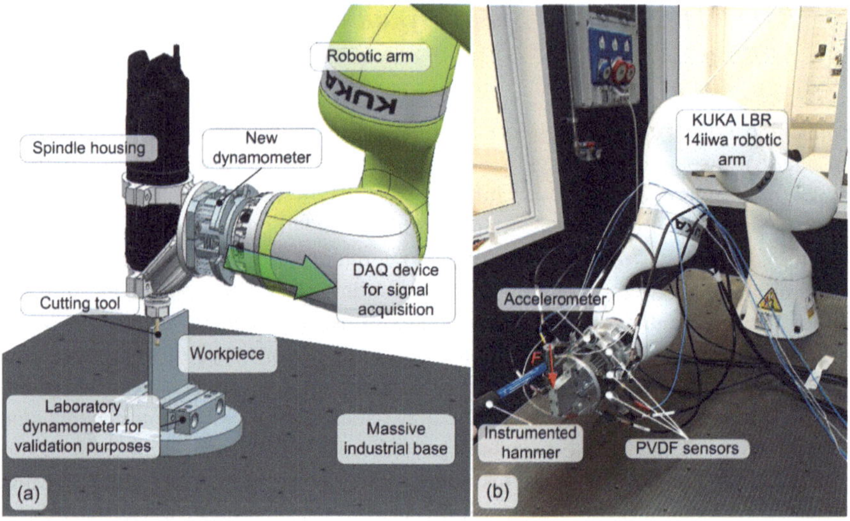

Fig. 9 **a** Concept of the new device; **b** preliminary modal analysis on the robotic arm (without the spindle system)

[128]. After an initial calibration by the manufacturer, it could be ideally used without further on-site recalibration. The numerical UIF filter does considerably attenuate all the inertial disturbances affecting the measured signals, both caused by cutting process or by exogenous vibration sources.

The novel device is composed of a stiff and lightweight (0.5 kg) structure made primarily from Ti6Al4V alloy, 3D printed via the L-PBF process. PVDF sensors are used as main sensing elements for the input cutting forces, while piezoelectric accelerometers are used for monitoring the 6 rigid body motions of dynamometer base, according to the principles of the Universal Dynamometer. The preliminary tests conducted on the single PVDF sensors taken separately and on the whole assembled device clamped at a rigid fixture and at the robotic arm (without the spindle) have been quite successful, as illustrated in [129]. Initial results showed that the usable frequency bandwidth is extended from 20–300 Hz up to 1400–7700 Hz along the direct directions. In conclusion, preliminary experimental validation was successful, indicating the proposed device has significant potential for both R&D and industrial applications. The development of this case study received funding from the Friuli Venezia Giulia Region of Italy through PR FESR 2021–2027, CALL A1.3.1—Non-Refundable Incentive Instrument for the Implementation of Proof of Concept (PoC), under the title "Development of a Universal Dynamometer for Process Monitoring and Control of Advanced Machining Operations Performed UsingRobotic Arms—SMARTROBODYN."

4.4 *Effect of Lubrication-Cooling Conditions on Surface Integrity of Inconel 718 Alloy*

Surface quality and integrity are paramount in aerospace component machining, encompassing surface finish (e.g., Ra), defect types, and the micro-hardness of the SPD layer. Aerospace materials must adhere to stringent standards like Aerospace Material Specification (AMS) and Metallic Materials Properties Development and Standardization (MMPDS) to ensure safety and performance under extreme conditions. Regulatory bodies like the European Union Aviation Safety Agency (EASA) and the Federal Aviation Administration (FAA), through standards, define critical performance criteria, material characteristics, and testing protocols for airworthiness certification. Compliance with these regulations is essential for ensuring the safety and efficiency of aerospace components. Nickel-based superalloys are among the most widely used materials in aerospace component manufacturing due to their superior mechanical properties, excellent chemical and oxidation resistance, and very low creep. Among these superalloys, Inconel 718 is extensively used for manufacturing turbine shafts, combustion cases, turbine discs, and turbine spools. Figure 10a shows a detail of a combustion case manufactured by FLY S.p.A., located in Grigno, TN, Italy. A primary challenge associated with the machining of these components lies in the development of significant tool wear. This wear can negatively impact the surface integrity of the workpiece, potentially leading to non-compliance with aeronautical certifications. This case study investigates the longitudinal turning of Inconel 718 using WC inserts with a triple-layer TiCN + Al_2O_3 + TiN coating, currently employed in industrial practice. The study evaluates the performance of these inserts under both dry and cryogenic cooling conditions, exploring viable alternatives to environmentally less sustainable flood machining. Longitudinal turning experiments were conducted on a Mori Seiki™ NL 1500 CNC lathe equipped with a cryogenic cooling system. Figure 10b illustrates the CNC lathe during cryogenic cooling, with magnified views of the cutting zone. Figure 10c presents the tool life curves for the three lubri-cooling conditions, with flood cooling serving as the current industrial benchmark. All curves exhibit a typical trend: an initial rapid increase in flank wear depth followed by a more gradual growth. Cryogenic cooling effectively decelerates flank wear progression throughout the tool's operational life. While dry cutting may initially appear efficient, it ultimately proves the most detrimental to tool life, exceeding the wear observed under flood cooling conditions after approximately 33,000 mm^3 of material removal.

In dry machining, initial heat accumulation can lead to thermal softening, reducing material resistance. However, prolonged cutting can cause the coating protecting the substrate to deteriorate. This can result in strong adhesion between the substrate and the workpiece due to the high chemical affinity between them, leading to a sudden increase in flank wear. The abrupt wear increase observed in Fig. 10c can be attributed to this adhesion phenomenon.

Measuring the average surface roughness (Ra) provides a direct assessment of surface finish quality. This parameter can be determined through non-destructive

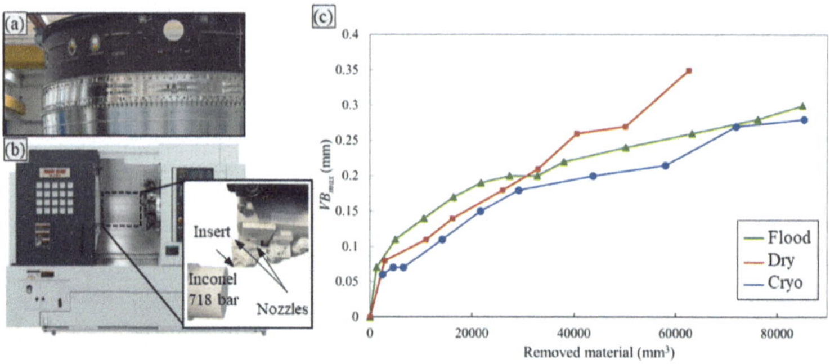

Fig. 10 **a** Detail of an Inconel 718 combustion case, **b** CNC lathe with magnified views of the cutting zone during cryogenic cooling, and **c** tool life curves of coated WC inserts under varying lubrication-cooling conditions

testing methods, making it easily accessible and safe for quality control. Table 9 presents Ra values obtained under varying lubrication-cooling conditions at the end of tool life. Results demonstrate that cryogenic turning offers a viable alternative to conventional flood turning in terms of surface finish, achieving a 15% reduction compared to flood machining. Conversely, dry turning does not consistently maintain the same level of surface finish due to significantly higher flank wear, resulting in a 120% increase compared to flood machining. Figure 11 provides a closer inspection of the workpiece's surface finish.

Cryogenic cutting yielded the most uniform surface with minimal observable defects. In contrast, both flood and dry turning exhibited surface defects, including adhered material and scratches. Adhesion of workpiece material to the tool increases

Table 9 Ra values of the machined surface under varying lubrication-cooling conditions

	Flood	Dry	Cryo
Ra (μm)	0.94 ± 0.03	2.07 ± 0.03	0.8 ± 0.02

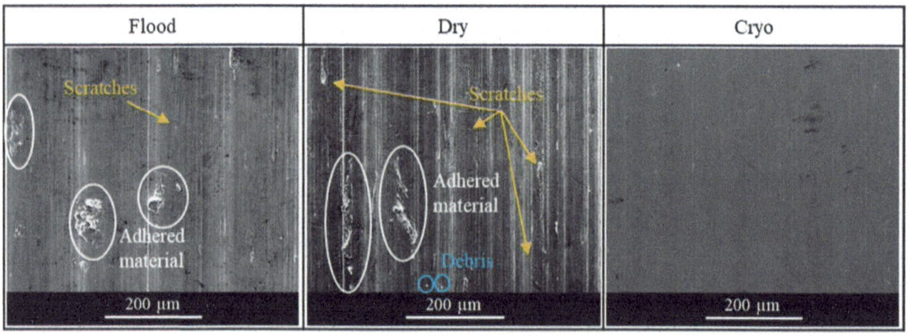

Fig. 11 SEM images of the machined surface under varying lubrication-cooling conditions

with rising cutting temperatures, as observed in dry cutting due to elevated thermal energy at the tool-workpiece interface.

This elevated temperature can soften the workpiece material, promoting adhesion and potentially leading to built-up edge formation on the cutting tool. The instability of the built-up edge increases the risk of surface scratches. Furthermore, in dry cutting, the absence of a cooling medium exacerbates the likelihood of debris accumulating in the cutting zone, leading to the formation of grooves on the machined surface.

5 Conclusions and Perspectives

The aerospace sector stands as a hub of innovation and a significant and growing contributor to the Italian and global economy. The sector relies on precision machining processes to meet the stringent requirements in terms of performance and components quality. Despite progress, trial-and-error methods are still common in industry, leading to inefficiencies, high scrap rates, and significant costs. This White Paper focuses on conventional machining processes (i.e., milling, turning, drilling) and investigates the manufacturing of three critical aerospace components: mirrors for space applications, aero-engine airfoils (i.e., blades), aerospace structures. The selection of these components was driven by their ability to accurately represent the intricate technological challenges inherent in aerospace machining. Their complexity aligns with the stringent requirements, advanced materials, and precision engineering that characterize this industry, making them particularly relevant for the scope of this work.

Manufacturing high quality mirrors for optical systems (e.g., telescopes) involves advanced machining processes, including ultra-precision machining (e.g., SPDT, UPFC) to achieve extremely tight surface roughness and tolerance. Key challenges in mirror production include deformations due to different factors (e.g., cutting forces, clamping, coupling surfaces, centrifugal forces, gravity) and vibrations (Table 2).

Machining of turbine and compressor blades is critical due to their complex geometries, high precision requirements, and the need for materials that withstand extreme conditions. After casting or forging to shape the initial blade blanks, 5-axis CNC milling is the most used process to achieve accurate dimensions and surface finish. Challenges include managing dimensional tolerances affected by distortions due to cutting forces and internal stress relaxation, avoiding unstable vibrations, and ensuring flawless surface integrity without post-polishing (Table 3).

Milling and drilling processes are the main conventional technologies used for aerostructures. These components are commonly composed of thin-walled structures making them prone to distortions and vibrations. In addition, aluminum alloys require high-speed machining with specific coatings to extend tool life, while titanium alloys pose challenges like poor thermal conductivity and high tool wear (Table 4).

By analyzing such critical components and focusing on quality of the product some common issues related to distortion, vibrations and surface integrity were identified.

Machining distortions in aerospace components are generally generated by residual stresses, cutting loads and clamping forces, leading to geometric errors. The extensive use of thin-walled components in the sector makes it critical for such an issue, requiring predictive approaches, measurement techniques, and mitigation strategies. While progress has been made, challenges remain in automating solutions, integrating methodologies, and addressing distortions in different materials and geometries (Table 5).

Machining vibrations, including forced vibrations and self-excited chatter, significantly impact surface quality, tool life, and machining efficiency, especially in milling, extensively used in the aerospace sector. While stability lobes diagrams (SLDs) guide optimal cutting parameters to prevent chatter, challenges remain due to uncertainties, case-specific dynamics, and the influence of material removal on workpiece behavior, especially in complex aerospace applications. Despite advancements in predictive, passive, and hybrid suppression methods, there is still a lack of reliable, general-purpose commercial software (Table 6).

Surface integrity is critical for the performance and reliability of aerospace components, especially those made of nickel-based superalloys and titanium alloys, which are challenging to machine due to high toughness, work-hardening, and poor thermal conductivity. Advancements in tooling, such as ceramic tools for roughing and carbide tools for finishing, along with sustainable cooling methods like cryogenic cooling and cryo-MQL, have improved surface quality and tool life. Innovative testing methods like the Pendulum-based Cutting Test (PbCT) and predictive simulations could enhance process understanding, while Additive manufacturing (AM) introduces new opportunities for aerospace applications but poses challenges in achieving desired surface integrity due to anisotropy and microstructural differences (Table 7).

This paper presented a practical implementation of the most interesting approaches developed by the authors in collaboration with industry, to show the potential solutions and advanced methodologies to tackle the analyzed issues: *(i)* simulation-based approaches to distortions mitigation in space component manufacturing, *(ii)* toolpath optimization for form error reduction in ultra-precision machining of space mirrors, *(iii)* advanced measurement technique for vibration monitoring in robotic milling and *(iv)* surface integrity assessment on aerospace materials, as shown in Table 8 and Fig. 1.

Future perspectives for the covered topics, highlighting critical components and issues involved, are outlined in Table 10.

Table 10 Future perspectives

Component	Issues	Future perspectives
Telescope mirrors	Distortions	The steadily increasing interest which is being put in the development of fully metallic mirrors. Ultra-precision machining of such components requires the implementation of advanced technologies to achieve the desired strict quality requirements, such as simulation-based or toolpath compensation approaches
Telescope mirrors	All	Space applications, often requiring very stringent tolerance and surface finish, increasingly demand for cost savings in the new space economy paradigm. Therefore, trial and error approaches are often obsolete, and manufacturing optimization techniques are still required
All	Surface integrity	Additive manufacturing which is increasingly used in the aerospace sector and requires investigations of the unique properties of the (often tailored) printed materials to optimize the additive-subtractive manufacturing integration
Aero-engine airfoils aerospace structures	Surface integrity	Investigating and applying advanced cooling strategies such as cryogenic cooling is crucial for enhancing surface integrity, reducing tool wear, and improving the machining efficiency of advanced materials used in the aerospace sector, such as nickel-based superalloys and titanium alloys
Aerospace structures	Distortions vibrations	The versatility of robotic milling, that could be a game changer in the machining of large and complex aerospace components. However, this technology needs dedicated methodologies to investigate and mitigate distortions and vibrations coming from the characteristic low-stiffness structure
Aero-engine airfoils aerospace structures	Vibrations	Machining of thin-walled structure, typically employed in the aerospace sector, suffering unstable vibrations occurrence, demanding for reliable chatter predictive and suppression methods that could consider workpiece dynamics modification during the process
All	All	The advancement of AI techniques, supported by growing data availability, which offers significant potential to address machining challenges. While data-driven approaches can leverage emerging trends, the complexity of these issues, influenced by various factors, necessitates the integration of physics-based models for more robust and effective solutions

References

1. PWC: Global aerospace and defense—annual industry performance and outlook (2023)
2. International air travel association: air passenger market analysis (2023)
3. US Department of Defense: defense budget overview: fiscal year 2024 budget request (2023)
4. Deloitte: aerospace and defense industry outlook—Deloitte Insights (2024)
5. ASD: facts & figures, aerospace and defense industries (2022)
6. European Commission: EU Aeronautics Industry (2024)
7. Grand view research: aerospace parts manufacturing market size report 2024–2030 (2024)
8. Flight global: top 100 aerospace companies ranked by revenue (2021)
9. Italian trade agency: aerospace industry report (2020)
10. Euroconsult: space economy report, 10th edition (2023)
11. Crippa P, Guidi A (2020) Investire nell'industria della difesa italiana: una garanzia per il mondo post-covid
12. Deloitte: global aerospace and defense industry outlook (2020)
13. Capasso S (2024) Stato dell'Arte nei settori Aeronautica-Aerospazio e Automotive (2024)
14. Ghidini T, Grasso M, Gumpinger J, Makaya A, Colosimo BM (2023) Additive manufacturing in the new space economy: current achievements and future perspectives. https://doi.org/10.1016/j.paerosci.2023.100959
15. Counterpoint: aerospace machining (2020)
16. Bowden DM, Halley JE (2001) Aluminium reliability improvement program-final report 60606. The Boeing Company, Chicago
17. Hatefi S, Abou-El-Hossein K (2020) Review of single-point diamond turning process in terms of ultra-precision optical surface roughness. Int J Adv Manuf Technol 106:2167–2187. https://doi.org/10.1007/s00170-019-04700-3
18. Zhang S, Bi Q, Ji Y, Wang Y (2019) Real-time thickness compensation in mirror milling based on modified Smith predictor and disturbance observer. Int J Mach Tools Manuf 144. https://doi.org/10.1016/j.ijmachtools.2019.103427
19. Zhang SJ, To S, Zhu ZW, Zhang GQ (2016) A review of fly cutting applied to surface generation in ultra-precision machining. https://doi.org/10.1016/j.ijmachtools.2016.01.001
20. Liu DF, Tang YJ, Cong WL (2012) A review of mechanical drilling for composite laminates. https://doi.org/10.1016/j.compstruct.2011.11.024
21. Pace E, Tozzi A, Abreu MA, Alonso G, Barroqueiro B, Bianucci G, et al (2022) The telescope assembly of the Ariel space mission. In: Proceedings of SPIE—the international society for optical engineering. https://doi.org/10.1117/12.2629432
22. He X, Jin H, Zhou C, Gao C, Zhang G, ES (2023) Modeling of material removal in magnetic finishing based on Maxwell's stress tensor theory and its experimental validation. J Mater Process Technol 312. https://doi.org/10.1016/j.jmatprotec.2022.117808
23. Chabot T, Brousseau D, Auger H, Thibault S (2021) Sub-assembly fabrication of diamond-turned aluminum image slicers. https://doi.org/10.1117/12.2601142
24. Yu DP, Hong GS, Wong YS (2012) Profile error compensation in fast tool servo diamond turning of micro-structured surfaces. Int J Mach Tools Manuf 52:13–23. https://doi.org/10.1016/j.ijmachtools.2011.08.010
25. Zhuang G, Zong W, Tang Y (2023) Statistical analysis and suppression of vibration frequency bifurcation in diamond turning of Al 6061 mirror. Mech Syst Signal Process 198. https://doi.org/10.1016/j.ymssp.2023.110421
26. Huang P, Lee WB (2016) Cutting force prediction for ultra-precision diamond turning by considering the effect of tool edge radius. Int J Mach Tools Manuf 109:1–7. https://doi.org/10.1016/j.ijmachtools.2016.06.005
27. Gottini D, Pace E, Tozzi A, Bianucci G, Bocchieri A, Brienza D, et al (2022) FEA testing the pre-flight ariel primary mirror. In: Proceedings of SPIE—the international society for optical engineering. https://doi.org/10.1117/12.2629815

28. Tozzi A, Brucalassi A, Canestrari R, Chioetto P, Del Vecchioi C, Carbonaro L, et al (2022) Toward ARIEL's primary mirror. In: Proceedings of SPIE—the international society for optical engineering. https://doi.org/10.1117/12.2628906
29. Moroni G, Petró S, Polini W (2014) Robust design of fixture configuration. In: Procedia CIRP. Elsevier B.V., pp 189–194. https://doi.org/10.1016/j.procir.2014.03.120
30. Yip WS, To S (2020) Preliminary investigation on ultra-precision diamond turning of titanium alloys using thermoelectric cooler fixture. J Manuf Process 58:187–192. https://doi.org/10.1016/j.jmapro.2020.08.020
31. Ji TG, Park W, Pak S, Jeong B, Kim S, Lee HI, et al (2021) Error compensation software to remove the low-frequency error of aluminum freeform mirror for an infrared off-axis telescope. J Korean Soc Precis Eng 38. https://doi.org/10.7736/JKSPE.020.117
32. Zhang J, Zhang X, Tan S, Xie X (2017) Design and manufacture of an off-axis aluminum mirror for visible-light imaging. Curr Optics Photonics 1:364–371. https://doi.org/10.3807/COPP.2017.1.4.364
33. Gottini D, Scimia G, Grossi N, Scippa A (2024) Deformation and optical aberration prediction in ultra-precision single point diamond turning of optical components. Precis Eng. https://doi.org/10.1016/J.PRECISIONENG.2024.10.010
34. Song P, Yang C, Bai Y, Ding J, Guo J, Li C, et al (2023) High-precision turning and ultra-smooth direct polishing of aluminum alloy mirrors. Opt Express 31. https://doi.org/10.1364/oe.495779
35. Zhang Y, Dong G, Zhou M (2016) Simulation and experiment study on cutting temperature in diamond ultra-precision cutting of aluminum alloy mirror. Key Eng Mater 667. https://doi.org/10.4028/www.scientific.net/KEM.667.136
36. Zhang Z, Yang X, Zheng L, Xue D (2017) High-performance grinding of a 2-m scale silicon carbide mirror blank for the space-based telescope. Int J Adv Manuf Technol 89. https://doi.org/10.1007/s00170-016-9107-1
37. Zhao Y, Zhang Y, Xu J, Rao M, Yin Y (2023) Error separation and compensation of arc wheel grinding for SiC segmented mirror considering wheel wear. International J Adv Manuf Technol 126. https://doi.org/10.1007/s00170-023-11194-7
38. Lin S, Jiang ZH, Yin YH (2019) Research on arc-shaped wheel wear and error compensation in arc envelope grinding. Int J Adv Manuf Technol 103. https://doi.org/10.1007/s00170-019-03460-4
39. Beaucamp A, Simon P, Charlton P, King C, Matsubara A, Wegener K (2017) Brittle-ductile transition in shape adaptive grinding (SAG) of SiC aspheric optics. Int J Mach Tools Manuf 115. https://doi.org/10.1016/j.ijmachtools.2016.11.006
40. Gorla RSR, Khan AA (2003). Turbomachinery: Design and Theory. https://doi.org/10.1201/9780203911600
41. Li Z, Zeng Z, Yang Y, Ouyang Z, Ding P, Sun J et al (2024) Research progress in machining technology of aerospace thin-walled components. J Manuf Process 119:463–482. https://doi.org/10.1016/j.jmapro.2024.03.111
42. Yan B, Hao Y, Zhu L, Liu C (2022) Towards high milling accuracy of turbine blades: a review. Mech Syst Signal Process 170:108727. https://doi.org/10.1016/j.ymssp.2021.108727
43. Zhang X, Chen Y, Hu J (2018) Progress in aerospace sciences recent advances in the development of aerospace materials. Prog Aerosp Sci 97:22–34. https://doi.org/10.1016/j.paerosci.2018.01.001
44. Dursun T, Soutis C (2014) Recent developments in advanced aircraft aluminium alloys. J Mater Design 56:862–871. https://doi.org/10.1016/j.matdes.2013.12.002
45. Boyer RR, Cotton JD, Mohaghegh M, Schafrik RE (2015) Materials considerations for aerospace applications. 40:1055–1065. https://doi.org/10.1557/mrs.2015.278
46. M'Saoubi R, Axinte D, Soo SL, Nobel C, Attia H, Kappmeyer G et al (2015) High performance cutting of advanced aerospace alloys and composite materials. CIRP Ann Manuf Technol 64:557–580. https://doi.org/10.1016/j.cirp.2015.05.002
47. Boyer RR (2010) Attributes, characteristics, and applications of titanium and its alloys. J Miner Metals Mater Soc (TMS) 62:21–24

48. Huda Z, Edi P (2013) Materials selection in design of structures and engines of supersonic aircrafts: a review. Mater Des 46:552–560. https://doi.org/10.1016/j.matdes.2012.10.001
49. Domínguez-monferrer C, Guerra-sancho A, Caggiano A, Nele L, Cantero JL (2024) Multiresolution analysis for tool failure detection in CFRP/Ti6Al4V hybrid stacks drilling in aircraft assembly lines 206
50. Caggiano A (2018) Machining of fibre reinforced plastic composite materials. Materials 11. https://doi.org/10.3390/ma11030442
51. Fleischer J, Teti R, Lanza G, Mativenga P, Möhring HC, Caggiano A (2018) Composite materials parts manufacturing. CIRP Ann 67:603–626. https://doi.org/10.1016/j.cirp.2018.05.005
52. Liu H, Wang C, Li T, Bo Q, Liu K, Wang Y (2022) Fixturing technology and system for thin-walled parts machining: a review. Front Mech Eng 17. https://doi.org/10.1007/s11465-022-0711-5
53. Cuesta M, Jose P (2023) Improving surface integrity when drilling CFRPs and Ti-6Al-4V using sustainable lubricated liquid carbon dioxide 36:129–146. https://doi.org/10.1016/j.cja.2022.09.004
54. Dornfeldl DA, Chen LJ, Technology U, Hartford E (1999) Drilling burr formation in titanium alloy, li-6AI-4V. Annals fo the CIRP. 48:73–76
55. Li R, Hegde P, Shih AJ (2007) High-throughput drilling of titanium alloys. Int J Mach Tools Manuf 47:63–74. https://doi.org/10.1016/j.ijmachtools.2006.02.012
56. Assemblies A (2022) A review on drilling of multilayer fiber-reinforced polymer composites and aluminum stacks: optimization of strategies aerospace assemblies
57. Dahnel AN, Fauzi MH, Raof NA, Mokhtar S (2022) Materials today: proceedings tool wear and burr formation during drilling of aluminum alloy 7075 in dry and with cutting fluid. Mater Today Proc. 59:808–813. https://doi.org/10.1016/j.matpr.2022.01.110
58. Melkote SN, Newton TR, Hellstern C, Morehouse JB, Turner S (2010) Interfacial burr formation in drilling of stacked aerospace materials. https://doi.org/10.1007/978-3-642-00568-8
59. Kondo Y, Miyake Y (2024) Drilling performance of aluminum plates under manual drilling conditions 13:18–19. https://doi.org/10.18178/ijmerr.13.1.18-25
60. Aurrekoetxea M, Llanos I, Zelaieta O, López de Lacalle LN (2022) Towards advanced prediction and control of machining distortion: a comprehensive review. https://doi.org/10.1007/s00170-022-10087-5
61. Akhtar W, Lazoglu I, Liang SY (2022). Prediction and control of residual stress-based distortions in the machining of aerospace parts: a review. https://doi.org/10.1016/j.jmapro.2022.02.005
62. Yang Y, Xia L, Zhao G, Meng L, He N (2018) Investigation of the coupled distribution of initial and machining-induced residual stress on the surface of thin-walled parts. Int J Adv Manuf Technol 98:213–222. https://doi.org/10.1007/s00170-017-1567-4
63. Liu Y, Xu D, Agmell M, Saoubi RM, Ahadi A, Stahl JE, et al (2021) Numerical and experimental investigation of tool geometry effect on residual stresses in orthogonal machining of Inconel 718. Simul Model Pract Theory 106. https://doi.org/10.1016/j.simpat.2020.102187
64. Gao H, Zhang Y, Wu Q, Li B (2018) Investigation on influences of initial residual stress on thin-walled part machining deformation based on a semi-analytical model. J Mater Process Technol 262:437–448. https://doi.org/10.1016/j.jmatprotec.2018.04.009
65. Aurrekoetxea M, de Lacalle LNL, Llanos I (2020) Machining stresses and initial geometry on bulk residual stresses characterization by on-machine layer removal. Materials 13. https://doi.org/10.3390/ma13061445
66. Chantzis D, Van-Der-Veen S, Zettler J, Sim WM (2013) An industrial workflow to minimise part distortion for machining of large monolithic components in aerospace industry. In: Procedia CIRP. Elsevier B.V. pp 281–286. https://doi.org/10.1016/j.procir.2013.06.103
67. Rodríguez-Sánchez AE, Ledesma-Orozco E, Ledesma S (2020) Part distortion optimization of aluminum-based aircraft structures using finite element modeling and artificial neural networks. CIRP J Manuf Sci Technol 31:595–606. https://doi.org/10.1016/j.cirpj.2020.08.011

68. Li X, Li L, Yang Y, Zhao G, He N, Ding X et al (2020) Machining deformation of single-sided component based on finishing allowance optimization. Chin J Aeronaut 33:2434–2444. https://doi.org/10.1016/j.cja.2019.09.015
69. Zhang Z, Luo M, Tang K, Zhang D (2020) A new in-processes active control method for reducing the residual stresses induced deformation of thin-walled parts. J Manuf Process 59:316–325. https://doi.org/10.1016/j.jmapro.2020.09.079
70. Wan M, Zhang WH, Qin GH, Wang ZP (2008) Strategies for error prediction and error control in peripheral milling of thin-walled workpiece. Int J Mach Tools Manuf 48:1366–1374. https://doi.org/10.1016/j.ijmachtools.2008.05.005
71. Morelli L, Grossi N, Campatelli G, Scippa A (2022) Surface location error prediction in 2.5-axis peripheral milling considering tool dynamic stiffness variation. Precis Eng 76:95–109. https://doi.org/10.1016/J.PRECISIONENG.2022.03.008
72. Grossi N, Scippa A, Croppi L, Morelli L, Campatelli G (2019) Adaptive toolpath for 3-axis milling of thin walled parts. MM Sci J 2019:3378–3385. https://doi.org/10.17973/MMSJ.2019_11_2019096
73. Regel J, Bräunig M, Putz M, Richter C (2018) Industrial relevance and causes of thermal issues in machine tools
74. Sousa VFC, Silva FJG, Fecheira JS, Lopes HM, Martinho RP, Casais RB et al (2020) Cutting forces assessment in CNC machining processes: a criticalreview review. Sensors (Switzerland). 20:1–26. https://doi.org/10.3390/s20164536
75. Bhirud NL, Gawande RR (2017). Measurement and prediction of cutting temperatures during dry milling: review and discussions. https://doi.org/10.1007/s40430-017-0869-7
76. Hussain A, Lazoglu I (2019) Distortion in milling of structural parts. CIRP Ann 68:105–108. https://doi.org/10.1016/J.CIRP.2019.04.053
77. Schindler S, Zimmermann M, Aurich JC, Steinmann P (2014) Thermo-elastic deformations of the workpiece when dry turning aluminum alloys—a finite element model to predict thermal effects in the workpiece. CIRP J Manuf Sci Technol 7:233–245. https://doi.org/10.1016/j.cirpj.2014.04.006
78. Huang N, Bi Q, Wang Y, Sun C (2014) 5-Axis adaptive flank milling of flexible thin-walled parts based on the on-machine measurement. Int J Mach Tools Manuf 84:1–8. https://doi.org/10.1016/j.ijmachtools.2014.04.004
79. Cherif I, Outeiro J, Cotton D, Poulachon G, Charrondiere G, Brosse A (2019) Effects of toolpath and clamping strategies in machining distortion of stainless-steel parts. In: Procedia CIRP. Elsevier B.V. pp 427–431. https://doi.org/10.1016/j.procir.2019.04.065
80. Cerutti X, Mocellin K, Hassini S, Blaysat B, Duc E (2017) Methodology for aluminium part machining quality improvement considering mechanical properties and process conditions. CIRP J Manuf Sci Technol 18:18–38. https://doi.org/10.1016/j.cirpj.2016.07.004
81. Wang H, Wang H (2019) Numerical and experimental investigation of bulk stress distribution in edge under different clamping sequence. Assem Autom 39:523–531. https://doi.org/10.1108/AA-10-2017-131
82. Gameros A, Lowth S, Axinte D, Nagy-Sochacki A, Craig O, Siller HR (2017). State-of-the-art in fixture systems for the manufacture and assembly of rigid components: a review. https://doi.org/10.1016/j.ijmachtools.2017.07.004
83. Brusa E (2015) Mechatronics principles, technologies and applications, chapter: vibrations in machining by M. Sortino, G. Totis. Mech Eng Theory Appl
84. Altintas Y, Weck M (2004) Chatter stability of metal cutting and grinding. CIRP Ann 53(2):619–642
85. Munoa J, Beudaert X, Dombovari Z, Altintas Y, Budak E, Brecher C et al (2016) Chatter suppression techniques in metal cutting. CIRP Ann Manuf Technol 65:785–808
86. Totis G, Insperger T, Sortino M, Stepan G (2019) Symmetry breaking in milling dynamics. J Mach Tools Manuf 139:37–59
87. Farahani ND, Altintas Y (2022) Chatter stability of serrated milling tools in frequency domain. ASME J Manuf Sci Eng 144:31011–31013

88. Marco S, Giovanni T (2023) Superior optimal inverse filtering of cutting forces in milling of thin-walled components. Measurement 206:112227
89. van Zyl D, Altintas Y, Ostling D (2022) Parametric design of boring bars with adaptive tuned mass dampers. CIRP J Manuf Sci Technol 38:491–499
90. Wan M, Qin X-B, Dai J, Zhang W-H, Sun X-L (2024) Parameters optimization for chatter-free milling tool path based on spindle speed variation method. CIRP J Manuf Sci Technol 50:65–79
91. Du J, Liu X, Long X (2023) Coupled LQG with robust control for milling chatter suppression. Int J Mech Sci 243:108051
92. Navarro-Devia JH, Chen Y, Dao DV, Li H (2023) Chatter detection in milling processes—a review on signal processing and condition classification. Int J Adv Manuf Technol 125:3943–3980
93. Totis G, Sortino M (2020) Polynomial chaos-kriging approaches for an efficient probabilistic chatter prediction in milling. Int J Mach Tools Manuf 157:103610
94. Wang Z, Liu S, Li H, Ou J, Peng D, Li Z (2024) A novel method for updating time-varying information of milling thin-walled components based on digital twin. IEEE Sens J 24:2531–2546
95. Field M, Kahles JF (1964) The surface integrity of machined-and ground high-strength steels (Surface integrity of machined and ground high strength steels). DMIC Report. 210:54–77
96. Chu MQ, Ding RG, Zhang SY, Zheng JP, Zhang N (2021) Surface integrity for machining aerospace parts. Materials Report 35:7183–7189
97. Ulutan D, Ozel T (2011) Machining induced surface integrity in titanium and nickel alloys: a review. Int J Mach Tools Manuf 51:250–280
98. Liao Z, Schoop JM, Saelzer J, Bergmann B, Priarone PC, Splettstößer A et al (2024) Review of current best-practices in machinability evaluation and understanding for improving machining performance. CIRP J Manuf Sci Technol 50:151–184. https://doi.org/10.1016/j.cirpj.2024.02.008
99. Liu J, Chen G, Zhao L, Yu Z, Jia X (2023) Research status and development trend of cutting surface integrity of aerospace alloy materials. Int J Adv Manuf Technol 127:45–63
100. Gong L, Bertolini R, Bruschi S, Ghiotti A, He N (2022) Surface integrity evaluation when turning Inconel 718 alloy using sustainable lubricating-cooling approaches. Int J Precision Eng Manuf-Green Technol 1–18
101. Yu W, Ming W, An Q, Chen M (2023) Wear behavior of SiAlON ceramic tool and its effects during high-speed cutting. Ceram Int 49:26694–26706
102. Molaiekiya F, Aliakbari Khoei A, Aramesh M, Veldhuis SC (2021) Machined surface integrity of inconel 718 in high-speed dry milling using SiAlON ceramic tools. Int J Adv Manuf Technol 112:1941–1950
103. Salur E (2022) Understandings the tribological mechanism of Inconel 718 alloy machined under different cooling/lubrication conditions. Tribol Int 174:107677
104. Padhan S, Das SR, Das A, Alsoufi MS, Ibrahim AMM, Elsheikh A (2022) Machinability investigation of Nitronic 60 steel turning using SiAlON ceramic tools under different cooling/lubrication conditions. Materials 15:2368
105. Pervaiz S, Anwar S, Qureshi I, Ahmed N (2019) Recent advances in the machining of titanium alloys using minimum quantity lubrication (MQL) based techniques. https://doi.org/10.1007/s40684-019-00033-4
106. Pereira O, Rodríguez A, Barreiro J, Fernández-Abia AI, de Lacalle LNL (2017) Nozzle design for combined use of MQL and cryogenic gas in machining. Int J Precision Eng Manuf Green Technol 4:87–95. https://doi.org/10.1007/s40684-017-0012-3
107. Supekar SD, Clarens AF, Stephenson DA, Skerlos SJ (2012) Performance of supercritical carbon dioxide sprays as coolants and lubricants in representative metalworking operations. J Mater Process Technol 212:2652–2658. https://doi.org/10.1016/j.jmatprotec.2012.07.020
108. Pereira O, Rodríguez A, Fernández-Abia AI, Barreiro J, López de Lacalle LN (2016) Cryogenic and minimum quantity lubrication for an eco-efficiency turning of AISI 304. J Clean Prod 139:440–449. https://doi.org/10.1016/j.jclepro.2016.08.030

109. Khochtali H, Ayed Y, Zemzemi F, Bensalem W (2021) Tool wear characteristics in rough turning of Inconel 718 with coated carbide tool under conventional and high-pressure coolant supplies. Int J Adv Manuf Technol 114:2371–2386
110. M'Saoubi R, Axinte D, Herbert C, Hardy M, Salmon P (2014) Surface integrity of nickel-based alloys subjected to severe plastic deformation by abusive drilling. CIRP Ann 63:61–64
111. Arrazola PJ, Özel T, Umbrello D, Davies M, Jawahir IS (2013) Recent advances in modelling of metal machining processes. CIRP Ann-Manuf Technol. 62:695–718
112. Imbrogno S, Rinaldi S, Umbrello D, Filice L, Franchi R, Del Prete A (2018) A physically based constitutive model for predicting the surface integrity in machining of Waspaloy. Mater Des 152:140–155
113. Rinaldi S, Imbrogno S, Rotella G, Umbrello D, Filice L (2019) Physics based modeling of machining Inconel 718 to predict surface integrity modification. Procedia CIRP. 82:350–355
114. Xu D, Liao Z, Axinte D, Hardy M (2020) A novel method to continuously map the surface integrity and cutting mechanism transition in various cutting conditions. Int J Mach Tools Manuf 151:103529
115. Blakey-Milner B, Gradl P, Snedden G, Brooks M, Pitot J, Lopez E et al (2021) Metal additive manufacturing in aerospace: a review. Mater Des 209:110008
116. Vlaeyen M, Sinico M, Witvrouw A, Yildiz S, Gallego Bordallo FJ, Milton S, et al (2023) End-to-end additive manufacturing for a structural aerospace component. In: Proceedings of the 23rd international conference of the european society for precision engineering and nanotechnology. Euspen; 33 Rothersthorpe Crescent Mews, Northampton, NN4 8JD, UK pp 103–106
117. Gradl PR, Greene SE, Protz C, Bullard B, Buzzell J, Garcia C, et al (2018) Additive manufacturing of liquid rocket engine combustion devices: a summary of process developments and hot-fire testing results. In: 2018 Joint propulsion conference
118. Zhang C, Zou D, Mazur M, Mo JP, Li G, Ding S (2023) The state of the art in machining additively manufactured titanium alloy Ti-6Al-4V. Materials 16:2583
119. Malakizadi A, Hajali T, Schulz F, Cedergren S, Ålgårdh J, M'Saoubi R et al (2021) The role of microstructural characteristics of additively manufactured Alloy 718 on tool wear in machining. Int J Mach Tools Manuf 171:103814
120. Lizzul L, Sorgato M, Bertolini R, Ghiotti A, Bruschi S (2020) Influence of additive manufacturing-induced anisotropy on tool wear in end milling of Ti6Al4V. Tribol Int 146:106200
121. Yeo W-J, Choi H-J, Jeon M, Kim M, Kim Y-J, Jeong BJ et al (2024) Enhancement of optical surface quality based on real-time compensation of temperature-driven thermal errors in diamond turning. J Manuf Process 110:424–433
122. Zhu Z, Tang X, Chen C, Peng F, Yan R, Zhou L et al (2022) High precision and efficiency robotic milling of complex parts: challenges, approaches and trends. Chin J Aeronaut 35:22–46
123. Makulavicius M, Petkevicius S, Rozene J, Dzedzickis A, Bucinskas V (2023) Industrial robots in mechanical machining: perspectives and limitations. Robotics 12:160
124. Wang W, Guo Q, Yang Z, Jiang Y, Xu J (2023) A state of the art review on robotic milling of complex parts with high efficiency and precision. Robot Comput Integr Manuf 79:102436
125. Mun CH, Rezvani S, Lee J, Park SS, Park HW, Lee J (2023) Indirect measurement of cutting forces during robotic milling using multiple sensors and a machine learning-based system identifier. J Manuf Process 85:963–976
126. Guo K, Zhang Y, Sun J (2022) Towards stable milling: principle and application of active contact robotic milling. Int J Mach Tools Manuf 182:103952
127. Totis G, Adams O, Sortino M, Veselovac D, Klocke F (2014) Development of an innovative plate dynamometer for advanced milling and drilling applications. Measurement 49:164–181
128. Totis G, Bortoluzzi D, Sortino M (2024) Development of a universal, machine tool independent dynamometer for accurate cutting force estimation in milling. Int J Mach Tools Manuf 198:104151

129. Totis G, Bordon A, Scalzo F, Vaglio E, Sortino M (2024) Development of an innovative, posture-independent dynamometer for robotic milling applications. In: Under publication on the Proceedings of the 18th CIRP conference on intelligent computation in manufacturing engineering

Challenges and Opportunities for Composite Manufacturing from a Sustainable Perspective

L. Boccarusso, V. Antonucci, I. Bianchi, P. Carlone, L. Donati, A. Langella, G. Parodo, F. Quadrini, L. Raimondi, M. R. Ricciardi, L. Santo, M. Simoncini, L. Sorrentino, and F. Tucci

Abstract The white paper aims to identify the challenges and opportunities in the manufacturing of fiber-reinforced polymer (FRP) composites. It provides an in-depth exploration of the evolving landscape of sustainable composite manufacturing, addressing the urgent need for environmentally friendly alternatives across various industrial sectors in terms of materials, processes, and applications. Beginning with an introduction to the context and vision of composite manufacturing, the paper emphasizes the growing importance of sustainable practices. A comprehensive overview of next-generation materials and their applications is presented, with a focus on the use of sustainable and bio-derived resources such as natural fibers,

L. Boccarusso (✉) · A. Langella
Department of Chemical, Materials and Production Engineering, University of Naples Federico II, Naples, Italy
e-mail: luca.boccarusso@unina.it

V. Antonucci · M. R. Ricciardi
Institute for Composite and Biomedical Materials Technology CNR, Naples, Italy

I. Bianchi · M. Simoncini
Department of Industrial Engineering and Mathematical Science, Polytechnic University of Marche, Ancona, Italy

P. Carlone · F. Tucci
Department of Industrial Engineering, University of Salerno, Fisciano, SA, Italy

L. Donati · L. Raimondi
Department of Industrial Engineering, Alma Mater Studiorum—University of Bologna, Bologna, Italy

G. Parodo · L. Sorrentino
Department of Engineering and Sciences, Faculty of Technological and Innovation Sciences, Universitas Mercatorum, Rome, Italy

F. Quadrini · L. Santo
Department of Industrial Engineering, University of Rome "Tor Vergata", Rome, Italy

L. Boccarusso
Department of Civil and Mechanical Engineering, University of Cassino and Southern Lazio, Cassino, Italy

L. Fratini et al. (eds.), *Selected Topics in Manufacturing*, Lecture Notes in Mechanical Engineering, https://doi.org/10.1007/978-3-031-99501-9_13

recycled materials, and green matrices. Similarly, it explores both traditional and emerging manufacturing processes, including circular economy-based production, out-of-autoclave processes, and recycling technologies. In line with the information presented, the white paper also includes recent studies conducted or currently being carried out by members of the SIMPCo group (Sustainable and Innovative Manufacturing Processes for Composites), presenting their main findings and highlighting the remaining open challenges. SIMPCo is a thematic group within A.I.Te.M. (The Italian Manufacturing Association). Finally, the paper concludes with a summary of the key points discussed, offering future perspectives and recommendations for sustainable composite manufacturing practices.

Keywords Manufacturing materials · Composites · Recycling · Repairing · Sustainable · Biocomposites

1 Introduction

The manufacturing sector is at a pivotal moment, with environmental sustainability moving from a niche concern to a primary driver of innovation, particularly in composite manufacturing. In recent years, fiber-reinforced polymer (FRP) composites have gained prominence across industries, including automotive, aerospace, energy, sport and construction, due to their strength-to-weight ratio, corrosion resistance, and durability [1]. However, the environmental impact of traditional composite materials and manufacturing processes is under increasing scrutiny. The reliance on petroleum-based resins, energy-intensive processes, and complex end-of-life disposal options poses significant challenges in terms of sustainability. As global regulations tighten and consumer awareness of environmental impacts grows, the composite manufacturing industry faces an urgent imperative to adopt more sustainable practices while meeting performance demands and cost requirements.

The shift towards sustainable manufacturing is driven by a combination of regulatory, economic, and social pressures. Government policies and regulations now emphasize reducing carbon emissions, minimizing waste, and promoting circular economy principles, while consumers and clients increasingly demand environmentally responsible products [2, 3].

For instance, in different fields of application:

- EU regulation 2023/851 sets strict limits on CO_2 emissions, pushing automotive manufacturers to use lighter materials, such as natural fibers and low-impact composites, to reduce weight and improve vehicle efficiency.
- UNI EN 15804:2021 standard requires that construction materials meet specific sustainability and transparency requirements, promoting the use of composite materials with natural fibers or bio-based matrices to minimize environmental impact.

- Directive (EU) 2018/2001 (RED II) encourages the adoption of sustainable and recycled materials in the production of renewable energy technologies, including wind turbines.

This situation presents a dual challenge for composite manufacturers: achieving sustainability without compromising on the technical performance that FRPs are valued for, such as high tensile strength, chemical resistance, and design flexibility. Meeting these expectations necessitates a comprehensive transformation of how materials are sourced, manufactured, and ultimately, disposed of or recycled. As a response, manufacturers, researchers, and policy-makers are working collaboratively to establish sustainable manufacturing standards and best practices tailored to the specific challenges of the composites industry [4].

One of the most promising areas in sustainable composite manufacturing is the development and adoption of alternative materials, innovation in composite manufacturing and recycling. Indeed, traditional FRPs rely heavily on non-renewable, synthetic fibers like carbon and glass, as well as petroleum-derived polymers, which are not biodegradable and generate substantial waste.

In contrast, sustainable material like natural fibers and bio-based polymers offer reduced environmental impact while trying to maintain performance characteristics.

These bio-derived materials not only decrease dependency on finite resources but also provide potential energy savings during production and disposal. Although these sustainable materials face technical challenges, such as variability in mechanical properties and compatibility with existing manufacturing processes, advancements in material science are trying to address these issues [2].

In addition to renewable raw materials, there is a growing emphasis on integrating recycled content into composite materials. This approach aligns with circular economy principles and has already shown promise in sectors like automotive, where manufacturers are incorporating post-industrial and post-consumer recycled fibers and resins. However, recycled composites face hurdles related to quality control, mechanical property retention, and regulatory compliance. Ongoing research and collaboration between industries, academic institutions, and governments aim to overcome these barriers, paving the way for broader adoption of recycled materials in high-performance applications.

Inevitably, this aspect is reflected in the selection and study of recycling technologies for FRPs. Recycling technologies, such as mechanical, chemical and thermal recycling methods, are evolving to meet the specific requirements of composite materials. These technologies aim to recover fibers and in some cases also resins from end-of-life components or manufacturing scraps, facilitating the reuse of materials in new applications. While the current recycling rate for composites remains low, advances in these technologies, coupled with supportive policy measures, are expected to increase the viability of composite recycling on an industrial scale.

Finally, the sustainability of composite manufacturing also hinges on innovations in production processes. Traditional manufacturing processes for composites, including autoclave curing process, are energy-intensive and have a large carbon footprint. Emerging techniques, such as out-of-autoclave (OOA) processes, automated layup, and additive manufacturing, are transforming the industry by enabling energy-efficient and low-waste production. OOA processes, for example, eliminate the need for high-temperature curing ovens, resulting in significant energy savings. Automated layup and digitalization further optimize material usage and reduce production time, making manufacturing both economically and environmentally viable [5].

This white paper provides a comprehensive exploration of the current state of sustainable composite manufacturing. It examines a variety of promising sustainable materials, innovative processes, and practical applications, all while considering the real-world challenges that arise in implementing these solutions. Recent research works conducted by SIMPCo section components, illustrate successful applications of these new technologies and materials and underscore the potential of sustainable practices in reshaping the future of composite manufacturing. Through a strategic combination of materials innovation, process efficiency, and circular economy principles, the composite industry can unlock new levels of environmental stewardship and economic resilience.

2 Current Challenges on Both Manufacturing and Materials

The concept of composite materials has been known to humanity since ancient times. The earliest examples of composites in construction date back to 3400 B.C., when the Mesopotamians created plywood by gluing wooden strips at different angles. Later, in 1200 A.D., evidence shows the use of a composite mixture for the production of bows.

However, when contextualizing modern fiber-reinforced composite materials, it is commonly agreed that their first historical appearance occurred in the 1930s. During this period, the combination of advancements brought about by the chemical revolution and the ability to produce glass fibers, pioneered by Corning Glass, led to the idea of using polymeric matrices combined with synthetic fibers as reinforcement. This marked the birth of glass fiber-reinforced polymers (GFRPs), which was widely used starting from World War II.

Since then, significant progress in terms of material diversity and production technologies has led to the development of various types of FRPs. These composites are manufactured in different ways and are used across numerous industrial sectors, such as automotive, marine, aerospace, and sports.

Figure 1 summarizes the chronological development of composite materials and their applications in several industries, including infrastructure, transportation, and communication, from 3400 B.C. to the present day [1].

Although there are various ways to classify different manufacturing technologies, it is common nowadays to categorize them into three main groups [2]:

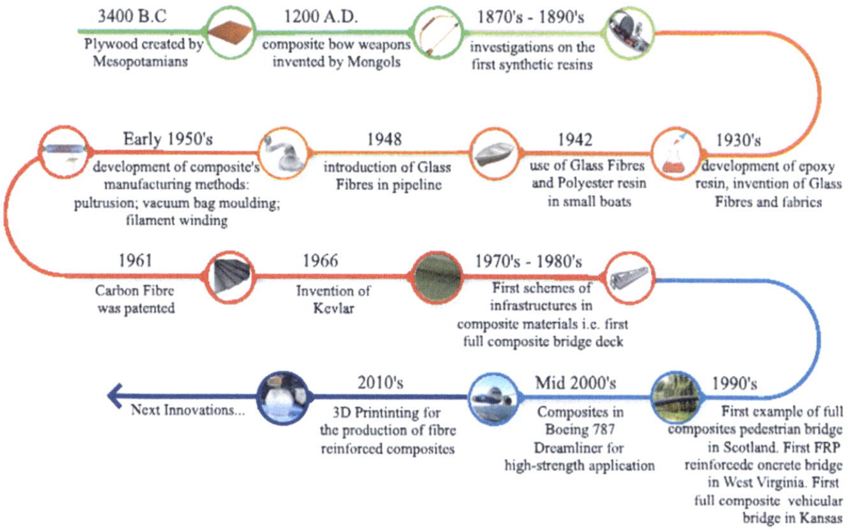

Fig. 1 Chronological evolution of composite manufacturing across various industries [1]

(i) **Conventional processes** (e.g., hand lay-up, spray-up, injection molding, compression molding, cold and hot press forming, autoclave, vacuum bagging, VARTM, LCM, filament winding, pultrusion, and so on).
(ii) **Automated processes** (e.g., automated tape placement, automated fiber placement, etc.).
(iii) **Additive manufacturing (AM) technologies** (e.g., stereolithography, selective laser sintering, continuous filament fabrication, and others).

Conventional and automated processes refer to manufacturing methods that rely on one or more tools or molds to achieve the final component shape [6]. In contrast, AM processes include all manufacturing methods where the part is built layer by layer directly within the manufacturing system, without requiring any additional tools, molds, or post-processing operations.

All three categories face ongoing challenges from both academic and industrial research perspectives, though these challenges exist at different levels.

For conventional technologies, the understanding of the processes is now fairly well-established. However, specific research areas remain, such as the exploration of next-generation materials, commonly referred to as eco-friendly materials, and the fine-tuning of related production processes and cycles. The use of materials such as natural reinforcements, thermoplastic matrices, biopolymers, vitrimeric matrices, and recycled materials presents a significant challenge for process optimization. Managing these materials, along with their chemical and physical differences compared to traditional materials, has a notable impact on production cycles and process parameters.

Another line of research aims to optimize the production process in terms of minimizing waste and energy consumption.

In the industry, the primary interest lies in the retrofit of industrial plants to achieve these objectives.

Differently from the first macro-category, in the case of automated deposition and additive manufacturing technologies, the maturity of the processes has not yet reached a level where basic research becomes marginal. In this scenario, research efforts are not limited to the aforementioned aspects but also include the refinement of the technologies themselves, which in some cases remain immature. Therefore, the main research areas focus on the refinement of emerging technologies, some of which are still underdeveloped, as well as on waste management and energy reduction. In this context, it should also be noted that the industry is pushing for the scalability of these technologies and for improving their productivity.

The true technological challenges, therefore, belong to the automated and additive manufacturing classes, while the challenges related to the use of next-generation materials remain open across all categories. It should be noted, however, that the use of innovative materials, while representing an emerging trend, is primarily studied and adopted by large companies, which have internal resources to fund research and development activities in this area. Conversely, small and medium-sized companies face issues related to the costs of such materials, the lack of established knowledge about their properties, and the objective difficulties in investing in new production facilities or modifying long-established production cycles.

Of great importance is the interest in life cycle assessment (LCA) studies to evaluate how much more sustainable composites truly are compared to conventional ones. In addition to addressing the challenges mentioned above, the composite materials sector is currently facing ambitious challenges, including the transformation of production and consumption models in line with the principles of the circular economy. In this context, the recovery and reuse of end-of-life products across various sectors and industrial processes is particularly significant. Moving away from the concept of waste or scrap in favor of reuse is central to achieving efficient resource utilization. Based on these premises, among the major challenges, it is important to include the study, and above all, the industrial-scale implementation of all processes related to the recovery and recycling of these materials, their management, and their utilization. This aspect becomes evident when observing Fig. 2, which clearly shows the continuous increase in demand for FRPs over the years, with further demand growth expected in the future.

Added to this, it is important to highlight the complexity of FRPs, which represents a significant obstacle to their recovery and reuse. Unlike monolithic materials, composites cannot be easily separated into their constituent components, making it difficult to recover the fibers or the matrix without compromising their mechanical properties and without incurring significant costs associated with obtaining recycled material. In fact, one of the main challenges lies in the absence of scalable and economically sustainable recycling technologies. Another issue is the lack of standardization in recycling and reuse processes. Given the variety of composite materials used across different industries, there is no single approach that can be

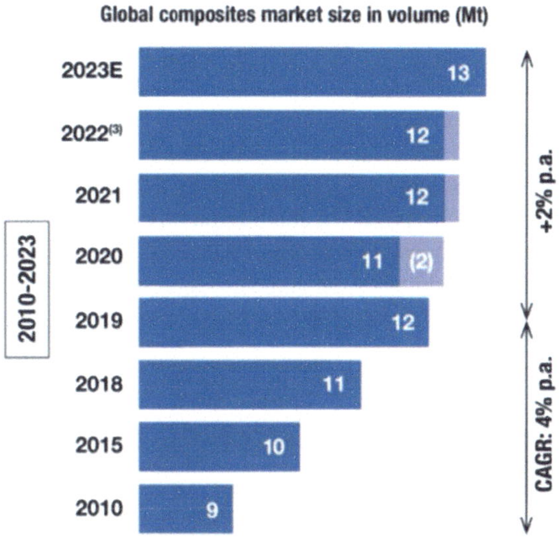

Fig. 2 Evolution of the composite materials market—1960-2023E(1)3—Global—In volume. *Source* JEC Observer 2024 [7]

universally applied. Furthermore, the limited availability of data on the performance and durability of recycled materials hinders their acceptance in industrial applications. This issue is further exacerbated by the lack of economic incentives and stringent regulations promoting the use of recycled materials in production sectors.

3 Towards Next Generation Composites

In the following sections, we discuss various classes of sustainable composite materials, highlighting their potential benefits as well as limitations. To effectively navigate this discussion and clearly interpret the terminology adopted, it is important to first clarify several sustainability-related concepts frequently used in literature, which can sometimes be ambiguous or overlapping. While the terminological distinction between synthetic fibres (such as glass and carbon) and natural fibres (animal, plant, or mineral) is fairly intuitive, some clarification is needed for polymer matrices.

Indeed, the term bio-based polymer refers to polymers derived wholly or partially from renewable biological resources, such as plant-based materials, agricultural biomass, or other living organisms. This definition clearly distinguishes them from conventional petroleum-based (fossil-derived) polymers. On the other hand, biodegradable polymers are defined by their capacity to decompose into simpler molecules through biological processes, typically involving microorganisms such as bacteria or fungi, under specific environmental conditions. Importantly, biodegradability does not depend on polymer origin. Both bio-based and petroleum-based polymers may exhibit biodegradability under suitable circumstances. Finally, the broader concept of green polymer matrices encompasses polymer matrices explicitly designed to minimize environmental impacts throughout their life cycle. This typically includes renewable raw material sourcing, reduced carbon emissions during production, enhanced recyclability, and biodegradability, whenever relevant to the specific application.

Establishing these distinctions clearly at the outset allows readers to better interpret the subsequent discussion, in which terms such as "bio-based", "biodegradable", and "green" polymers will be frequently encountered. Such clarification is essential to accurately evaluate polymer sustainability, emphasizing renewable sourcing and reduced dependency on finite fossil resources as key criteria guiding sustainable development and ecological transition strategies.

3.1 Sustainable Composite Materials

The concept of sustainability, particularly "sustainable development", has gained significant traction over the past four decades, stemming from the environmental movement of the late 1960s and 1970s. A pivotal moment arrived in 1987 with the publication of the landmark report "Our Common Future," which defined sustainable development as "development that meets the needs of the present without compromising the ability of future generations to meet their own needs" [1]. This Brundtland definition ignited a global surge of academic discourse and policy discussions, culminating in the United Nations Conference on Environment and Development (UNCED) in Rio de Janeiro in 1992.

The scope of sustainability broadened to encompass a harmonious balance between economic, social, and environmental considerations to ensure the availability of resources for future generations. Consequently, key environmental sustainability parameters include water, energy, and chemical usage, greenhouse gas emissions, and the generation of wastewater, waste materials, and chemicals throughout the entire supply chain. Currently, Life Cycle Assessment (LCA) methodologies, encompassing various approaches such as cradle-to-cradle, cradle-to-grave, and cradle-to-gate, are employed to comprehensively assess environmental impacts.

In recent decades, this principle has fortunately permeated various sectors, including the composite materials industry, whose demand, as discussed above, is continuously increasing. Within the specific context of materials, sustainability implies minimizing environmental impact throughout the entire product lifecycle, from production to disposal, prioritizing the use of renewable resources, waste reduction, and the optimization of production processes.

Regarding materials, it is essential to consider those that fall into the category of sustainable composites, better known as bio-composites. However, the use of the term "bio" or "eco-friendly composite materials" is often subject to speculation and abuse, also due to the absence of a single regulation that provides a clear definition of bio-composite. At most, the concept is supported by international standards such as ISO 14021 for recycled materials, ASTM D6866/ISO 16620-2 for biobased content and the work of CEN/TC 411 on bio-based products.

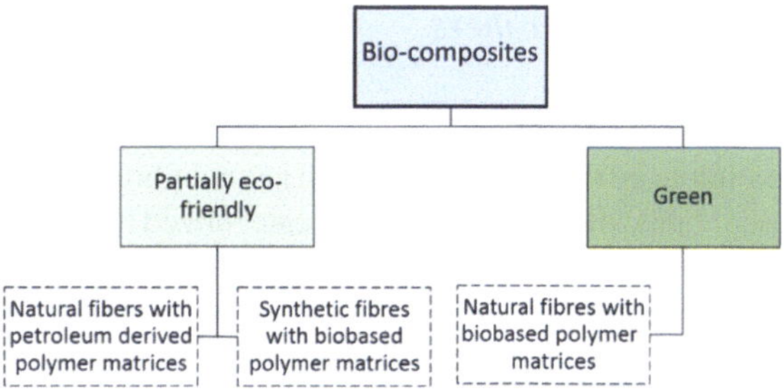

Fig. 3 Classification of bio-composites based on material constituents

Despite the absence of a regulatory definition, in the scientific literature the term bio-composites is generally recognized as a material in which at least one of the two phases (matrix or reinforcement) is of renewable or recycled origin [8]. However, the level of environmental sustainability of a bio-composite varies depending on the constituent materials and their ability to biodegrade. According to the literature, bio-composites can be divided into several categories, distinguishing between those that are truly biodegradable and those that, although of renewable origin, do not easily decompose in the environment. Their actual eco-compatibility therefore depends both on the choice of raw materials and on the end of the material's life cycle. In Fig. 3 their classification is schematized. As a result of this non-uniform and global classification, it is currently really complicated to present an estimate of the global or European market assessment of the subject. Most of the available studies focus on what are the natural fiber composites (NFCs).

One of the most recent studies reports that their global market is experiencing significant growth. In 2022, the market was valued at $19.27 billion and is projected to reach $46.85 billion by 2028, with a compound annual growth rate (CAGR) of 16% during the 2022–2028 period. In Europe, considerable growth is expected, with the market estimated at about $11 billion by 2028 [9].

For Italy, reliable data are not currently available, but certainly the growing use is evidenced by the growing interest from companies such as:

Alkivio®: Specializing in the production of bio-composites and bio-pellets derived from paper waste, Alkivio® offers sustainable alternatives to plastic, formulating biodegradable, compostable and eco-sustainable materials [10].

Novamont: A leading company in the production of biodegradable and compostable bioplastics that has registered the MATER-BI brand for a family of biodegradable and compostable bioplastics usable as matrix for bio-composites [11].

MAIP Group: Through the BIOntier project, the MAIP Group, in collaboration with Bio-on, aims to develop new biobased bio-composites with advanced properties, intended for multiple applications, including the automotive sector [12].

SAPA and Stellantis: Within the European project Life Biobcompo, SAPA, in collaboration with FCA Italy (now part of Stellantis) and the Fiat Research Center, has developed low-density bio-composites with a thermoplastic matrix and cellulose fiber reinforcement for weight reduction applications in the automotive sector [13].

3.2 Natural Fibers

Considering the previously discussed points, the bio-composites that attract the most attention are those reinforced with natural fibres [14], regardless of the type of matrix used. Logically, natural fibres are categorized into vegetal, mineral, and animal fibres. Typically, vegetable fibres are the most used reinforcements for polymer matrices in engineering applications. These fibres are primarily composed of cellulose, hemicellulose, and lignin and can be derived from different parts of the plant, including the blast, leaf, fruit, or seeds. The differences between vegetable fibres are mainly due to variations in their chemical composition, as well as their physical and mechanical properties.

Blast fibres (like hemp, jute, flax, etc.) are particularly suitable for composite reinforcement, as their primary function in plants is to provide structural support. Leaf fibres (like sisal, banana, agave, etc.), on the other hand, are known for their superior toughness, while fruit (like coir) and seed (like cotton) fibres contribute to the composite's elastic toughness. Table 1 [15] summarizes some properties of some vegetable fibers.

The interest in vegetable fibres arises when comparing the specific properties of some of these fibres with those of glass fibres. Notably, their density is approximately 1.5 g/cm^3, significantly lower than that of glass fibres, which is around 2.5 g/cm^3. This, combined with their respective mechanical properties, makes the specific properties of certain vegetable fibres particularly attractive. Consequently, research is increasingly focused on evaluating the feasibility of replacing glass fibres in composite materials for specific applications.

Table 1 was adapted from Pickering et al. [15], who compiled data from multiple sources to illustrate the variability in the mechanical properties of natural fibres compared to E-glass.

Table 1 Mechanical properties of vegetable and E-Glass fibers

Fibre	Density (g/cm^3)	Length (mm)	Failure strain (%)	Tensile strength (MPa)	Young's modulus (GPa)	Specific tensile strength (MPa/ g cm^{-3})	Specific young's modulus (GPa/ g cm^{-3})
Ramie	1.5	900–1200	2.0–3.8	400–938	44–128	270–620	29–85
Flax	1.5	5–900	1.2–3.2	345–1830	27–80	230–1220	18–53
Hemp	1.5	5–55	1.6	550–1110	58–70	370–740	39–47
Jute	1.3–1.5	1.5–120	1.5–1.8	393–800	10–55	300–610	7.1–39
Sisal	1.3–1.5	900	2.0–2.5	507–855	9.4–28	362–610	6.7–20
Cotton	1.5–1.6	10–60	3.0–10	287–800	5.5–13	190–530	3.7–8.4
Coir	1.2	20–150	15–30	131–220	4–6	110–180	3.3–5
Wool	1.3	38–152	13.2–35	50–315	2.3–5	38–242	1.8–3.8
E-glass	2.5	Continuous	2.5	2000–3000	70	800–1400	29

Several vegetable fibres, particularly hemp and flax, offer promising mechanical properties, making them strong candidates for composite reinforcement. Their appeal is growing, with companies like Bcomp Ltd, Porcher Industries, Saertex, and Terre de Lin supplying flax textiles and prefabricated panels. Even traditional reinforcement suppliers like Hexcel are incorporating woven and stitched flax into bio-based prepregs.

In the automotive sector, hemp and flax are used in race car body panels, hoods, and commercial vehicle interiors to reduce noise and vibration. The marine industries are also adopting these materials, for example flax composites are emerging as a sustainable alternative to glass composites in marine applications, as seen in racing yachts like the Class40 CrossCall [12].

Naturally, there are both advantages and disadvantages associated with these fibres, as described below. The key approach is to maximize their intrinsic benefits to fully exploit their potential in the production of optimized bio-composites for specific applications while simultaneously investigating ways to mitigate their drawbacks.

Among the advantages, it is certainly possible to mention:

Low Carbon Footprint: Vegetable fibres come from renewable resources, reducing environmental impact compared to synthetic fibres. Their production generally requires less energy and results in lower CO_2 emissions. For example, the production of flax and hemp fibres has a significantly lower carbon footprint than glass fibres and requires up to 10 times less energy.

Additionally, cultivating these fibre crops contributes to carbon sequestration. A hectare of flax can capture up to 3.7 tons of CO_2 per year, helping to reduce greenhouse gas emissions. Hemp further enhances sustainability due to its high CO_2 absorption capacity, ranging from 8 to 22 tons per year. It grows rapidly (requiring only 100–130 days for harvesting) and, thanks to its dense foliage, effectively suppresses weeds, reducing or eliminating the need for herbicides and pesticides. Moreover, hemp is highly effective in phytoremediation, absorbing heavy metals such as lead, cadmium, and nickel from the soil, contributing to the decontamination of polluted lands [16–18].

However, it is important to recognize that here, as well as in most papers available in the literature, the focus has been placed solely on the carbon footprint. If one aims to address the broader topic of environmental sustainability for vegetable fibres, it must be clarified that this is a multifaceted concept. In fact, while the carbon footprint is a key indicator, other relevant factors such as land use (especially in competition with food production), water consumption, and potential social impacts should also be considered for a more comprehensive view.

Low Density: vegetable fibres have a density of approximately 1.5 g/cm^3, enabling the production of lighter components, which is advantageous across various industries. Besides leveraging this intrinsic property for lightweight composite production, current research also focuses on maximizing this benefit by designing structures inherently characterized by low density, for example by producing grid structures used as cores for sandwich structures [19–23]. Boccarusso et al. [16, 24] have demonstrated the feasibility of producing such structures through tailored manufacturing processes, achieving composite structures with a density as low as 0.31 g/cm^3, which serve as high-performance sandwich cores with excellent impact and flexural resistance, while also exploring the possibility of obtaining bio-inspired gradient structures [24]. Figure 4 illustrates the manufacturing process proposed in [16, 24], the used hemp fabrics and the corresponding sandwich structures made with hemp fibres and CFRP skins after impact tests. Meanwhile, Fig. 5 shows the application developed by McLaren, in collaboration with BComp, where flax fibres were incorporated into the seat shell of their Formula 1 car [25].

Vibration Damping Properties: Natural fibres exhibit excellent vibration damping capabilities. This characteristic can be leveraged both acoustically, as studied by Dragonetti et al. [26], and mechanically for vibration absorption, as highlighted by Pinto et al. [27], whose main results on damping tests of hybrid hemp/carbon specimens are presented in Fig. 6. Given these properties, it is not surprising that bio-composites reinforced with vegetable fibres are widely used in automotive, marine, and construction industries.

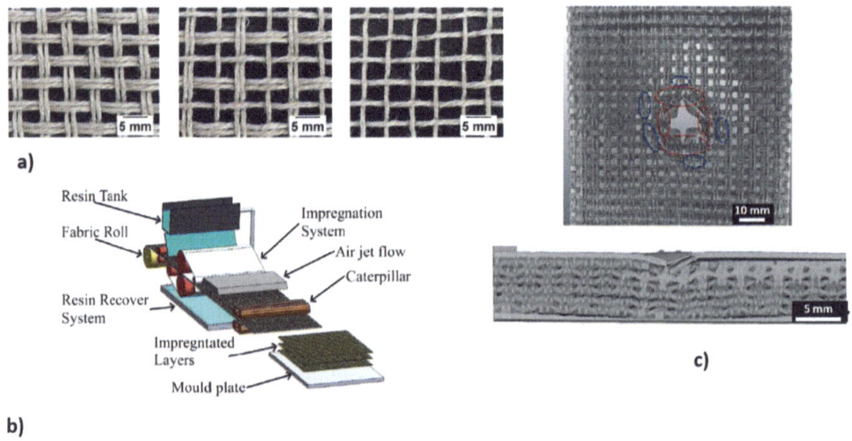

Fig. 4 Typologies of hemp fabrics (**a**); manufacturing process proposed (**b**) and sandwich structures impact tests (**c**) [14, 22]

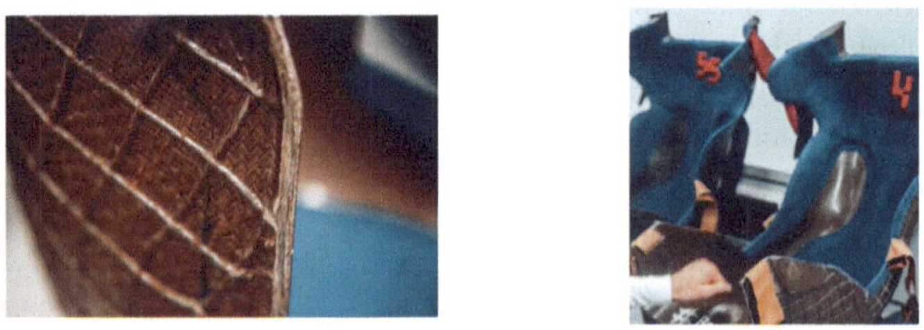

Fig. 5 McLaren seat section made from a three-dimensional grid structure placed on one side of the seat, utilizing Bcomp's proprietary PowerRibs technology [25]

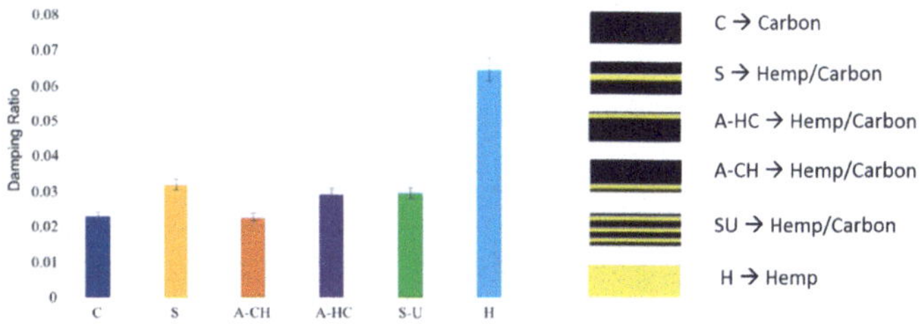

Fig. 6 Damping ratio histograms of all sample typologies studied in [27]

Safety and Health: Unlike synthetic fibres, vegetable fibres do not release irritating particles, reducing health risks for workers during fibre production, composite fabrication, and machining processes such as drilling. Furthermore, production waste from natural fibre composites, if uncontaminated, does not qualify as hazardous waste but rather as biodegradable waste, which can be composted rather than requiring specialized disposal.

Non-Abrasive: Another key advantage of vegetable fibre is their ease of processing, particularly in drilling and machining operations carried out on composites reinforced by these fibres. Unlike glass or carbon fibre composites, which tend to be highly abrasive and accelerate the tool wear, vegetable fibres do not cause excessive wear, leading to economic and operational benefits. These materials also generate less heat during machining, helping preserve tool integrity, and they do not produce harmful dust or debris. Naturally, this aspect must also be evaluated in relation to the type of matrix used in the composite.

Nonlinear stress–strain behaviour: Vegetable fibres exhibit a nonlinear mechanical response under load, influenced by their cellular structure and the intrinsic variability of their mechanical properties. This results in a stress–strain curve that does not follow a linear relationship, typically displaying an initial elastic deformation phase, followed by a visco-plastic phase before failure. This characteristic can be beneficial for applications where material ductility is required, as well as for the predictive modelling of mechanical behaviour, such as creep performance. In this context, analytical models traditionally used for polymeric materials can be applied, as demonstrated by Durante et al. [28] in their study on Hemp/PLA composites.

Among the disadvantages, it is certainly possible to mention:

Poor Adhesion to Polymer Matrix: The weak chemical compatibility between hydrophilic vegetable fibers and hydrophobic polymer matrices can lead to poor interfacial adhesion, compromising the mechanical properties of the composite. Several studies [14, 29] have demonstrated that adhesion can be improved through chemical treatments, such as alkalization, silane treatment, acetylation, or mechanical treatments like plasma treatment. However, while these treatments enhance mechanical performance, they also impact the overall sustainability of the product, as discussed by Das et al. in [30].

Low Thermal Resistance: Vegetable fibers begin to degrade at relatively low temperatures (~ 200 °C), limiting both processing methods and high-temperature

Fig. 7 Specimens at the end of the vertical burning test, with an increasing percentage of APP from left to right

applications [31]. This restricts the selection of thermoplastic matrices to those that can be processed within the thermal stability range of natural fibers, such as HDPE, LDPE, PP, PLA, and PVC.

Low Fire Resistance: vegetable fibers are combustible and offer lower fire resistance than synthetic fibers, which limits their use in fire-critical applications, such as transportation sectors. However, this drawback can be mitigated by incorporating flame retardants, either as additives in the matrix or as surface treatments on fibers. For example, in [17], it was demonstrated that hemp/epoxy composites containing ammonium polyphosphate (APP) at specific weight fractions achieved stringent fire resistance standards, meeting aerospace regulations outlined in FAR25.853A standard. Figure 7 shows test samples after vertical burning tests. Other studies, such as [32], have also shown that fire resistance can be enhanced using silica treatments.

Variability in Mechanical Properties: The mechanical properties of vegetable fibers can vary due to multiple factors.

By analysing Table 1, it becomes evident that there is a significant variation in mechanical properties due to differences in fibre geometry, density, and chemical composition, as well as plant species, growth conditions, soil quality, harvesting and processing methods, measurement techniques for tensile properties, and other influencing factors. The mechanical properties of vegetable fibres are highly dependent on their batch, variety, and agricultural conditions, including temperature, humidity, and soil mineral content, all of which significantly impact the final performance of natural fibre-reinforced polymer (NFRP) composites.

Furthermore, the selection of vegetable fibres for composite manufacturing is often influenced by their country of origin. For instance, flax and hemp are widely cultivated in Europe, making them readily available for industrial applications. Meanwhile, fibres such as coir, jute, kenaf, and ramie are primarily grown in Southeast and East Asia, where they are commonly used in textiles, including reinforcement fabrics for composites. Sisal, originally native to Mexico, is now predominantly cultivated in Brazil and several African countries such as Kenya, Tanzania, and Tunisia [31].

Addressing this issue is highly challenging and may never be fully resolved. However, strategies such as standardized agricultural practices and strict control over harvesting and processing conditions can improve fiber uniformity. Additionally, fiber engineering techniques—such as controlled alignment and morphological modifications can enhance mechanical properties and reduce variability [33].

Moisture Absorption: Similar to the adhesion issue, the hydrophilic nature of natural fibers makes them prone to moisture absorption, which can degrade mechanical properties and accelerate material deterioration. This issue can be mitigated through chemical treatments, as mentioned for adhesion enhancement [34, 35]. Additionally, this characteristic necessitates an extra processing step in composite manufacturing: fiber drying before molding. Typically, a drying treatment at 60 °C for 12 h is enough [17].

Cost Considerations: The prices of vegetable fibers fluctuate due to various factors:

- Raw material costs and harvest productivity
- Trade regulations, such as tariffs and import/export restrictions, can impact prices, especially for fibers sourced from specific regions
- Environmental policies promoting sustainable materials can drive demand, influencing costs.

While many sources [36] claim that natural fibers are cheaper than synthetic ones, these comparisons often overlook long fiber fabrics specifically designed for composites. In reality, the price of woven natural fiber fabrics can sometimes exceed that of synthetic alternatives. For example, a 200 g/m^2 flax fabric costs approximately € 40/m^2, whereas a 200 g/m^2 satin weave glass fiber fabric (UTE 200P) costs only € 9/m^2 [37].

3.3 Green Matrices

In the field of Fiber-Reinforced Polymer composites, the term "green matrices" generally refers to polymer matrices that are specifically designed and developed with the goal of reducing environmental impact compared to traditional polymer matrices, which often derive from non-renewable fossil resources and can be associated with significant environmental issues, such as pollution and disposal challenges. In fact, green matrices can represent an innovative and sustainable approach in the world of composite materials, characterized by the use of renewable resources and the implementation of production processes that minimize greenhouse gas emissions and energy consumption. These matrices are generally derived from biomass or natural sources, such as vegetable oils, starch, lignin, and other bio-based polymers, which are not only renewable but are often also biodegradable or compostable, contributing to a more sustainable lifecycle of the material [38]. However, can a petroleum-based polymer truly be considered green? The answer to this question may be more complex than it seems at first glance. For example, using biodegradable matrices in applications that require long-lasting components may not be feasible with the polymer formulations currently available. As a result, even some traditional polymer matrices can be seen as green if they are designed to be recyclable or engineered to minimize their environmental impact throughout their lifecycle [39]. For example, conventional polymers that can be effectively recycled, thereby reducing the need for new materials and limiting waste generation, can also be labeled as green. In these cases, the emphasis is not only on the material's origin but also on its ability to be reused or recycled in a closed-loop system, minimizing resource consumption and waste production. The analysis presented in this section primarily emphasizes greenhouse gas emissions reduction associated with sustainable composite materials. However, it is important to acknowledge that sustainability encompasses broader ecological and societal dimensions (such as land use, water footprint, eutrophication potential, biodiversity impacts, and social factors) which are beyond the scope of the current discussion.

3.3.1 Bio-based Thermoset Matrices

The transition toward using green matrices in polymeric composites is motivated by the growing need to mitigate the environmental impact of materials across various industries. These matrices are generally derived from renewable sources such as vegetable oils, cellulose derivatives, and other bio-based polymers [40]. The primary appeal of green matrices lies in their ability to reduce the carbon footprint of composite materials compared to traditional synthetic matrices. The main bio-based thermosetting resins include bio-epoxy resins (derived from vegetable oils and lignin) [41], bio-phenolic resins (based on lignin and tannins) [42], furan resins (from furfuryl alcohol sourced from biomass) [43], bio-based polyesters and vinyl esters (synthesized from bio-derived acids), and bio-polyurethanes (produced from vegetable oil-based polyols). In [44], a bio-based structural adhesive based on epichlorohydrin/cardanol is analyzed, comparing its environmental, economic, and mechanical performance with a conventional epoxy resin. The life cycle analysis (LCA) highlights a significant reduction in the carbon footprint of the bio-based adhesive, with a 73% decrease in Global Warming Potential (GWP), along with 62%–73% reductions in abiotic depletion potential (ADP), acidification potential (AP), and ozone depletion potential (ODP). However, eutrophication potential (EP) increases by 44%, mainly due to the use of plant-derived raw materials, whose cultivation and processing contribute to nitrogen and phosphorus emissions into aquatic ecosystems. This case demonstrates that bio-based thermosetting resins can significantly reduce greenhouse gas emissions while requiring careful assessment of their indirect ecological impacts. Economically, the production cost of the bio-based adhesive is 77% lower than that of the reference resin, reinforcing its industrial competitiveness. Mechanical characterization indicates that the bio-based adhesive exhibits suitable properties for non-structural applications, albeit with lower stiffness compared to conventional epoxy systems.

The main bio-based epoxy resins currently available on the market are characterized by an increasing content of renewable carbon, supported by the use of bio-based diluents and curing agents (Table 2). These systems, developed by leading companies such as Cardolite, Entropy Resins, Sicomin, and Resoltech, offer a range of bio-based content in the mixtures, varying from 19% to over 40%.

The biodegradability of these matrices is a critical advantage, as it allows the material to decompose naturally at the end of its lifecycle, thereby reducing the accumulation of plastic waste and environmental pollution [45]. For example, furan-based epoxy resins, as explored by Marotta et al. [46], exhibit notable biodegradability, making them ideal for applications prioritizing sustainability. These resins are derived from furfuryl alcohol, a by-product of agricultural processes, and offer the dual benefit of waste reduction and the production of sustainable materials. The resulting epoxy resins not only possess commendable mechanical and thermal properties but also maintain a chemical structure conducive to biodegradation under appropriate environmental conditions.

Green matrices also hold promise for their adaptability to various processing techniques. For instance, the use of bio-based resins in composite manufacturing processes like vacuum infusion or compression molding has shown that these matrices can be tailored to meet the specific needs of different applications, ranging from lightweight automotive parts to sustainable building materials. Their compatibility with natural fibers, such as flax and hemp, further enhances their appeal in environmentally conscious design [47]. However, the higher viscosity compared

Table 2 Summary of some bio-based epoxy resins available on the market [41]

Supplier	Epoxy name	Bio-based diluent?	Curing agent name	Curing agent bio-based?	Mixture bio-based percentage (%)
Cardolite (Belgium)	Formulite 2500A	Yes	LITE2002LP	Yes	46.8
Cardolite (Belgium)	Formulite 2500A	Yes	Formulite 2401B	Yes	36.6
Entropy Resins (USA/Spain)	SuperSap INR	No	SuperSap INS	No	19
Gurit (UK)	AMPRO BIO	Yes	AMPRO BIO Slow	Yes	> 40
Resoltech (France)	1800 ECO	Yes	1804 ECO	No	33
Sicomin (France)	Greenpoxy 56	Yes	SD8822	No	43
Sicomin (France)	Greenpoxy 56	Yes	SD8824	No	42
Sicomin (France)	Infugreen 810	Yes	SD8822	No	29
Sicomin (France)	Infugreen 810	No	SD8824	No	31

to synthetic ones could generally represent a limit for their application [48]. In the context of automotive applications, Fantuzzi et al. [49] explored the use of sustainable composites for electric vehicles, demonstrating that bio-based matrices, when combined with natural fibers, can provide sufficient mechanical strength while significantly reducing the vehicle's overall carbon footprint. This application highlights the potential for green matrices to meet both performance and environmental requirements in demanding applications.

Despite the clear advantages in terms of reduced carbon footprint and renewable sourcing, the sustainability of bio-based thermoset matrices involves more complex considerations. Factors such as land-use changes due to biomass cultivation, potential competition with food production, increased water use, and impacts on local biodiversity are critical aspects often overlooked when focusing solely on greenhouse gas emissions. For instance, cultivating biomass feedstocks could lead to deforestation or loss of biodiversity if not managed sustainably. Moreover, socioeconomic issues, including labor conditions and land rights, may also influence the overall sustainability profile of these materials. These trade-offs highlight the importance of comprehensive sustainability evaluations beyond carbon accounting alone, as recently emphasized in literature [47].

Moreover, several technical and commercial limitations hinder the widespread adoption of green matrices. One of the primary challenges is the mechanical performance and durability of these matrices. Natural polymers, while environmentally friendly, often exhibit lower strength and stiffness compared to their synthetic counterparts, which can limit their applicability in scenarios where high-performance materials are required. Moreover, bio-based matrices are generally more sensitive to moisture than synthetic polymers, a factor that can significantly compromise their mechanical properties over time. Protective coatings are often required to ensure adequate durability, especially in environments where the material is exposed to varying humidity levels [47].

Another significant barrier is the cost of production. Although green matrices offer environmental benefits, they are often more expensive to produce than traditional polymers. This is due not only to the cost of renewable raw materials but also to the complexity of the synthesis processes required to achieve polymers with mechanical properties comparable to those of synthetic materials. In addition to production costs, the compatibility of green matrices with existing industrial processes poses a challenge. Many manufacturing facilities are optimized for processing synthetic polymers and introducing bio-based matrices may necessitate substantial modifications to equipment and processes, further increasing costs and complexity. Maiti et al. [1] discuss this issue, noting that the successful integration of green matrices into mainstream production will require significant advancements in processing technology.

Furthermore, the long-term performance of green matrices remains a concern. While laboratory tests have shown promising results, extensive field testing is needed to fully understand their behavior over time, particularly in harsh environmental conditions. Factors such as UV exposure, temperature fluctuations, and prolonged mechanical loading can all affect the performance of bio-based matrices, and more research is needed to establish their reliability for critical applications like structural components in construction and transportation [50].

3.3.2 Thermoplastic Matrices

The use of thermoplastic matrices in composites has grown significantly in recent years, both due to their potential for recyclability and sustainability, and for the possibility of being produced from renewable sources. Depending on their origin and sustainability characteristics, thermoplastic matrices can be grouped into two main categories: recyclable thermoplastics (regardless of their origin) and bio-based thermoplastics, derived from renewable resources.

Recyclable Thermoplastic Matrices

The future of composite matrices is increasingly being shaped by the growing demand for sustainability and compliance with new environmental regulations. As industries seek materials that can meet rigorous recyclability standards, thermoplastic polymers are emerging as the preferred choice for composite matrices. These polymers, whether derived from synthetic or bio-based sources, offer unique advantages in terms of recyclability, cost-effectiveness, and ease of processing, making them ideal for high-production-rate applications across various sectors.

Thermoplastic polymers are distinguished by their ability to be remelted and reshaped multiple times without significant degradation of their mechanical properties. This characteristic makes them highly recyclable, aligning with the increasing regulatory demands for sustainable materials, particularly in industries where large volumes of composite materials are produced. The automotive and aerospace sectors, for instance, are increasingly adopting thermoplastic matrices for components that require both high performance and environmental compliance.

As regulatory bodies worldwide tighten their environmental standards, the ability to recycle materials at the end of their life cycle has become a critical factor in material selection. Thermoplastic matrices are uniquely positioned to meet these requirements, as they not only facilitate easier recycling but also reduce the overall environmental impact through energy-efficient processing and the potential for closed-loop recycling systems. This is in stark contrast to thermosetting polymers, which, once cured, cannot be remelted and thus pose significant challenges for recycling and disposal.

Among synthetic thermoplastics, polypropylene (PP) stands out as a particularly advantageous option for composite matrices. PP offers an excellent balance of mechanical strength, chemical resistance, and cost-effectiveness. Its widespread use is supported by well-established recycling infrastructure, making it a practical choice for industries that require high production volumes with a focus on sustainability. PP's durability and resistance to environmental degradation ensure a longer lifespan for composite parts, which reduces the frequency of replacements and, consequently, the environmental impact associated with manufacturing and disposal. Moreover, the ability to recycle PP at the end of its lifecycle without significant loss of properties makes it a cornerstone of sustainable materials management in the short term. This aligns with the industry's current need for materials that are both economical and environmentally responsible, particularly as companies strive to meet new regulations on material recyclability [50]. Recent studies have demonstrated the effectiveness of PP as a matrix material in natural fiber composites, particularly when combined with flax fibers, due to its favorable mechanical properties and recyclability. In this context, a research conducted in the SLightWeight project (ref. A0375-2020-36531, funded under the POR FESR Lazio 2014-2020-Progetto T0002E0001, "Progetti di Gruppi di Ricerca 2020"), aims of exploring sustainable composite solutions for lightweight applications. It was investigated the mechanical performance of Flax/PP laminates manufactured via the vacuum bagging prepreg process, highlighting their potential for structural applications in the automotive sector. However, the study also emphasizes the lack of standardized testing protocols and specific regulations for these novel eco-friendly composites, which currently limit their adoption in industrial applications. While traditional materials such as metals and thermosetting composites benefit from well-defined qualification procedures, the introduction of Flax/PP composites in the automotive sector requires the development of appropriate evaluation methodologies to ensure compliance with industry standards. The results of the activities from this project confirm the feasibility of producing polypropylene/flax fiber components using aerospace-grade manufacturing processes such as autoclave molding and vacuum bagging [51]. The project demonstrated that these techniques allow for effective consolidation of thermoplastic composites, ensuring optimal fiber-matrix adhesion and structural integrity. This validates the potential of PP/flax composites for lightweight applications in sectors requiring high-performance and sustainable materials. Further studies on the durability of hybrid composite panels for the railway and automotive sectors are currently underway as part of the GreenMOB project (A0613-2023-078308, funded under the POR FESR Lazio 2021–2027). These investigations focus on the barrier effect of aluminum sheets in enhancing panel durability, while also addressing the significant challenges associated with the one-shot forming and bonding of aluminum sheets in composite structures. Additionally, the feasibility of manufacturing sandwich panels for railway applications is being explored. This final phase is crucial for the industrial adoption of these materials, particularly for large-area surface components. Ensuring the feasibility of manufacturing processes at scale is essential to meet the stringent structural and durability requirements demanded especially by the railway industries.

An emerging and promising class of thermoplastic matrices for composite applications is represented by cyclic oligomers, such as cyclic polybutylene terephthalate (CBT) oligomers and cyclic polycarbonate oligomers. Unlike conventional linear thermoplastics, these materials polymerize rapidly in situ upon heating during composite processing, resulting in low-viscosity monomers that effectively impregnate reinforcing fibers and achieve high-quality matrix distribution. This feature significantly simplifies manufacturing processes, including resin transfer molding (RTM), vacuum-assisted infusion, and compression molding, while maintaining the recyclability and reprocessability characteristic of thermoplastics. Moreover, cyclic oligomer-based thermoplastics offer substantial potential for enhancing composite sustainability, thanks to their ability to facilitate closed-loop recycling and reduced processing energy consumption compared to traditional thermosetting matrices. Recent studies have demonstrated that composites manufactured from cyclic oligomers exhibit mechanical properties comparable or superior to conventional thermoplastic matrices, thereby broadening their potential in structural and semi-structural applications [52].

Bio-based Thermoplastic Matrices

As environmental concerns intensify and the shift towards a circular economy accelerates, bio-based thermoplastics are gaining significant traction in the market. They maintain the advantageous properties of traditional thermoplastics, such as moldability upon heating and solidification upon cooling, while contributing to a reduced carbon footprint across various industries. In the realm of bio-based thermoplastics, polylactic acid (PLA) is gaining attention as a potential long-term solution for sustainable composite matrices [53]. PLA is derived from renewable resources such as corn starch and sugarcane, making it an environmentally friendly alternative to fossil-fuel-based polymers. Its biodegradability and compostability under industrial conditions add to its appeal, particularly for applications where end-of-life disposal is a critical consideration.

However, while PLA offers significant long-term potential, it currently faces challenges that limit its widespread adoption in high-performance applications. One of the main issues with PLA is its susceptibility to environmental degradation, particularly in the presence of moisture and high temperatures. This limits its use in applications that require long-term durability and stability, such as automotive or aerospace components [1].

Despite these challenges, ongoing research in polymer chemistry aims to develop PLA formulations that are less prone to environmental degradation while retaining their biodegradable properties. Advances in this area could make PLA a viable option for a broader range of applications, particularly as the infrastructure for recycling and industrial composting of bio-based materials continues to improve [46]. From a forward-looking perspective, PLA represents one of the most promising polymer formulations; however, its use for semi-structural applications is not yet feasible.

In practical applications, PLA-based composites have demonstrated notable mechanical improvements when reinforced with natural fibers [53]. For instance, PLA reinforced with 20 wt% sisal fibers exhibited a tensile strength of 80.6 MPa, a flexural strength of 249.8 MPa, and an impact strength of 106.06 kJ/m^2. Similarly, jute fiber-reinforced PLA composites, manufactured via compression molding, achieved a tensile strength of 64.13 MPa with a 30 vol% fiber content. From a thermal perspective, the degradation behavior of PLA biocomposites varies depending on the reinforcement used. PLA composites reinforced with 30 wt% pineapple fibers exhibited an initial degradation temperature of 325 °C, reaching a maximum decomposition temperature of 415 °C, with a weight loss of 80%. In contrast, hemp fiber-reinforced PLA composites (20 wt%) showed a slightly lower thermal stability, beginning degradation at 303 °C and peaking at 398 °C, with a weight loss of 95%. Moreover, PLA's environmental footprint remains significantly lower compared to traditional petroleum-based polymers. The biodegradation of PLA emits approximately 1600 kg of CO_2 per metric ton, which is considerably lower than polypropylene (1850 kg), polyethylene terephthalate (4150 kg), and polystyrene (2740 kg). Furthermore, the energy required for PLA production is 75.4 MJ/kg, with a 25–55% reduction in fossil energy consumption compared to petroleum-derived plastics.

One of the most national and recent applications of PLA as a matrix for bio-composites is being studied in the project PRIN 2022 (CUP E53D23005190006), which explores the feasibility of producing PLA-impregnated hemp yarns through a pultrusion process. The goal is to use these impregnated yarns for manufacturing specific laminated components or as a filler in filament-based additive manufacturing processes.

3.4 Hybrid Fiber Composites

The design flexibility of FRPs is not only evident in the selection of the type and size of fibrous reinforcement and the choice of matrix, but also in the orientation of the fibers, their content percentage, and the stacking sequence. With this in mind, the logical progression is to integrate two or more types of fibers within the same matrix system to combine their attributes and achieve a final product with distinctive properties. The materials resulting from this combination are referred to as fiber-hybrid composites (FHCs), and their primary advantage lies in mitigating the inherent limitations of each fiber type while retaining the benefits of both, allowing for synergistic effects that can produce properties that neither fiber type alone can provide.

Because of new generation of composite systems will have to minimize the environmental impact and considering that the most common hybridization of CFRPs materials, i.e. glass coupled with carbon, a very interesting approach that has attracted the attention of the scientific community over the last decade, is to replace glass fibres, with vegetable ones. Carbon/vegetable fiber hybrid composites offer potential benefits by combining properties, particularly in damage propagation and energy dissipation. Vegetable fibers, with their nonlinear response to external loads, enhance energy absorption and may help reduce the brittleness of carbon fiber composites. This can improve performance in applications requiring effective vibration damping. For example, Le Guen et al. [54] pointed out that the damping coefficient of carbon/flax hybrid composites was four times higher than composites reinforced with carbon fiber alone, but to the detriment of the elastic modulus and strength. Pinto et al. [27] studied the possibility to obtain hemp/carbon hybrid composites and they pointed out that the hybridization process affects the flexural behavior of carbon composites by changing the failure mechanism due to the combination of the brittle nature of

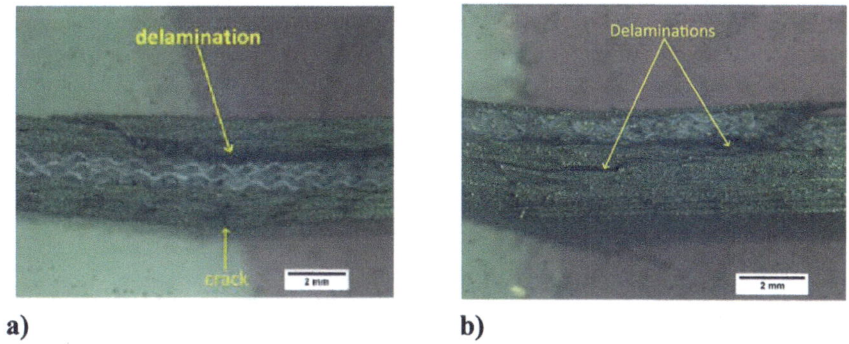

Fig. 8 Representative cross-section of hemp/carbon hybrid samples at the end of the bending test, with hemp layers positioned near the neutral axis (**a**) or close to the top surface (**b**) [27]

carbon with the gradual deformation of hemp fibers. By positioning hemp layers near the neutral axis, the flexural modulus remains similar to pure carbon, while placing them in different thickness positions enhances energy dissipation, increasing absorbed energy by 20% when located on the outer layers Fig. 8. Similar trends were observed in low-velocity impact (LVI) tests. When hemp fibers were near the neutral axis, impact energy absorption remained unchanged, as damage was confined to the hemp layers. However, positioning hemp on the impacted side allowed greater deformation around the contact area, improving energy transfer and reducing internal damage by nearly 40% for a 10 J impact.

Other examples of hybrid structures include sandwich panels, where a core made of natural fibers is combined with high-performance CFRP skins. An example of this is proposed in [16] and shown in Fig. 4. Additionally, another interesting solution to balance sustainability and mechanical performance is hybrid composites that combine virgin and recycled fibers. For instance, recently Mabalane et al. [55] highlighted the potential of hybrid panels consisting of one layer of recycled carbon fibers and two layers of virgin carbon fibers, comparing them to panels made entirely of virgin carbon fibers.

4 Innovations in Manufacturing for a Sustainable Prospective

Recent advances in composite manufacturing are increasingly driven by the need to enhance sustainability throughout the entire production chain. This section provides an overview of modern processing techniques designed to improve performance, reduce environmental impact, and enable recyclability. It also presents experimental research conducted by the SIMPCo group (Sustainable and Innovative Manufacturing Processes for Composites), a thematic group within A.I.Te.M. (Italian Association of

Manufacturing Technologies). In addition, the section discusses tools for evaluating sustainability, with a focus on Life Cycle Analysis (LCA). It concludes with a review of the latest developments in material regeneration, reuse, and recycling technologies for high-performance composites.

4.1 Modern Techniques in Composite Manufacturing

Sustainable components in fiber-reinforced polymeric materials are achievable via sustainability-oriented design strategies. This approach is based on a careful choice of materials, as described in the previous section, and on the optimization of the topology and reinforcement orientation [56]. The latter two factors are necessary to obtain components that properly respond to the specific solicitations during their in-service life while minimizing environmentally impactive characteristics, such as their weight [57]. In general, whatever the process is considered, manufacturing of fiber-reinforced polymers is based on three fundamental actions, namely raw materials combination, shaping, and consolidation. These actions can occur in different chronological orders or eventually can overlap depending on the manufacturing process adopted. The combination of the raw materials occurs by forcing the polymeric matrix to flow through the fibrous reinforcement, which must be oriented and well-constrained to avoid alteration of the properties [58]. The combination can occur online, whereas the mixing of the phases is part of the process, or offline in the manufacturing of pre-impregnated fibrous bundles or textiles [59]. The achievement of complex and well-oriented reinforcing architecture is fundamental to optimizing the specific mechanical and physical behaviors of the composite parts [60]. The shaping process must be carefully planned to avoid misorientation or displacement of the reinforcement and the consequent alteration of the stiffness matrix or the generation of poor-of-fibers weak spots [61]. Finally, the consolidation is based on the provision of heat in the case of thermoset matrices to complete the curing process, or on the absorption of thermal energy in the case of thermoplastics matrices to harden the softened or melted polymeric phase [61]. The consolidation action impacts mostly on the energy request, while combination and shaping mainly affect the specific properties, and therefore, the possibility to optimize the shape and the weight of the component. In synthesis, high specific performances and good selection of the process parameters are the key drivers for the sustainable manufacturing of fiber-reinforced polymers. The increasing social and government push towards the increase of sustainability and the reduction of the environmental impact determined the need to develop analytical methodologies to measure the sustainability performance and the impact relative to pollution of the atmosphere, contamination of sea, rivers, and ground, and depauperating of the natural resources [62]. Most analytical methodologies are based on energy considerations, evaluating the embodied energy of the raw materials adopted and the specific energy consumption of the manufacturing process [63]. Each raw ingredient adopted holds the energy necessary for the extraction of the elementary substances and for the previous processes carried out. This embodied energy assessment approach considers also the beneficial effects of recycling. According to this approach, in composite materials, the embodied energy of the reinforcing materials depends also on the efficiency of the manufacturing process of fibers, matrix, and any other ingredient included [64]. Consequently, for instance, the recent technological advancement in carbon fiber realization determined a relevant decrease in their embodied energy.

From this point of view, carbon fibers present values of embodied energy per unit of mass more than one order of magnitude higher than glass fibers (200/1500 MJ/kg for CF; 10/100 MJ/kg for GF), while natural fibers are the lowest impactive with an embodied energy lower than 10 MJ/kg [65–68]. Of course, the form of the ingredients adopted is another fundamental variable for the evaluation of the embodied energy. Indeed, preimpregnated textiles (prepregs) require complex and elaborate processing, and, consequently, hold higher embodied energy if compared to dry textiles or fiber rovings [2]. On the other hand, processing prepregs guarantees high specific properties, high control of the process, and easy handling and manufacturing. Due to these reasons, most modern technologies for composite materials manufacturing use prepregs as raw materials. In these cases, even if the embodied energies of the raw materials are sensibly higher, the specific energy consumption at the manufacturing level is significantly lower, and, at the same time, the geometry and the specific properties are dramatically optimized. The highest average value of the specific energy consumption is related to the autoclave curing of composites (400/600 MJ/kg), which, despite the high installation and operative costs and the environmental impact, remains the main technology to manufacture primary structures in composites in aeronautics, due to the high reliability and quality achieved [63]. Due to the high installation costs, energy demand, manufacturing management complexity, and autoclave volume occupation, numerous out-of-autoclave manufacturing processes have been developed [69]. The most widespread out-of-autoclave processes are liquid composite molding (LCM), compression molding, pultrusion, and filament winding. LCM is a wide class of processes based on the application of a gradient of pressure in a sealed environment to force a liquid resin system to flow through a dry fibrous preform laminated inside the mold. Depending on the driving pressure, the presence of heating systems, the dimension and shape complexity the specific energy consumption ranges among 15 and 100 MJ/kg [62, 63]. Processes based on compression obtain complex shapes, with high control of the geometries and consistent volume fraction by compacting and heating preimpregnated or partially impregnated layers between two temperature-controlled molds connected to a press [70]. In this case, thermal energy is the main factor determining the specific energy consumption, which lies in the range 10/80 MJ/kg [2].

As reviewed by Lunetto et al. [2], this significant range is due to the heterogeneity of data in the literature, dependence on sub-optimized processing parameters, scatters in materials characteristics, and final component size. In the compression process used for Sheet Molding Compounds (SMCs) and Bulk Molding Compounds (BMCs), the specific energy consumption values range from 3 to 18 MJ/kg, indicating that compression processes can be a compelling alternative to autoclave methods, particularly when dealing with high production volumes. Proof of concepts for the compression molding of SMCs/BMCs from recycled carbon fibers with both thermosetting and thermoplastic matrices has been demonstrated by several works [71, 72]. Primary challenges associated with the scalability of this material to industrially relevant applications include the lack of information about their mechanical properties, the lack of validated numerical tools to predict their processability [73], and the lack of standards to characterize their formability and flowability [74]. Besides the possibility of being molded into complex shapes, a valuable advantage of BMC and SMC compression molding is the possibility of integrating metallic inserts directly during the forming process [75], avoiding subsequent operations and minimizing waste. Prepreg compression molding (PCM) is a sustainable technology for producing natural fiber-reinforced composites, effectively managing continuous fiber preforms while minimizing waste [76]. The process is relatively simple and has lower energy consumption, making it particularly suitable for medium production. To achieve superior properties in natural fiber-reinforced composites during the compression molding process, it is essential to control and optimize three key parameters: temperature, pressure, and holding time. Each of these dimensions is equally important and must be effectively managed to develop the desired characteristics of the composite, as documented by several reviews [15].

In pultrusion fibers impregnated online or offline are guided through a heated die to produce constant cross-section profiles in fiber-reinforced thermoset or thermoplastic. Pultrusion is the only one continuous process to manufacture polymer-matrix composites. This aspect is extremely advantageous since it permits superior control, ease in automation, and optimized use of thermal energy. Indeed, the specific energy consumption is remarkably lower than the aforementioned processes, varying between 0.5 and 10 MJ/kg [2, 77]. Filament winding is based on the deposition of impregnated filaments on a rotating mandrel, to achieve convex hollow shells in composite. The filament deposition involves long cycle time, but, on the other hand, it guarantees the optimization of the reinforcement orientation at a local scale and high automation. The specific energy consumption in this case ranges between 3 and 8 MJ/kg [65]. The recent evolutions experienced by the manufacturing sector during the last two decades have led to the development of robotic deposition and additive techniques. Automated fiber placement, automated tape laying, and 3D printing of continuous fiber-reinforced composites are examples of this industrial ferment towards automation, high control of the local properties, and additive approach in composite manufacturing. In these cases, the values of specific energy consumption are dramatically lower than the ones reported until this point, and the constant push towards quality improvement and optimization promises further decreases in the environmental impact of manufacturing of composites [78]. The management of the end-of-life is the last crucial point in the sustainability of composites [79]. Traditionally, polymer-composites manufacturing has been based on thermoset matrices since they are characterized by low viscosity, and therefore they flow through the reinforcement and impregnate it efficiently. Nevertheless, after completion of the curing reaction, thermosets cannot be softened or remelted. This aspect makes challenging the management of the end-of-life, and the composites cannot be recycled without a relevant loss in their value and functionality. Recently, thermoplastic matrices have gained increasing relevance in composite manufacturing, thanks to technical development and synthesis of new molecules which are improving their applicability. In this case, apart from recycling, several technological processes permit the post-processing of the composite, by healing, reshaping, and/or assembling components by welding [80, 81].

4.1.1 Some SIMPCo' Innovations

In addition to the energy required for the production of raw materials and the final product, it is also necessary to consider the amount of material contaminated by matrices, especially when dealing with thermosetting resins, and the possibility of reducing related waste. This aspect is particularly relevant for all processes characterized by high waste, the use of disposable materials, and the application of thermosetting resins, even if they are bio-resins. A striking example is the Resin Infusion under Flexible Tools (RIFT) process, which belongs to the LCM processes and is typically associated with a high production of contaminated waste material.

The RIFT process has been analysed because it is highly flexible and widely used for various applications, offering the possibility to design the reinforcement in advance and to produce high-quality parts with high reproducibility. In the case of RIFT, in fact, the composite consolidation occurs after the impregnation of a proper selected dry net-shape fiber preform by a polymeric thermoset resin. Therefore, the quality of the realized part depends on the complete impregnation of the fibers and the complete cure cycle. In particular, the conventional RIFT basically involves the following steps: lay up of a fiber preform onto a one-sided rigid mold, vacuum application after the assembly sealing by a formable vacuum bag material, fiber impregnation by thermoset resin and cure of the resin. Vacuum is applied through a single or multiple injection points in order to remove the air from the fiber preform and to drive the fiber impregnation of the part by resin. Further, to ease the resin flow and to avoid voids and dry spots, a resin distribution net medium is placed onto the reinforcement. This kind of technology is characterized by low investment costs due to the use of a one-sided mold and low operating costs, related to the low temperature and pressure. However, RIFT process has some drawbacks, that have to be considered. It is a manual process, strongly related to the operator skill and experience. Due to the large use of several ancillary materials, such as peel ply, vacuum bag, distributing net, it has a high amount of waste. Further, large parts produced by vacuum infusion can have a not uniform thickness. Thus, researchers constantly aim to the development of more efficient and sustainable vacuum infusion techniques in order to improve the properties of the final composite part, reduce waste and enhance the automation [82]. For example, Ricciardi et al. [83] developed Pulsed Infusion technology, that is a double-bag vacuum infusion process, based on the use of a proper designed reusable pressure distributor and able to better control the vacuum pressure in pulsed way by an automated system. The adoption of a reusable distributor instead of a distributing net significantly reduces waste materials and resin consumption, lowering the material costs. In addition, since the transverse resin flow is promoted, the flexural properties of the final composites resulted higher than those of composites produced by conventional vacuum infusion. In addition, these findings were confirmed also by the adoption of Pulse Infusion (see Fig. 9a) to realize hybrid composites based on nanotubes modified thermoset resin. Due to resin permeation enhancement and better distribution of nanotubes through the fiber tows, the composites showed better interlaminar and fracture properties (see Fig. 9b) [84]. Other authors have implemented new automated vacuum infusion processes. In particular, Wang et al. [85] automated the process by a vacuum infusion workstation able to monitor processing parameters, like, laminate thickness, resin mass, mold temperature and resin infusion pressure and assessed higher fiber volume fractions for the produced composites thanks to the reduce material and process variability. Further, Grisin et al. [86] have proposed the use of a vacuum chamber infusion, similar to the chamber sealing machine used for packaging to perform the resin impregnation. The resin and the dry fibers are positioned in a vacuum bag, but in a way that they do not touch and a complete evacuation is possible. The filled bag is then evacuated in the vacuum chamber and sealed. After ventilation of the chamber, the impregnation occurs. The realized composites were characterized by void content similar to that of composites produced by conventional vacuum infusion, but low fiber volume fraction due to not performing process optimization. On the other hand, the main advantage of the vacuum chamber infusion approach was the reduction of consumables, set-up assembly and manufacturing time. It's important to note that the use of a closed chamber has limitations on the size of the composite part that has to be produced.

Thus, the above analysis evidences that extensive research has been taking place and needs to be performed to improve the sustainability of RIFT not only in terms of using natural or green raw materials, but also in terms of cost-efficient automation.

In the context of liquid composite molding, high efforts have been devoted to developing resin flow-cure optimizing strategies, allowing an increase in quality, a reduction in the amount of scrap and waste, and a sensible decrease in energy consumption and volatile compound emissions. The microwaves systems for resin preheating demonstrated a sensible improvement in filling and saturating the fibrous preform [87]. The resin flow is guided by a pressure gradient through the microwaves

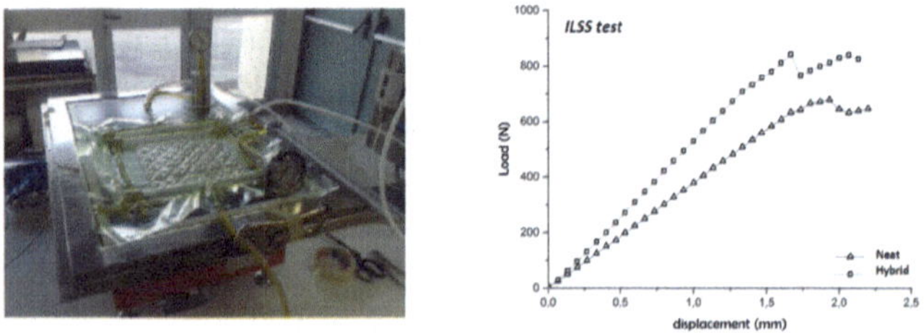

a) b)

Fig. 9 Pulse infusion system (**a**) and ILSS results of [84] (**b**)

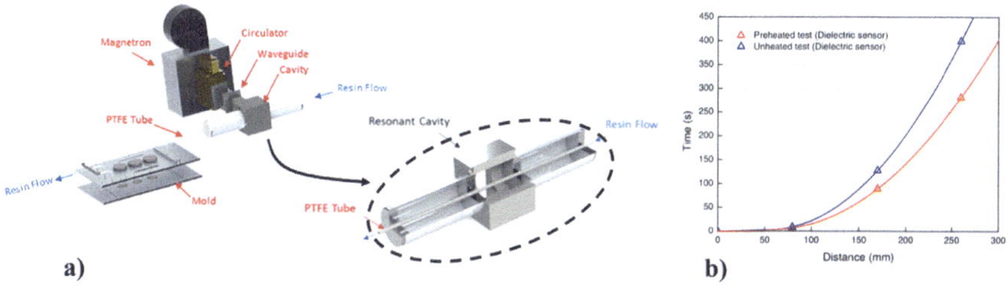

Fig. 10 Microwaves preheated liquid composite molding setup (**a**) and flow enhancement compared to a conventional infusion setup (**b**) [88]

resonant cavity in controlled flow conditions. The length of the resonant cavity and the resin flow must be carefully calibrated to control the exposure to microwave of the resin and avoid premature cure of the mixed thermoset system [88]. The preheating is planned on the base of previous cure kinetics and rheological characterization of the resin system and sensibly decreases the cycle time. Moreover, the microwaves technology maximizes heat transfer with a visible improvement in the energy efficiency of the process. Finally, as shown in Fig. 10a, this setup is completely sealed and minimizes the emissions of volatile compounds. The results achieved in terms of reduction of filling time in the vacuum resin infusion in a closed mold are illustrated in Fig. 10b [88].

The recent developments in thermoplastic polymers allowed industry and academics to pursue sustainable manufacturing by substituting conventional thermoset materials with innovative re-shapeable/remeltable thermoplastic polymers to realize composite matrices. A refulgent example is the adoption of the pultrusion process to manufacture thermoplastic matrix composites. It is worth noting that pultrusion is considered a highly efficient process in terms of specific energy consumption if compared to the other processes to manufacture composites even in its conventional version [2]. Several approaches have been proposed in the literature to convert the process towards the thermoplastic matrices [89]. A promising strategy is the tape-pultrusion, in which preconsolidated tapes constituted of fiber rovings preimpregnated with thermoplastic polymer are pultruded to realize a constant cross-section composite [90]. An example of tape-pultrusion dies is shown in Fig. 11a, b [91]. The tapes are pulled through a heating-forming die where the thermoplastic phase is softened and melted while compacted in the heating-forming die, presenting a convergent cavity to compress the advancing material. Afterward, a cooling die consolidates and decreases the temperature of the pultruded composite. This system, which has been widely experimented for polypropylene [5] (Fig. 11c), has been recently adopted to realize recycled PET matrix composites (Fig. 11d) [92].

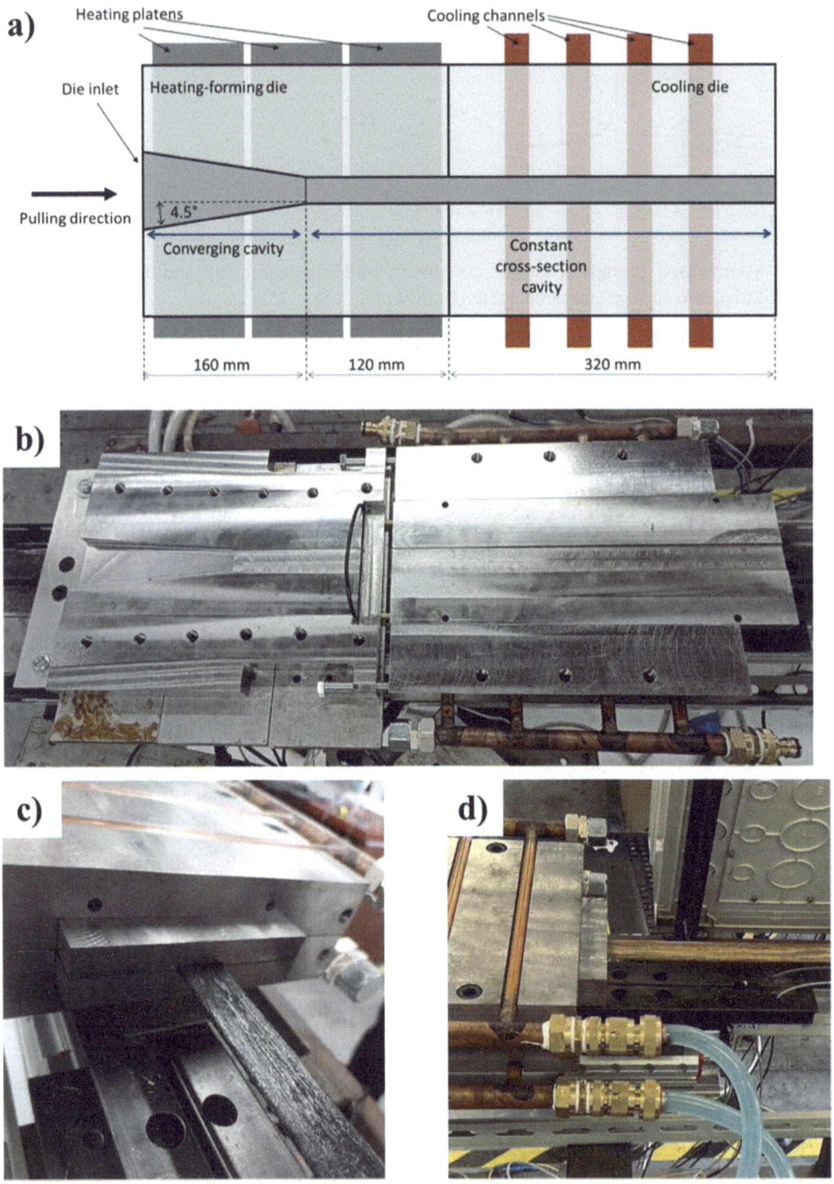

Fig. 11 Schematic of the heating/forming die and cooling die (**a**); **b** view of the internal cavities of the dies; **c** obtained pultruded profile; **d** Pultrusion setup while processing glass/polypropylene tapes

4.2 *Environmental Impact of Manufacturing Processes*

4.2.1 Sustainability in Composite Manufacturing

Despite their benefits in weight reduction and lower fuel consumption in transport, composites pose significant environmental burdens, especially in raw material production and component manufacturing. Their impact reduction is widely studied, with various sustainability solutions emerging. Given the high global demand, their environmental footprint is a critical concern. Scientific literature highlights that the production of polymer matrix composites significantly contributes to their carbon footprint [93]. Depending on the matrix type (thermosetting or thermoplastic), different processes are used, all requiring high temperatures for melting or curing, often with autoclave pressure, leading to high energy consumption. Additionally, raw material extraction and processing, mainly from petroleum-based sources, further increase environmental impacts due to intensive energy use. The sustainability of composite products depends on various life cycle factors, including design, material selection, production, use, and end-of-life management, all closely interconnected. To assess their footprint, studies typically use a Life Cycle Assessment (LCA) to quantify environmental impacts across categories like Cumulative Energy Demand (CED) and Global Warming Potential (GWP). LCA can also be combined with Life Cycle Costing (LCC) to evaluate economic impacts. These analyses follow a cradle-to-grave approach, covering raw material extraction, manufacturing, use, and disposal or recycling. The standardized LCA framework includes four iterative phases: (i) goal and scope definition (defining the system and study objectives); (ii) life cycle inventory (collecting data on materials, energy, and waste flows); (iii) life cycle impact assessment (calculating potential environmental impacts); (iv) results interpretation (identifying conclusions, critical issues, and improvement opportunities). LCA and LCC studies highlight challenges in composite production, guiding research toward more sustainable and efficient manufacturing processes.

4.2.2 Potential Issues

Considering composites manufacturing, sustainability can strongly vary depending on the specific materials and processes considered.

For what concerns raw materials production, at an industrial level, composites are typically constituted of synthetic fibers (such as glass or carbon fibers) and a petroleum-based polymer matrix. Hence, high-energy consumption is typically required for the production of both reinforcement fibers and polymeric matrix. According to recent studies, raw materials production can account for the majority of CFRP life cycle impacts (i.e. up to 90% of the total). Especially the production of the reinforcing fibers is highly energy intensive and can determine the majority of raw materials impact. Despite the low densities of both fiber and reinforcement constituents with respect to traditional materials such as metals, their high unitary impacts (i.e. impacts per kg of material) makes their use a potential issue on an environment level. For instance, epoxy resin have an embodied energy ranging from 76 to 80 MJ/kg, with impacts equal to about 3 kg of CO_2 eq per kg in terms of Global Warming Potential [94]; for what concerns the reinforcement, for example, virgin carbon fibers have a primary energy demand between 198 and 595 MJ/kg, with GWP value as high as 54 kg of CO_2 eq per kg [95]. These impacts derives from the energy intensive production processes such as fiber spinning and carbonisation that are performed on a synthetic precursor (polyacrylonitrile). On the other hand, glass fibers have much lower embedded energy and GWP (7–16 MJ/kg and about 2.5 kg of CO_2 eq per kg, respectively). However, their lower mechanical properties and higher density compared to carbon fibers make them unsuitable for specific high-performance applications. The embodied energy of raw materials also depends on the form in which they are supplied. For example, fabric manufacturing and prepregging require energy use equal to 2.6 MJ/kg and 40 MJ/kg respectively [67]. The additional processes required to obtain this feedstocks can also require high temperatures and electric energy use for the machines, resulting in increased carbon footprint. Indeed, the impacts value sensibly vary depending on the analyses (e.g. due to data uncertainty) and specific production conditions; despite that, it is clear that composite feedstock production is a relevant issue and improvements strategies are required to reduce virgin material use and develop more sustainable production routes.

Manufacturing processes for composite components often require considerable energy inputs and generate waste of raw materials constituents and consumables. Polymeric composites manufacturing processes can be classified considering the matrix material or considering the level of automation of the process. Depending on each technology, impacts can strongly vary and are associated with different materials, energy and waste flows. Regardless of the specific technology, literature Life Cycle analyses focusing on composite manufacturing processes typically consider the following elements: raw materials, molding machines and their associated energy consumption, tools such as mold and countermolds, consumables materials such as release agents and vacuum bag layers and personal protection equipment, and possible waste and scrap production and disposal.

In addition, LCC analysis may also include items such as human labour cost.

Traditional processes such as autoclave molding require high temperature and pressure to ensure complete impregnation of the fibers and reduction in void content; this typically results in energy intensive processes.

Literature studies reported specific energy consumption between 20 and 600 MJ/kg [2] for autoclave molding. Great energy consumption variability is observed also in this case due to different processing conditions; higher values are typically obtained for the production of high performance parts (e.g. for the aeronautic components) for which low production volumes and low saturation of the autoclave capacity are considered; on the other hand, the specific consumption can be reduced in the case of large mass production. This impacts range highlights the improvement possibilities that are possible for the composite sector. Even though autoclave molding is preferred as high performance parts are required, other molding technologies are usually characterized by lower specific energy use and carbon footprint. This depends on different impregnation and curing systems used that lead to lower energy demand; some examples of traditional molding processes are Vacuum Assisted Resin Infusion (VARI), with specific energy demand typically between 3 and 11 MJ/kg, and Resin Transfer Molding, with specific energy demand typically between 12 and 13 MJ/kg [2]. Hence, a correct selection of the process and process conditions are a feasible way to reduce the impacts of the manufacturing stages. It is worth noting that the impacts associated with the production of molding machines are usually not included within literature LCA analyses irrespective of the technology. In fact, since their service life is much longer than the temporal system boundaries of the analyses (e.g. related to the production of one component or a batch), they would have negligible impacts associated with the functional unit.

Tools production can be another relevant contributor the components life cycle emissions. For traditional processes, rigid molds are used for the lay-up phase and give the part its shape. Flexible or rigid counter molds can also be employed in some processes (e.g. in RTM or hot press forming). Metals or composites molds can be employed; typically, these tools are heavy and associated with relevant costs and environmental impacts [96].

Indeed, the production of a single mold can have significantly higher environmental impacts than other stages of composite manufacturing (e.g., energy consumption, use of composite raw materials). For this reason, the appropriate design and material selection of the tools have great importance from a sustainability perspective. These considerations are closely related to the expected production volume, as the environmental burden of the mold must be distributed over the number of parts produced. Since the tools can be reused multiple times before relevant wear phenomena arise, their impacts can be allocated on numerous composites parts (i.e. from few hundreds to thousands depending on the mold materials). Long tools service life and sufficiently high production volume is important to guarantee low impacts associated with mold production. As a matter of fact, in low-volume production scenarios (e.g. for specific or customized components) the tooling phase can represent a significant contribution to the overall environmental impact and may become a critical issue. This is the case, for instance, of production processes such as compression and autoclave molding, for which heavy and robust molds are typically required. However, in medium to high production volumes typical of industrial manufacturing, the impact of tooling can become almost negligible when allocated over the entire production batch. In addition, recycling metal tools at the end of their service life further contributes to reduce tooling phase relevance. Depending on the process, different tools may be used; some examples are the rotating mandrel in Filament Winding, the die in pultrusion and nozzles in 3D printing. Considerations similar to those for traditional processes can be made. Consumables materials play a crucial role in the manufacturing of composites and they vary depending on the process. Some examples for manual lay-up, VARI and autoclave molding processes are vacuum bagging materials such as polyamide vacuum bags, sealant adhesive tapes, breather, bleeder and release films [97]. Other examples include infusion materials such as flow mesh fabrics and resin distribution channels. Consumables are a significant source of waste in composite production as they are typically disposed of after just one production process. These materials have only a fraction of the weight of the composite parts; however, their use can be associated with relevant environmental impacts according to specific categories. For instance, PTFE release film can be responsible of the majority of CFRP parts life cycle according to "ozone depletion" impact category [98]. Solutions to minimize the use of consumables, reuse them or recycle the materials are needed to improve this aspect of manufacturing.

Finally, a relevant issue associated with composites manufacturing is represented by raw materials and process waste. Scrap materials are often generated during cutting and shaping processes of laminates (i.e. either dry or preimpregnated fabrics) due to wasteful cutting techniques and low nesting efficiency. This waste can account for 30–50% of raw materials used in laminates production [99]. Cured resin waste can be related to resin excess within mixing containers or distribution tubes (which in turns became waste material). In addition, components that do not meet the quality standards due to defects such as voids, delamination or incomplete curing became waste as well. Due to the limited adoption and efficiency of recycling systems for composites, much of this waste typically ends up in landfills or incineration facilities. This is an even more relevant issue considering the high impacts associated with the raw materials that are disposed of without ever being used. To reduce the carbon footprint associated with these processes, optimization strategies and minimization of produced waste are required.

4.2.3 Solutions for Improved Sustainability in Composites Manufacturing

Improving the sustainability of composite manufacturing is essential to address the environmental challenges associated with these materials. This requires an integrated approach that combines advancements in materials, technology, and effective waste management.

For what concerns composite constituents, the use of bio based fibers and resin for the production of composite parts represents a valuable solution. As reported in many literature studies, vegetable fibers are associated with much lower impacts with respect to synthetic ones. For instance, flax, hemp and jute fibers have GWP values between 1.4 and 1.9 kg CO_2 eq (i.e. about half the values for glass fibers and an order of magnitude lower than carbon fibers) [100]. As a result, the use of bio-based composites is under active investigation, with ongoing research focusing on their mechanical properties, the impact of various treatments, and the potential environmental implications [101]. Figure 12 compares synthetic fibers (carbon and glass) with bio-based alternatives (flax and hemp). The two natural fibers, which exhibit similar overall properties, offer mechanical characteristics almost comparable to glass fibers but with lower density and reduced environmental impact. While their properties are still significantly lower than those of carbon fibers, they provide substantial advantages in terms of cost reduction and environmental sustainability.

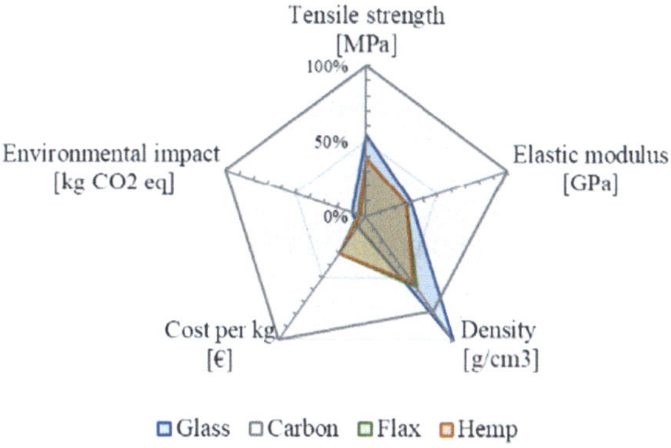

Fig. 12 Comparison between synthetic (carbon and glass fibers) and bio based (hemp and flax) reinforcements

Implementing advanced manufacturing techniques to substitute traditional processes can potentially enhance composites sustainability. As shown in literature, advanced systems can provide noteworthy reduction in manufacturing energy consumption.

Hence, more efficient machines with reduced heat and mechanical losses will represent a significant improvement. Similarly, the use of more sustainable energy mixes (e.g. with higher percentage of renewable energy with respect to fossil fuels) can provide important reductions in greenhouse gas emissions. Process parameters optimization is also an interesting strategy that can lead to lower resources consumption. Precise manufacturing processes can first reduce the possibility of producing defected parts and the generation of waste. For example, nesting software can reduce scarps produced during laminates cutting phases. Furthermore, as shown in literature, correct selection of parameters can minimize the process energy requirements whilst maintaining the same products quality and performance [98]. Systematic evaluations of the correlation between process parameters and sustainability are required to improve both innovative and traditional manufacturing processes.

Waste minimization and End of Life management are also essential to limit the composites sector impacts. Currently, there are different possible recycling methods for FRP including mechanical, thermal and chemical processes [102]. From an LCA perspective, effective end-of-life management translates into environmental credits and impact reductions equivalent to those associated with the avoided production of new materials. However, there are still relevant issues related to the recycling of polymer matrix composites; recycled fibers are typically reduced in size and show a strong reduction in mechanical properties with respect to virgin material. The investigation of efficient and cheap recycling process is an ideal solution to reduce both the impacts of raw materials (using secondary feedstock) and possible manufacturing waste production. In addition, the use of recyclable materials such as vitrimer and thermoplastic polymers can contribute to solve this industrial problem [103]. The minimization of production waste is also a possible route; zero-waste approaches have been developed to avoid manufacturing waste and to reuse possible scraps obtained during virgin component production [98].

These options are viable pathways and should be carefully evaluated throughout of the entire life cycle of composite products. A holistic life cycle approach is essential to assess all aspects together and ensure a sustainable and effective improvements.

4.3 Recent Developments in Material Regeneration and Reuse

The recycling and reuse of composite materials are emerging as crucial aspects of sustainability and the circular economy. Composite recyclability depends on its matrix and the possibility that this matrix can be re-processed, by the same processes of the virgin materials (primary recycling) or by different technologies. For polymer matrix composites, recyclability is associated to the matrix nature, if thermoplastic or thermosetting, as thermosets cannot be easily re-processed. End-of-life (EOL) thermoplastic matrix composites can be ground and processed by the typical technologies of thermoplastic polymers such as injection molding or extrusion. For this fact, these materials are assumed to be recyclable, apart from the real diffusion of recycling routes. Instead, thermoset matrix composites, such as fiberglass (GFR) of carbon fiber reinforced (CFR) epoxies, are considered difficult-to-recycle materials and destined to energy recovery or landfills.

At the state-of-the-art technologies, three different recycling routes are identified for thermoset composites: mechanical, thermal and chemical [104]. Mechanical recycling mainly consists of grinding and separating composite materials into chips of various sizes, from powders to coarse fibers. Powders can be used as fillers in thermoplastics or thermosets, but their economic value is very low. Coarse fibers can also be used as a reinforcement phase. Chemical processes, instead, tend to remove the matrix and recovery the fibers but these processes require high energies and produce large environmental impacts. Moreover fibers partially degrade during their extraction. In thermal recycling methods, fibers are released from the matrix using high temperatures that deteriorate the polymeric matrix. These recycling methods are usually classified in pyrolysis and fluidized beds.

All these technologies have been studied and proposed for the recycling of both glass and carbon fibers, aiming to overcome technological and cost challenges. For example, glass fiber is inherently inexpensive, so recycled fiber can end up costing more than virgin fiber.

For fiberglass, mechanical processing remains the most effective recycling method. In this context, innovation lies in the potential reuse of ground powders, either by exploiting their residual reactivity or by reactivating them, rather than relegating them to filler applications [105]. Powders from fiberglass machining can be also processed in the same way, or mixed with those from grinding of waste fiberglass [106]. The recycling technology is known as "direct molding" and is a compression molding process of powders in absence of virgin raw materials or any other linking

Fig. 13 Recycled products from direct molding of fiberglass powders

agent. Part agglomeration occurs under the combined action of pressure and temperature, by the combination of several mechanisms such as residual reactivity of the EOL products, powder re-activation (because of broken chemical bonds on particle surfaces), and incipient degradation during molding. In Fig. 13, such recycled products from direct molding of fiberglass powders are shown together with the initial powder. These parts are easy to paint because of surface porosity which helps the paint grabbing and it is also possible to put metal inserts during molding. Results from 4-point bending test show an average bending strength about 30 MPa and an elongation break under 2%, therefore the material is brittle because of its agglomerated nature. The bending modulus if 1.9 GPa, lower than the virgin fiberglass but enough for many applications.

Another example, which also includes several Italian partners in the Consortium, is the European project FiberEUse [3] which demonstrated at large scale the possibility to implement circular value-chains in the composites sector and their and their potentiality. The project developed highly controlled mechanical shredding processes through cyber-physical system, with an application focus on the recycling of fiberglass, and exploited thermochemical processes for the recovery of fibers and resin. The reuse of this materials has been successfully demonstrated in several sectors as automotive, aerospace, design, construction and sports equipment, leading to the ongoing European project DeremCo (GA 101084037), that aims to bring these solutions to the market, based on the inter-relation between the technical and social eco-systems at local and interregional level to bring benefit to the environment, industry, consumers and the European society, to unlock the cost-effective reuse of post-use composite materials and components in new high-added value products. The interest in glass fiber stems from its widespread use in the past and the large number of structures that are now being decommissioned, such as wind turbine blades, which make extensive use of GFRP materials. For instance, in [107] it was studied the innovative application of High Voltage Fragmentation (HVF) process to EoL wind blades. Experimental results demonstrate that this technology can act as a feasible recycling solution, both to remove impurities and to recycle GFRP, as a highly-controlled shredding process as shown in Fig. 14. This opens up interesting prospects for the reuse of the material itself in high-quality, high-added value products, following a demand-driven approach.

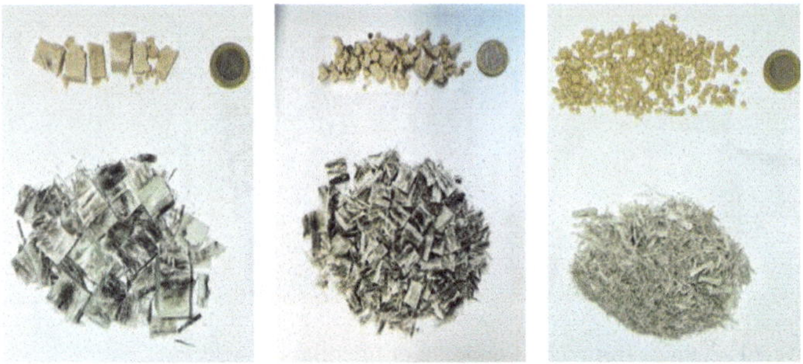

Fig. 14 Material obtained after recycling in using different pulses (from the left) 50 pulses, 250 pulses, 1000 pulse

Fiber recovery seems to be a valuable solution for EOL CFR parts because of the high residual value of the carbon fibers, instead of fiberglass. An important example is the full-scale experimental activity currently being conducted by Leonardo SPA and Gruppo Hera SPA in Emilia-Romagna, Italy, for the recovery of carbon fibers from the aerospace sector through a pyro-gasification process [108]. Regarding research, while initial studies primarily focused on thermal recycling processes, there is now growing interest in mechanical and chemical recycling. For instance, an innovative approach to composite recycling has been defined to recover the maximum residual value from waste composites and composite production waste. In this approach, depending on the residual value of the laminate constituents, different routes are proposed.

For example, Proietti et al. [109] proposed an innovative approach for CFR laminates in which single plies are delaminated by thermo-mechanical disassembly. Fibers do not degrade as they are not separated from the matrix and the full recycled ply is used for manufacturing new parts. Consolidation of the disassembled plies occurs under compression molding, thanks to the insertion of thermoplastic interlayers during lamination. In Fig. 15, this innovative approach is depicted. Hybridization occurs at the interface between the old thermoset and new thermoplastic matrix. Final hybrid laminates have unique properties such as thermoformability, repairability, damping and enhanced recyclability, with remarkable residual mechanical properties of the EOL laminates. Moreover, 100% recycled composites can be manufactured by using EOL thermoplastic sheets such as waste packaging or spent vacuum bags. Mechanical testing, under bending, of hybrid CFR laminates has shown values of strength over 380 MPa, and modulus over 35 GPa, 47% and 54% of the virgin laminate, respectively.

In Fig. 16, the optimal thermoformability of the hybrid panel is shown for a laminate with 10 CFR plies and 9 polyamide interlayers. In the same figure, thermoformability of a hybrid fiberglass laminate is also shown after disassembly of scraps. However, not all the fiberglass has enough residual value to allow this recycling technology and final laminate performances are less interesting than CFR composites.

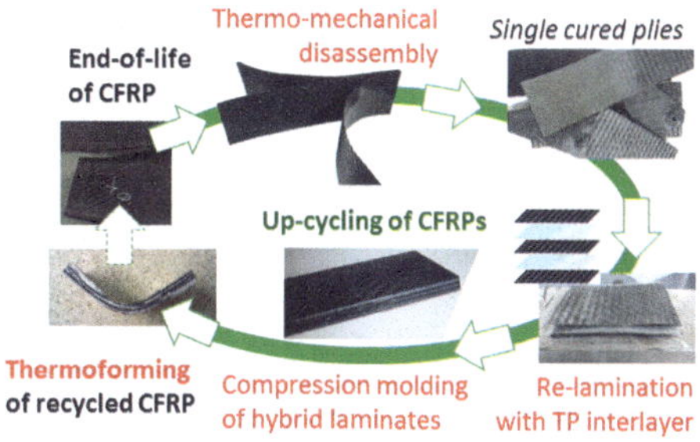

Fig. 15 Production of hybrid composite laminates by thermo-mechanical disassembly of waste laminates and subsequent compression molding with thermoplastic interlayers

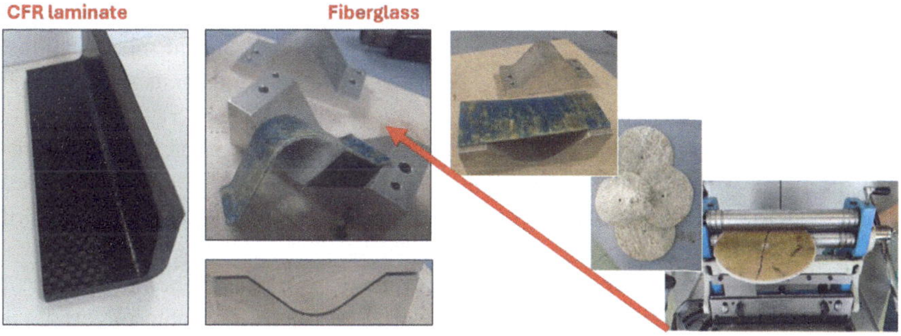

Fig. 16 Thermoformability of hybrid laminate from fiberglass recycling

Another innovative approach for recycling CFR has been proposed in [110, 111] where the aim is to overcome one of the major limitations of mechanical recycling, namely the use of machines dedicated to this operation, such as grinding machines, hammer milling machines, etc., in which it is also difficult to control the size of the recycled chips obtained. Indeed, the use of a milling process with dedicated tools is proposed, where it is possible to obtain CFRP chips of different sizes, which can then be reused for the production of new laminates. To remove the resin residues from the obtained chips [112] proposes the use of a soft chemical treatment using a peroxide and acetone solution (H_2O_2 20 wt% and C_3H_6O 80 wt%) for 1 h at temperatures ranging from 100 to 200 °C. The effectiveness of the treatment was highlighted by flexural tests and adhesion tests. Indeed, the flexural strength of specimens reinforced with fibers subjected to chemical treatment at 100 °C and 200 °C respectively experienced an increase of approximately 55 and 100% compared to samples reinforced with untreated fibers. Similarly, the flexural modulus showed an increase of about 75 and 155%, depending on the treatment temperature. Regarding adhesion, the pull-out force increased by approximately 152% compared to untreated fibers when a chemical treatment at 100 °C was performed and by about 252% for the chemical treatment at 200 °C.

5 Conclusions and Perspective

The transition toward sustainable composite manufacturing requires a multi-faceted approach, balancing material innovation, process efficiency, and circular economy principles. Based on the key insights presented in this white paper, future efforts should focus on:

Challenges in Adopting Green Composites: While the interest in sustainable composites is growing, several challenges still hinder their widespread adoption. These include higher variability in natural fiber properties, limited fire resistance, moisture sensitivity, lack of appropriate regulations, and, in some cases, higher production costs due to additional treatments or process modifications. Overcoming these limitations requires further research on material modifications, improved fiber-matrix compatibility, and process standardization to ensure industrial feasibility.

Enhancing Recycling and Reuse Strategies: Improving fiber recovery techniques, especially for thermoset composites, remains a crucial challenge. Advanced recycling methods, such as chemical and mechanical recycling or a combination of both could enable more effective material recovery.

Development of Hybrid and Multifunctional Composites: The integration of hybrid materials, including a mix of virgin, recycled fibers and green materials, is an effective strategy to improve both mechanical performance and sustainability. Future research should focus on optimizing the fiber/matrix interface and tailoring these materials for high-performance applications.

Energy Efficient of Manufacturing Processes: Reducing the carbon footprint of composite production through out-of-autoclave (OOA), automated, and additive manufacturing processes will be essential. Digitalization and artificial intelligence-driven optimization could further enhance production efficiency.

Lifecycle Assessment (LCA) and Standardization: Establishing clear LCA metrics for bio-composites and recycled materials is crucial to evaluating their true environmental benefits. Additionally, new industry standards and certification frameworks will help accelerate their adoption.

AITeM through its collaborative research initiatives, plays a vital role in addressing these challenges. Current and future activities could focus on:

- Experimental validation of new bio-composites and hybrid structures combining natural fibers with synthetic and recycled reinforcements and addressing production processes to correctly use next-generation materials as starting materials.
- Process optimization for sustainable composite manufacturing, including the study of novel low-energy processing techniques.
- Industry partnerships aimed at scaling up innovative solutions and facilitating their adoption in several applications.

By integrating these strategies, the composite manufacturing industry can unlock new opportunities for sustainability, efficiency, and innovation, ensuring its alignment with future environmental and regulatory frameworks.

References

1. Maiti S, Islam MR, Uddin MA, Afroj S, Eichhorn SJ, Karim N (2022) Sustainable fiber-reinforced composites: a review. Adv Sustain Syst. https://doi.org/10.1002/adsu.202200258
2. Lunetto V, Galati M, Settineri L, Iuliano L (2023) Sustainability in the manufacturing of composite materials: a literature review and directions for future research. J Manuf Process 85:858–874
3. Colledani M (2022) Engineering and construction systemic circular economy solutions for fiber reinforced composites.https://doi.org/10.1007/978-3-031-22352-5
4. Rajak DK, Pagar DD, Menezes PL, Linul E (2019) Fiber-reinforced polymer composites: manufacturing, properties, and applications. Polymers (Basel). https://doi.org/10.3390/polym11101667
5. Tucci F, Rubino F, Esperto V, Pasquino G, Carlone P (2023) Investigation on the pultrusion of thermoplastic preimpregnated polypropylene-glass tapes. Mater Res Proc 28:393–398
6. Goh GD, Yap YL, Agarwala S, Yeong WY (2019) Recent progress in additive manufacturing of fiber reinforced polymer composite. Adv Mater Technol. https://doi.org/10.1002/admt.201800271
7. Witten E, Mathes V (2024) The European market for fiber-reinforced plastics. In: Composites 2023 market developments, trends, challenges and prospects, vol 42
8. Fowler PA, Hughes JM, Elias RM (2006) Biocomposites: technology, environmental credentials and market forces. J Sci Food Agric 86:1781–1789
9. https://www.theinsightpartners.com/it/reports/biocomposites-market?utm_source=chatgpt.com
10. https://www.alkivio.com
11. https://www.novamont.com/eng/mater-bi
12. https://www.maipsrl.com/avviato-il-progetto-biontier-per-rivoluzionare-il-settore-dei-biocompositi-il-gruppo-maip-con-maip-compounding-e-bio-on-tra-i-partner-strategici?utm_source=chatgpt.com
13. https://www.polimerica.it/articolo.asp?id=29387&utm_source=chatgpt.com
14. Andrew JJ, Dhakal HN (2022) Sustainable biobased composites for advanced applications: recent trends and future opportunities—a critical review. Compos Part C: Open Access. https://doi.org/10.1016/j.jcomc.2021.100220
15. Pickering KL, Efendy MGA, Le TM (2016) A review of recent developments in natural fibre composites and their mechanical performance. Compos Part A Appl Sci Manuf 83:98–112
16. Boccarusso L, Pinto F, Cuomo S, De Fazio D, Myronidis K, Durante M, Meo M (2022) Design, manufacturing, and characterization of hybrid carbon/hemp sandwich panels. J Mater Eng Perform 31:769–785
17. Boccarusso L, Carrino L, Durante M, Formisano A, Langella A, Memola Capece Minutolo F (2016) Hemp fabric/epoxy composites manufactured by infusion process: improvement of fire properties promoted by ammonium polyphosphate. Compos B Eng 89:117–126
18. Iucolano F, Boccarusso L, Langella A (2019) Hemp as eco-friendly substitute of glass fibres for gypsum reinforcement: Impact and flexural behaviour. Compos B Eng. https://doi.org/10.1016/j.compositesb.2019.107073
19. Thomsen OT (2009) Sandwich materials for wind turbine blades—present and future. J Sandwich Struct Mater 11:7–26
20. Boccarusso L, Durante M, Langella A (2018) Lightweight hemp/bio-epoxy grid structure manufactured by a new continuous process. Compos B Eng 146:165–175
21. Abou-Ali AM, Al-Ketan O, Rowshan R, Abu Al-Rub R (2019) Mechanical response of 3D printed bending-dominated ligament-based triply periodic cellular polymeric solids. J Mater Eng Perform 28:2316–2326
22. Boccarusso L, Durante M, De Fazio D, Myronidis K, Meo M, Pinto F (2023) Hybridization of thin laminates with natural fibers: a comparison study between hemp and flax. In: Lecture notes in mechanical engineering, pp 254–260

23. Cicala G, Recca G, Oliveri L, Perikleous Y, Scarpa F, Lira C, Lorato A, Grube DJ, Ziegmann G (2012) Hexachiral truss-core with twisted hemp yarns: out-of-plane shear properties. Compos Struct 94:3556–3562
24. Boccarusso L, De Fazio D, Myronidis K, Durante M, Pinto F (2024) Bio-inspired optimisation design strategy of graded hemp core for hybrid carbon/hemp sandwich laminates. Int J Adv Manuf Technol. https://doi.org/10.1007/s00170-024-14604-6
25. https://www.formulapassion.it/motorsport/formula-1/f1-team/formula-1-mclaren-esplora-nuove-frontiere-materiali-seidl-lando-norris-carlos-sainz
26. Dragonetti R, Napolitano M, Boccarusso L, Durante M (2020) A study on the sound transmission loss of a new lightweight hemp/bio-epoxy sandwich structure. Appl Acoust. https://doi.org/10.1016/j.apacoust.2020.107379
27. Pinto F, Boccarusso L, De Fazio D, Cuomo S, Durante M, Meo M (2020) Carbon/hemp bio-hybrid composites: effects of the stacking sequence on flexural, damping and impact properties. Compos Struct. https://doi.org/10.1016/j.compstruct.2020.112148
28. Durante M, Formisano A, Boccarusso L, Langella A, Carrino L (2017) Creep behaviour of polylactic acid reinforced by woven hemp fabric. Compos B Eng 124:16–22
29. Sepe R, Bollino F, Boccarusso L, Caputo F (2018) Influence of chemical treatments on mechanical properties of hemp fiber reinforced composites. Compos B Eng. https://doi.org/10.1016/j.compositesb.2017.09.030
30. Das PP, Chaudhary V (2020) Environmental impact and effect of chemical treatment on bio fiber based polymer composites. Mater Today Proc 49:3418–3422
31. El Hawary O, Boccarusso L, Ansell MP, Durante M, Pinto F (2023) An overview of natural fiber composites for marine applications. J Mar Sci Eng. https://doi.org/10.3390/jmse11051076
32. Branda F, Malucelli G, Durante M, Piccolo A, Mazzei P, Costantini A, Silvestri B, Pennetta M, Bifulco A (2016) Silica treatments: a fire retardant strategy for hemp fabric/epoxy composites. Polymers (Basel). https://doi.org/10.3390/polym8080313
33. Rayyaan R, Kennon WR, Potluri P, Akonda M (2020) Fibre architecture modification to improve the tensile properties of flax-reinforced composites. J Compos Mater 54:379–395
34. Abdollahiparsa H, Shahmirzaloo A, Blok R, Teuffel P (2023) Influence of moisture absorption on tensile and compressive properties of natural fiber-reinforced thermoplastic composites. Polym-Plast Technol Mater 62:2138–2142
35. Bachchan AA, Das PP, Chaudhary V (2020) Effect of moisture absorption on the properties of natural fiber reinforced polymer composites: a review. Mater Today Proc 49:3403–3408
36. Elfaleh I, Abbassi F, Habibi M, Ahmad F, Guedri M, Nasri M, Garnier C (2023) A comprehensive review of natural fibers and their composites: an eco-friendly alternative to conventional materials. Results Eng. https://doi.org/10.1016/j.rineng.2023.101271
37. https://www.castrocompositesshop.com/it/rinforzo-in-fibre-di/1190-200-gm2-tessuti-di-vetro-satin-ute-200p.html
38. Garrison TF, Murawski A, Quirino RL (2016) Bio-based polymers with potential for biodegradability. Polymers (Basel). https://doi.org/10.3390/polym8070262
39. Korol J, Burchart-Korol D, Pichlak M (2016) Expansion of environmental impact assessment for eco-efficiency evaluation of biocomposites for industrial application. J Clean Prod 113:144–152
40. Xie Y, Gao S, Zhang D, Wang C, Chu F (2023) Bio-based polymeric materials synthesized from renewable resources: a mini-review. Resour Chem Mater 2:223–230
41. Terry JS, Taylor AC (2021) The properties and suitability of commercial bio-based epoxies for use in fiber-reinforced composites. J Appl Polym Sci 138:1–12
42. Sarika PR, Nancarrow P, Khansaheb A, Ibrahim T (2020) Bio-based alternatives to phenol and formaldehyde for the production of resins. Polymers (Basel) 12:1–24
43. Ipakchi H, Shegeft A, Rezadoust AM, Zohuriaan-Mehr MJ, Kabiri K, Sajjadi S (2020) Bio-resourced furan resin as a sustainable alternative to petroleum-based phenolic resin for making GFR polymer composites. Iran Polym J 29:287–299

44. Burelo M, Martínez A, Hernández-Varela JD, Stringer T, Ramírez-Melgarejo M, Yau AY, Luna-Bárcenas G, Treviño-Quintanilla CD (2024) Recent developments in synthesis, properties, applications and recycling of bio-based elastomers. Molecules 29:387
45. Claassen W, Zarifis G (2021) First Bio-based composite movable bicycle bridge. Struct Eng Int 31:227–232
46. Marotta A, Faggio N, Brondi C (2022) Curing kinetics of bioderived furan-based epoxy resins: study on the effect of the epoxy monomer/hardener ratio. Polymers (Basel). https://doi.org/10.3390/polym14235322
47. McCracken A, Sadeghian P (2018) Green sandwich composites fabricated from flax FRP facings and corrugated cardboard cores. Non-Conventional Mater Technol 7:285–294
48. Sorrentino L, Turchetta S, Parodo G, Papa R, Toto E, Santonicola MG, Laurenzi S (2022) RIFT process analysis for the production of green composites in flax fibers and bio-based epoxy resin. Materials. https://doi.org/10.3390/ma15228173
49. Fantuzzi N, Bacciocchi M, Benedetti D, Agnelli J (2021) The use of sustainable composites for the manufacturing of electric cars. Compos Part C: Open Access. https://doi.org/10.1016/j.jcomc.2020.100096
50. Koczoń P, Bartyzel B, Iuliano A, Klensporf-Pawlik D, Kowalska D, Majewska E, Tarnowska K, Zieniuk B, Gruczyńska-Sękowska E (2022) Chemical structures, properties, and applications of selected crude oil-based and bio-based polymers. Polymers (Basel). https://doi.org/10.3390/polym14245551
51. Parodo G, Sorrentino L, Turchetta S (2023) Study of autoclave process to manufacture thermoplastic composites constituted by PP/flax fibers. Mater Res Proc 35:385–392
52. Miranda Campos B, Bourbigot S, Fontaine G, Bonnet F (2022) Thermoplastic matrix-based composites produced by resin transfer molding: a review. Polym Compos 43:2485–2506
53. Trivedi AK, Gupta MK, Singh H (2023) PLA based biocomposites for sustainable products: a review. Adv Ind Eng Polym Res 6:382–395
54. Le Guen MJ, Newman RH, Fernyhough A, Emms GW, Staiger MP (2016) The damping-modulus relationship in flax-carbon fibre hybrid composites. Compos B Eng 89:27–33
55. Mabalane PN, Molnar K, Molnar K, Khoathane C (2024) Structural integrity of hybrid panel of recycled and virgin carbon fibre. In: Lecture notes in mechanical engineering, pp 12–19
56. Pokkalla DK, Garg N, Paramanathan M, Kumar V, Rencheck ML, Nandwana P, Kunc V, Hassen AA, Kim S (2024) Design optimization of lightweight automotive seatback through additive manufacturing compression overmolding of metal polymer composites. Compos Struct. https://doi.org/10.1016/j.compstruct.2024.118504
57. Bravo A, Vieira D (2017) Design and manufacturing of sustainable composites. Handb Compos Renew Mater 1–8:533–601
58. Gomes D, Amorim L, Santos RM, Gonçalves ND (2023) Numerical simulations of carbon-fibre impregnation with a polymer as an anisotropic permeability medium. Materials. https://doi.org/10.3390/ma16206627
59. Galos J (2020) Thin-ply composite laminates: a review. Compos Struct. https://doi.org/10.1016/j.compstruct.2020.111920
60. Haj-Ali R, Kilic H (2002) Nonlinear behavior of pultruded FRP composites. Compos B Eng 33:173–191
61. Li J, Huang Q, Yin Y, Xu J, Jiang Y, Du W (2024) The technology and current applications of continuous fiber-reinforced thermoplastic composites. Polym Compos. https://doi.org/10.1002/pc.28764
62. Katsiropoulos CV, Loukopoulos A, Pantelakis SG (2019) Comparative environmental and cost analysis of alternative production scenarios associated with a helicopter's canopy. Aerospace. https://doi.org/10.3390/aerospace6010003
63. Witik RA, Gaille F, Teuscher R, Ringwald H, Michaud V, Månson JAE (2012) Economic and environmental assessment of alternative production methods for composite aircraft components. J Clean Prod 29–30:91–102
64. Bontempi E (2017) A new approach for evaluating the sustainability of raw materials substitution based on embodied energy and the CO_2 footprint. J Clean Prod 162:162–169

65. Oliveux G, Dandy LO, Leeke GA (2015) Current status of recycling of fibre reinforced polymers: review of technologies, reuse and resulting properties. Prog Mater Sci 72:61–99
66. Howarth J, Mareddy SSR, Mativenga PT (2014) Energy intensity and environmental analysis of mechanical recycling of carbon fibre composite. J Clean Prod 81:46–50
67. Song YS, Youn JR, Gutowski TG (2009) Life cycle energy analysis of fiber-reinforced composites. Compos Part A Appl Sci Manuf 40:1257–1265
68. Joshi SV, Drzal LT, Mohanty AK, Arora S (2004) Are natural fiber composites environmentally superior to glass fiber reinforced composites? Compos Part A Appl Sci Manuf 35:371–376
69. Ekuase OA, Anjum N, Eze VO, Okoli OI (2022) A review on the out-of-autoclave process for composite manufacturing. J Compos Sci. https://doi.org/10.3390/jcs6060172
70. Taylor T, Zhang J, Yanagimoto J (2020) Evaluation of a concept out-of-autoclave process for manufacturing carbon fibre reinforced polymer automotive parts. Procedia CIRP 86:162–166
71. Austermann V, Neuhaus J, Schneider D, Dahlmann R, Hopmann C (2022) Investigation of the mechanical properties of Vinylester-based sheet moulding compound (SMC) subject to particle recycling. J Compos Sci. https://doi.org/10.3390/jcs6030084
72. Qian C, Yuan H, Begum H, Hill T, Groves R, Joesbury A, Harper L (2024) Experimental characterisation for compression moulding of hybrid architecture composites using reclaimed prepreg manufacturing waste. Mater Res Proc 41:631–640
73. Rothenhäusler F, Meyer N, Wehler S, Hohberg M, Gude M, Henning F, Kärger L (2022) Experimental and numerical analysis of SMC compression molding in confined regions—a comparison of simulation approaches. J Compos Sci. https://doi.org/10.3390/JCS6030068
74. Yuan H, Khan M, Peijs T, Qian C (2024) Experimental process characterisation for high-volume compression moulding of hybrid-architecture composites. Compos Part A Appl Sci Manuf. https://doi.org/10.1016/j.compositesa.2024.108137
75. Raimondi L, Tomesani L, Donati L, Zucchelli A (2021) Lattice material infiltration for hybrid metal-composite joints: manufacturing and static strenght. Compos Struct. https://doi.org/10.1016/j.compstruct.2021.114069
76. Li W, Feng T, Lu T, Zhao F, Zhao J, Guo W, Hua L (2024) Optimization of compression molding parameters and lifecycle carbon impact assessment of bamboo fiber-reinforced polypropylene composites. Polymers (Basel). https://doi.org/10.3390/polym16233435
77. Khan WA, Methven J (2014) Correlation of the heater's duty cycle and specific energy consumption, and reduction in energy consumption in the pultrusion process. Sci Eng Compos Mater 21:295–302
78. Brecher C, Schmitt R, Lindner F, Peters T, Emonts M, Böckmann MG (2016) Increasing cost and eco efficiency for selective tape placement and forming by adaptive process design. Procedia CIRP 57:769–774
79. Tolio T, Bernard A, Colledani M, Kara S, Seliger G, Duflou J, Battaia O, Takata S (2017) Design, management and control of demanufacturing and remanufacturing systems. CIRP Ann Manuf Technol 66:585–609
80. Jia C, Zhang P, Seraji SM et al (2022) Effects of BN/GO on the recyclable, healable and thermal conductivity properties of ENR/PLA thermoplastic vulcanizates. Compos Part A Appl Sci Manuf. https://doi.org/10.1016/j.compositesa.2021.106686
81. Rubino F, Parmar H, Mancia T, Carlone P (2023) Ultrasonic welding of glass reinforced epoxy composites using thermoplastic hybrid interlayers. Compos Struct. https://doi.org/10.1016/j.compstruct.2023.116980
82. Zhao J, Yang Y, Kobir MH, Faludi J, Zhao F (2024) Driving additive manufacturing towards circular economy: state-of-the-art and future research directions. J Manuf Process 124:621–637
83. Ricciardi MR, Antonucci V, Durante M, Giordano M, Nele L, Starace G, Langella A (2014) A new cost-saving vacuum infusion process for fiber-reinforced composites: pulsed infusion. J Compos Mater 48:1365–1373
84. Ricciardi MR, Martone A, Borriello A, Zarrelli M, Giordano M, Langella A, Antonucci V (2017) Mechanical behavior of hybrid fiber-reinforced composites manufactured by pulse infusion. Polym Compos 38:2254–2260

85. Wang T, Huang K, Guo L, Zheng T, Zeng F (2023) An automated vacuum infusion process for manufacturing high-quality fiber-reinforced composites. Compos Struct. https://doi.org/10.1016/j.compstruct.2023.116717
86. Grisin B, Carosella S, Middendorf P (2024) Vacuum chamber infusion for fiber-reinforced composites. Polymers (Basel). https://doi.org/10.3390/polym16192763
87. Rubino F, Tucci F, Esperto V, Carlone P (2022) Filling time reduction in liquid composite molding processes. J Compos Sci 6:1–12
88. Rubino F, Esperto V, Tucci F, Carlone P (2020) Flow enhancement in liquid composite molding processes by online microwave resin preheating. Polym Eng Sci 60:2377–2389
89. Minchenkov K, Vedernikov A, Safonov A, Akhatov I (2021) Thermoplastic pultrusion: a review. Polymers (Basel) 13:1–36
90. Vedernikov A, Minchenkov K, Gusev S, Sulimov A, Zhou P, Li C, Xian G, Akhatov I, Safonov A (2022) Effects of the pre-consolidated materials manufacturing method on the mechanical properties of pultruded thermoplastic composites. Polymers (Basel). https://doi.org/10.3390/polym14112246
91. Tucci F, Rubino F, Pasquino G, Carlone P (2023) Thermoplastic pultrusion process of polypropylene/glass tapes. Polymers (Basel). https://doi.org/10.3390/polym15102374
92. Tucci F, Rubino F, Esperto V, Carlone P (2024) Thermoplastic pultrusion of recycled PET matrix composites. Mater Res Proc 41:422–429
93. Duflou JR, Deng Y, Van Acker K, Dewulf W (2012) Do fiber-reinforced polymer composites provide environmentally benign alternatives? A life-cycle-assessment-based study. MRS Bull 37:374–382
94. Tserpes K, Tzatzadakis V (2022) Life-cycle analysis and evaluation of mechanical properties of a bio-based structural adhesive. Aerospace. https://doi.org/10.3390/aerospace9020064
95. Ramachandran K, Gnanasagaran CL, Vekariya A (2023) Life cycle assessment of carbon fiber and bio-fiber composites prepared via vacuum bagging technique. J Manuf Process 89:124–131
96. Forcellese A, Marconi M, Simoncini M, Vita A (2020) Life cycle impact assessment of different manufacturing technologies for automotive CFRP components. J Clean Prod. https://doi.org/10.1016/j.jclepro.2020.122677
97. Shahid AT, Silvestre JD, Hofmann M, Garrido M, Correia JR (2024) Life cycle assessment of an innovative bio-based unsaturated polyester resin and its use in glass fibre reinforced bio-composites produced by vacuum infusion. J Clean Prod. https://doi.org/10.1016/j.jclepro.2024.140906
98. Bianchi I, Forcellese A, Mancia T, Simoncini M, Vita A (2022) Process parameters effect on environmental sustainability of composites FFF technology. Mater Manuf Processes 37:591–601
99. Nilakantan G, Nutt S (2015) Reuse and upcycling of aerospace prepreg scrap and waste. Reinf Plast 59:44–51
100. Le Duigou A, Davies P, Baley C (2011) Environmental impact analysis of the production of flax fibres to be used as composite material reinforcement. J Biobased Mater Bioenergy 5:153–165
101. Bozaci E, Sever K, Sarikanat M, Seki Y, Demir A, Ozdogan E, Tavman I (2013) Effects of the atmospheric plasma treatments on surface and mechanical properties of flax fiber and adhesion between fiber-matrix for composite materials. Compos B Eng 45:565–572
102. Ateeq M (2023) A state of art review on recycling and remanufacturing of the carbon fiber from carbon fiber polymer composite. Compos Part C: Open Access. https://doi.org/10.1016/j.jcomc.2023.100412
103. Zhang H, Cui J, Hu G, Zhang B (2022) Recycling strategies for vitrimers. Int J Smart Nano Mater 13:367–390
104. Pakdel E, Kashi S, Varley R, Wang X (2021) Recent progress in recycling carbon fibre reinforced composites and dry carbon fibre wastes. Resour Conserv Recycl. https://doi.org/10.1016/j.resconrec.2020.105340

105. Quadrini F, Bellisario D, Santo L (2016) Molding articles made of 100% recycled fiberglass. J Compos Mater 50:2959–2969
106. Bellisario D, Iorio L, Proietti A, Quadrini F, Santo L (2024) Recycling of thermoset fiberglass by direct molding of ground powders. Mater Res Proc 41:2410–2417
107. Diani M, Torvi S, Colledani M (2024) Exploiting high voltage fragmentation to enable demand-driven recycling of end-of-life wind blades. Procedia CIRP 122:837–842
108. https://www.compositimagazine.it/il-riciclo-della-fibra-di-carbonio-rivoluziona-il-mondo-aerospaziale/
109. Proietti A, Iorio L, Gallo N, Regi M, Bellisario D, Quadrini F, Santo L (2022) Recycling of carbon fiber laminates by thermo-mechanical disassembly and hybrid panel compression molding. Mater Plast 59:44–50
110. Boccarusso L, De Fazio D, Durante M, Formisano A, Langella A (2023) Unconventional method for recycling CFRPs by using a milling process. Mater Res Proc 28:1683–1688
111. Durante M, Boccarusso L, De Fazio D, Formisano A, Langella A (2023) Investigation on the mechanical recycling of carbon fiber-reinforced polymers by peripheral down-milling. Polymers (Basel). https://doi.org/10.3390/polym15040854
112. Boccarusso L, de Fazio D, Durante M, Formisano A, Langella A (2024) Mechanical and chemical combined recycling process for CFRP scraps. Mater Res Proc 41:568–575

The manufacturer's authorised representative in the EU is Springer
Nature Customer Service Centre GmbH, Europaplatz 3, 69115 Heidelberg,
Germany. If you have any concerns regarding our products, please
contact ProductSafety@springernature.com

Printed and bound by CPI Group (UK) Ltd, Croydon, CR0 4YY
01/10/2026
02225961-0013